T0309088

Modern Production Technology in Food and Agriculture

Modern Production Technology in Food and Agriculture

Editor: Brandon Corbyn

www.callistoreference.com

Callisto Reference,
118-35 Queens Blvd., Suite 400,
Forest Hills, NY 11375, USA

Visit us on the World Wide Web at:
www.callistoreference.com

ISBN: 978-1-64116-817-5 (Hardback)

Cataloging-in-Publication Data

Modern production technology in food and agriculture / edited by Brandon Corbyn.
p. cm.
Includes bibliographical references and index.
ISBN 978-1-64116-817-5
1. Food industry and trade. 2. Agricultural innovations. 3. Agricultural productivity.
4. Food. 5. Agriculture. I. Corbyn, Brandon.
HD9000.5 .M63 2023
338.19--dc23

Table of Contents

Preface

This book has been a concerted effort by a group of academicians, researchers and scientists, who have contributed their research works for the realization of the book. This book has materialized in the wake of emerging advancements and innovations in this field. Therefore, the need of the hour was to compile all the required researches and disseminate the knowledge to a broad spectrum of people comprising of students, researchers and specialists of the field.

Food technology refers to a subfield of food science that deals with the invention, production, preservation and quality control of food products. It uses modern technologies to create and enhance the food production and distribution systems. It is categorized into agricultural technology, food science and food delivery. Agricultural technology refers to the application of technology within the field of agriculture in order to improve efficiency, yield and profitability. Technological solutions within this field can be in the form of software solutions, sensors, satellite imagery and agriculture machines. Agricultural technology can be in the form of products, services or applications which have been derived from agriculture. This book unfolds the innovative aspects of modern production technology in food and agriculture. It elucidates new techniques and their applications in a multidisciplinary manner. This book is a resource guide for experts as well as students.

At the end of the preface, I would like to thank the authors for their brilliant chapters and the publisher for guiding us all-through the making of the book till its final stage. Also, I would like to thank my family for providing the support and encouragement throughout my academic career and research projects.

Editor

Preface

1

Precision and Digital Agriculture: Adoption of Technologies and Perception of Brazilian Farmers

Édson Luis Bolfe [1,2,*], Lúcio André de Castro Jorge [3], Ieda Del'Arco Sanches [4], Ariovaldo Luchiari Júnior [1], Cinthia Cabral da Costa [3], Daniel de Castro Victoria [1], Ricardo Yassushi Inamasu [3], Célia Regina Grego [1], Victor Rodrigues Ferreira [5] and Andrea Restrepo Ramirez [5]

1 Embrapa Informática Agropecuária, Brazilian Agricultural Research Corporation, Campinas 13083-886, Brazil; ariovaldo.luchiari@embrapa.br (A.L.J.); daniel.victoria@embrapa.br (D.d.C.V.); celia.grego@embrapa.br (C.R.G.)
2 Department of Geography, Graduate Program in Geography, University of Campinas (Unicamp), Campinas 13083-885, Brazil
3 Embrapa Instrumentação, Brazilian Agricultural Research Corporation, São Carlos 13560-970, Brazil; lucio.jorge@embrapa.br (L.A.d.C.J.); cinthia.costa@embrapa.br (C.C.d.C.); ricardo.inamasu@embrapa.br (R.Y.I.)
4 Divisão de Sensoriamento Remoto, National Institute for Space Research (INPE), São José dos Campos 12227-010, Brazil; ieda.sanches@inpe.br
5 Unidade de Competitividade do Sebrae Nacional, Brazilian Micro and Small Business Support Service (Sebrae), Brasília 70770-900, Brazil; victor.ferreira@sebrae.com.br (V.R.F.); andrea.ramirez@sebrae.com.br (A.R.R.)
* Correspondence: edson.bolfe@embrapa.br

Abstract: The rapid population growth has driven the demand for more food, fiber, energy, and water, which is associated to an increase in the need to use natural resources in a more sustainable way. The use of precision agriculture machinery and equipment since the 1990s has provided important productive gains and maximized the use of agricultural inputs. The growing connectivity in the rural environment, in addition to its greater integration with data from sensor systems, remote sensors, equipment, and smartphones have paved the way for new concepts from the so-called Agriculture 4.0 or Digital Agriculture. This article presents the results of a survey carried out with 504 Brazilian farmers about the digital technologies in use, as well as current and future applications, perceived benefits, and challenges. The questionnaire was prepared, organized, and made available to the public through the online platform LimeSurvey and was available from 17 April to 2 June 2020. The primary data obtained for each question previously defined were consolidated and analyzed statistically. The results indicate that 84% of the interviewed farmers use at least one digital technology in their production system that differs according to technological complexity level. The main perceived benefit refers to the perception of increased productivity and the main challenges are the acquisition costs of machines, equipment, software, and connectivity. It is also noteworthy that 95% of farmers would like to learn more about new technologies to strengthen the agricultural development in their properties.

Keywords: agriculture 4.0; smart farming; farmer's attitudes; Brazil

---

1. Introduction

Digital innovation in agriculture represents, according to the United Nations Food and Agriculture Organization [1], a great opportunity to eradicate poverty and hunger and mitigate the effects of climate change. Through digitalization, all parts of the agri-food production chain will be modified, since connectivity and the processing of large amounts of information in an instant allows for more efficient work, greater economic return, greater environmental benefits, and better working conditions in the field. However, implementing these changes will require governments to increasingly strengthen rural infrastructure and promote the development of rural communities [1] and small rural businesses, so that they can adopt and implement innovative solutions.

In this context of innovation, Digital Agriculture (DA) is part of the so-called "fourth industrial revolution", and its conceptual bases address aspects associated with Agriculture 4.0 [2], which derives from the Industry 4.0, and refers to the use of cutting-edge technology in food production. More recently, the term "Smart Farming" has also been used from the perspective of a development that emphasizes the use of information and communication technologies in the digital farm management cycle, through the intensive use of new technologies such as the Internet of Things, cloud computing, artificial intelligence, and big data [3]. In general, the conceptual basis of "Smart Farming" or "Digital Agriculture" comes from scientific knowledge, techniques, and equipment from Precision Agriculture [4] started from 1990s decade.

The digital agriculture can be understood and encompasses communication, information, and spatial analysis technologies that allow rural producers to plan, monitor, and manage the operational and strategic activities of the production system. In addition to the technologies already consolidated, such as field sensors [5–7], orbital remote sensors [8,9] and also embedded in UAV-Unmanned Aerial Vehicle [10], global positioning systems [11], telemetry and automation [12], digital maps-soil relief, production, productivity [12], digital agriculture also involves the Internet and connectivity in crops [13,14], cloud computing, big data, blockchain and cryptography [3,15,16], deep learning [17–19], Internet of Things (IoT) [20], mobile applications and digital platforms [21,22], and artificial intelligence [23]. All these technologies support pre and post-production decisions and greater sustainability of production systems [24,25], in addition to access to a differentiated market benefiting short marketing chains.

The perspectives of American rural producers were assessed for benefits of using precision agriculture technologies, observing that the perceptions of the derived benefits are heterogeneous and differentiated according to the agricultural culture [26]. The authors point out that in order to better understand farmers' adoption decisions, or lack of, it is important to first understand their perceptions of the benefits that technologies provide. Reference [27] emphasize that digital agriculture supports better decision-making based on consistent analyzes of agricultural systems, supporting the farmer in the form of digital solutions associated with robotics and artificial intelligence. However, they stress that it is necessary to coordinate more solid user training, especially for young farmers eager to learn and apply modern agricultural technologies and to grant a generational renewal yet to come. They consider the right time for society to advance in a modern and sustainable agriculture, capable of presenting all the power of agricultural management based on data to face the challenges posed to food production in the 21st century.

Projections from Brazil's Ministry of Agriculture, Livestock, and Supply [28] indicate that grain production may rise from the current 250.9 million tons in 2019/2020 (65.6 million hectares) to 318.3 million tons in the 2029/2030 harvest (76.4 million hectares), corresponding to an increase of 26.9%; that is, a 2.4% growth per year. For meat production (beef, pork, and poultry), projections indicate that it should rise from 28.2 million tons to 34.9 million tons by the end of the next decade, representing a 23.8% increase in the period. Precision agriculture and the digital transformation that has occurred in rural areas can contribute to Brazil reaching or exceeding these expectations, strengthening the country's position as one of the World leaders in food production and export, based on increased productivity and a sustainable use of natural resources.

Reference [29] investigated the adoption of precision agriculture technologies in sugarcane production in the State of São Paulo, through a questionnaire sent to all companies that operate in the sugar and alcohol sector in the region. The authors concluded that companies that adopted and used these technologies have proven to reap benefits, such as management improvement, higher productivity, lower costs, minimization of environmental impacts, and improvements sugarcane quality. Reference [30] investigated the use and adoption by producers and service providers on precision agriculture technologies in different Brazilian agricultural regions and found that the growth in technology adopted was linked to economic gains in agriculture. Economic aspects combined with the difficulty in using software and equipment provided by the lack of technical training by field teams were highlighted as the main factors that limit the expansion in the use of these technologies in the field.

Precision agriculture is already a reality for Brazilian technicians and rural producers. The knowledge that there is a variability in the production areas according to soil variations, vegetation, and the history of land use, is spreading progressively [31]. However, there are still important gaps in studies on a national scale to support strategic decisions in the development of new research, innovation, and the market. It is necessary to conduct studies in the context of digital agriculture in Brazil on aspects such as which technologies and applications are used, the perception of benefits, and the main challenges and expectations.

The future of using decision support systems in Agriculture 4.0 lies in the researchers' ability to better understand the challenges of this decision-making, including its applications in planning agricultural activities, managing water resources, adapting to climate change and food waste control [32]. The literature review shows different thematic clusters of extant social science on digitalization in agriculture: (i) adoption, uses, and adaptation of digital technologies on farm; (ii) effects of digitalization on farmer identity, farmer skills, and farm work; (iii) power, ownership, privacy and ethics in digitalizing agricultural production systems and value chains; (iv) digitalization and agricultural knowledge and innovation systems; and (v) economics and management of digitalized agricultural production systems and value chains [33]. This future research agenda provides ample scope for future interdisciplinary and transdisciplinary science on precision farming, digital agriculture, smart farming, and agriculture 4.0. Thus, this work aimed to gather information through an online consultation with rural producers about the digital technologies used in Brazil today, their applications, challenges, and future perspectives.

2. Materials and Methods

For the researcher's methodological definition, it was based on aspects applied: (i) in the evaluation of the factors of adoption of remote sensing images in the precision management of cotton producers in the United States [34]; (ii) in assessing the adoption and future prospects of precision agriculture in Germany [35]; (iii) in who evaluated the adoption of precision agriculture technologies in the sugarcane sector in Brazil [29], (iv) in who evaluated the adoption of precision agriculture under the perception of farmers and service providers in Brazil [30], and (v) in who evaluated different perspectives of American rural producers regarding the benefits of precision agriculture [26]. From the bibliographic references and the experience of the project team in the Brazilian context, specific questions were established, addressing aspects with the possibility of multiple choice answers on the technologies in precision agriculture and digital agriculture used in different complexity of applications (Table 1), among them in different sectors of agriculture and profiles of rural producers, added to perceptions of benefits, challenges and future expectations in Brazilian agriculture.

Table 1. Technologies in precision agriculture and digital agriculture in the context of the present study.

Technologies	Reference	Complexity of Applications
Internet and Connectivity/Wireless	[14,36]	Low
Mobile APPs, Digital Platforms and Software	[21,22]	
Global Positioning Systems	[11]	
Digital Maps	[12]	

Table 1. *Cont.*

Technologies	Reference	Complexity of Applications
Proximal and Field Sensors	[5–7]	Medium
Remote Sensors	[8–10]	
Embedded Electronics, Telemetry, and Automation	[12,37]	
Deep Learning and Internet of Things	[17–20]	High
Cloud Computing, Big Data, Blockchain and Cryptography	[3,15,16]	
Artificial Intelligence	[23]	

The questionnaire was prepared, organized, and made available to the public through the online platform LimeSurvey under the registration number 25889/2020 [38]. The system was available from 17 April to 2 June 2020, allowing answers to be collected based on specific questions previously defined (Table 2). The based used for the sample the research, was the Mailing List of Embrapa (Brazilian Agricultural Research Corporation, Campinas, Brazil) and Sebrae (Brazilian Micro and Small Business Support Service, Brasília, Brazil), agricultural cooperatives and associations of rural producers. A similar approach to sample respondents was applied in [30]. An "invitation message" was sent explaining the objectives with an access link and other guidelines to those interested in voluntarily participating in this online survey.

Table 2. Questions asked in precision agriculture and digital agriculture in the context of the research.

Questions
What is the identification and location of the responding farmer?
In which productive sector does he/she operate in?
For how many years have you been working in this job?
What is the cultivated area?
What techniques and inputs are used in the agricultural production?
What technologies in digital agriculture were used?
What are the main functions of the technologies used in digital agriculture?
How do you access the technology used by digital agriculture in on your property?
Is it by cell phone apps, machines, equipment, data or images?
What are your perceptions on the advantages enabled by using technologies from digital agriculture?
What are the difficulties in accessing and using technologies from digital agriculture?
In what applications would you like to start or strengthen the use of digital technologies in the future?

The primary data obtained for each question (single choice, multiple choice, and matrix questions) and its respective complete answers were consolidated in a LimeSurvey platform report and later exported in csv and included in a spreadsheet. Subsequently, statistics were generated based on absolute frequency data and graphs representative of the relative frequency represented by the percentages of each of the variables associated with the survey questions. Considering the size of the sample population of rural properties in Brazil approximately 1.5 million farmers in the survey profile, [39], it was stipulated to reach at least 385 questionnaires answered in full to obtain up to 5% margin of error with a 95% reliability level. However, results were gathered for 504 farmers (0.03% of population), a higher response rate than originally expected.

3. Results and Discussion

The consultation was nationwide and a total of 504 questionnaires were answered in full, including 154 (30.6%) from the South; 150 (29.8%) were from the Southeast region; 137 (27.2%) from the Northeast; 39 (7.7%) from the Midwest; and 24 (4.8%) in the North (Figure 1). The five states with the highest number of respondents were: Rio Grande do Sul (18.9%), Minas Gerais (13.9%), São Paulo (11.9%),

Bahia (11.1%), and Paraná (7.9%), which represent 61.7% of the respondents. These States are part of consolidated agricultural regions and, together with Mato Grosso, Goiás, and Mato Grosso do Sul, are part of the eight States with the highest gross agricultural production value in Brazil [40].

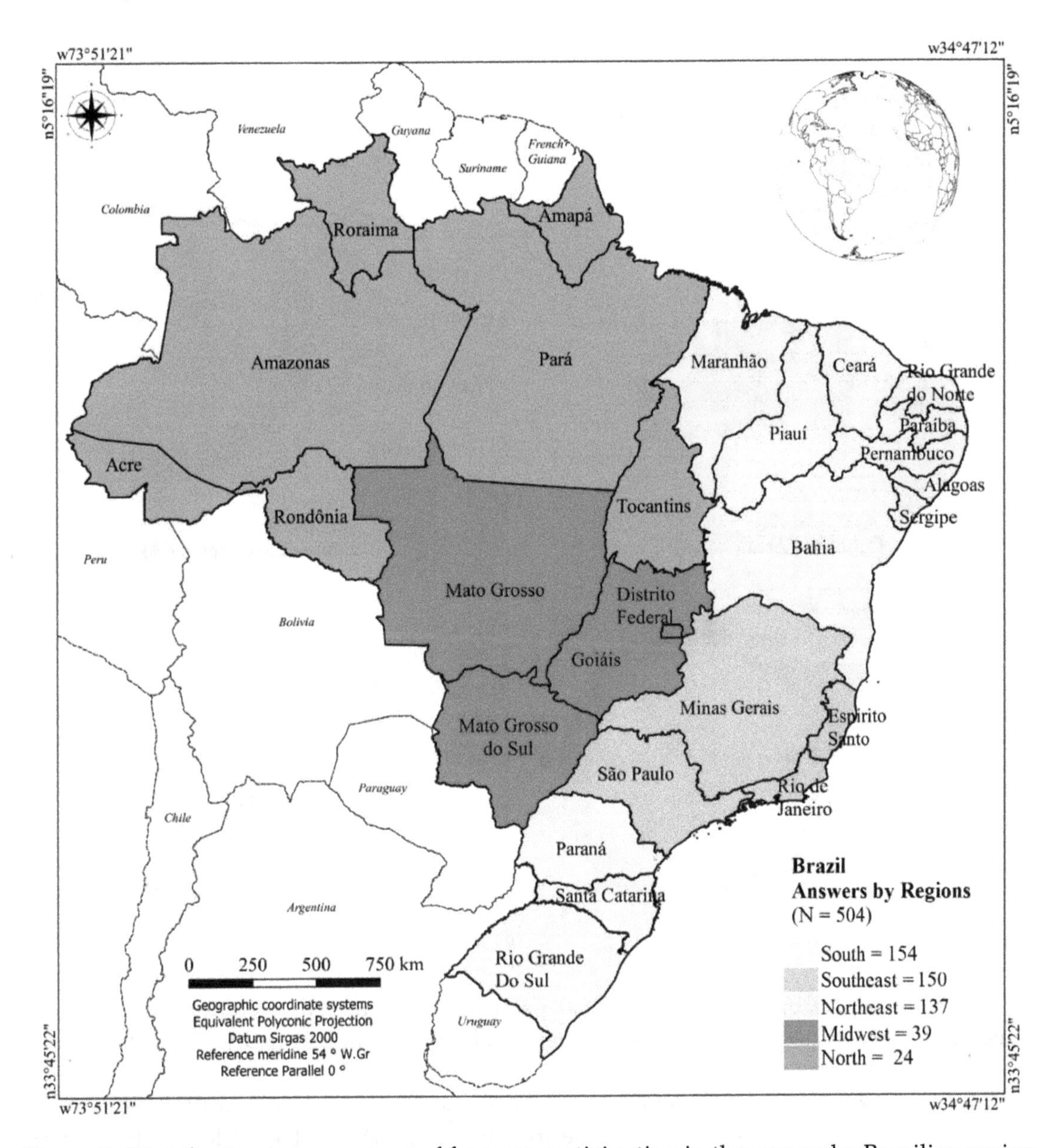

Figure 1. Distribution in percentage of farmers participating in the survey by Brazilian region.

Among the rural producers who participated in the survey, 74% work in agriculture (grains, fruit, horticulture, etc.); 54% with livestock (beef, pork, poultry, etc.); 6% with forestry (eucalyptus, pine, native, etc.) and 14% with other activities such as beekeeping, floriculture, aquaculture, and fish farming—not exclusively (Figure 2a); 72% cultivate areas of up to 50 hectares (Figure 2b); and 69% declared that they have more than 10 years of experience (Figure 2c). The percentage of respondents with areas of up to 50 hectares is in line with the distribution in size of properties in Brazil. Data from the last Brazilian Agricultural Census shows that 81.5% of rural establishments have less than 50 hectares, 15% between 50 and 500 hectares, and 2% above between 500 and 10,000 hectares of area, of which 46.7% work exclusively with agriculture, 48.8% with livestock, and 3.8% with forestry [39].

Half of the farmers participating in the survey use chemical inputs and controls; 43% crop rotation or pasture; 37% organic inputs and biological controls; and 24% intercropped or integrated systems, such as crop–livestock–forest integration systems or agroforestry systems (Figure 2d). The cultivation of 53 types of perennial and temporary crops was reported by farmers, the main ones being corn, beans, soybeans, coffee, sugar cane, cassava, wheat, rice, vegetables, and fruits, especially banana, orange, grape, papaya, and mango. These crops represent the main Brazilian commodities, according to crop data for the years 2019/2020 [28].

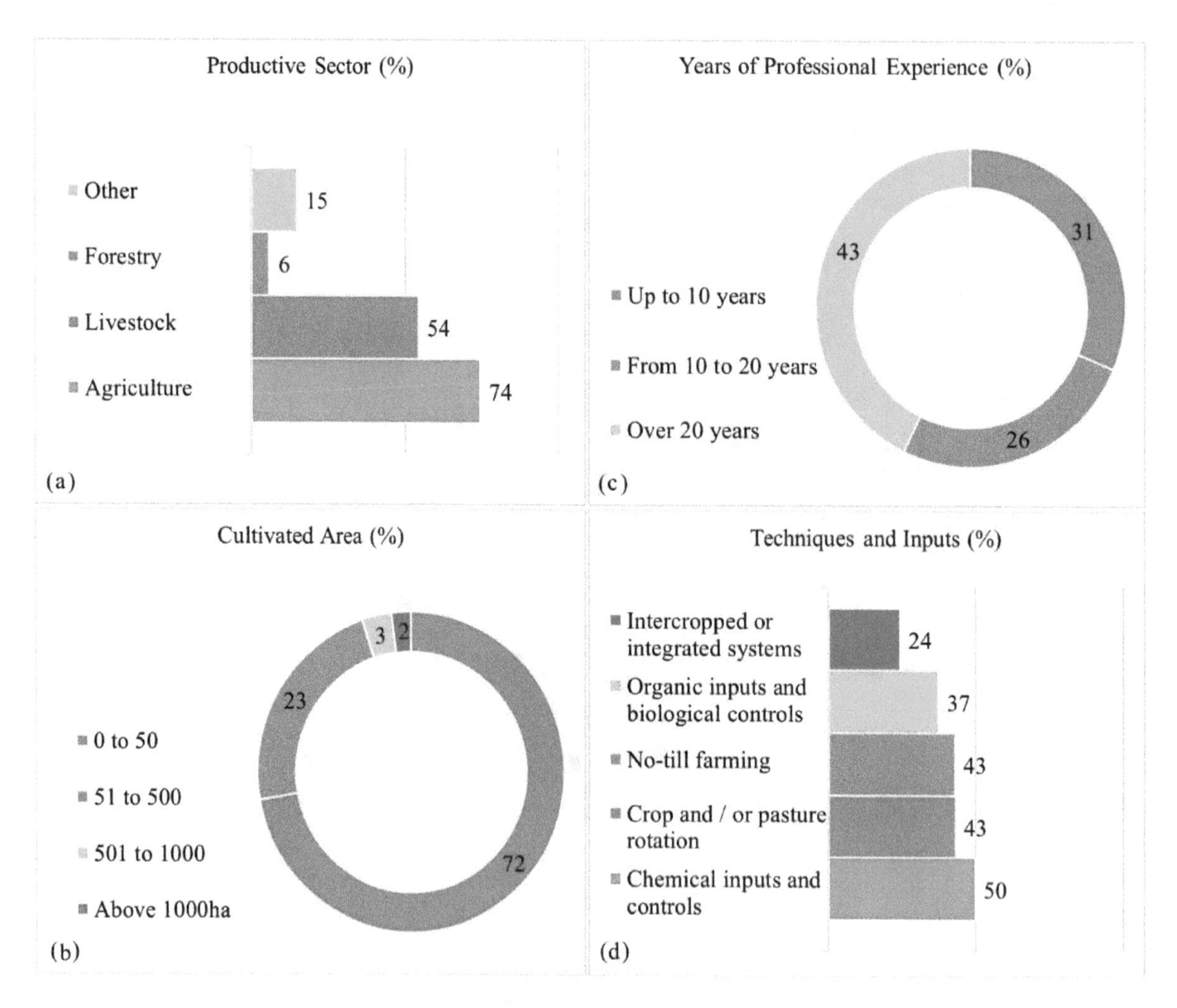

Figure 2. Profile of the Brazilian farmers that participated in the survey. Productive Sector (**a**); Cultivated Area (**b**); Years of Professional Experience (**c**); and Techniques and Inputs (**d**).

When interpreting the results presented, it is important to keep in mind that they are associated with a research by sample, which means it represents the different profiles of Brazilian farmers participating in the research. This is particularly important, considering the regional differences in relation to the socioeconomic and structural profile, where properties with larger dimensions and with a higher level of technification tend to show a higher level in the use of digital technologies. Therefore, caution must be taken when extrapolating or generalizing the results presented here to all Brazilian profiles and regions.

In reference to the digital technologies used, 15.9% of rural producers indicated that they still do not use any of these technologies; that is, 84.1% use at least one of the technologies listed in their production process (Figure 3). Among the technologies used, mainly low complexity ones, involving Internet access and connectivity/wireless on the property (70.4%), mobile applications, digital platforms, and software to obtain general information (57.5%), stand out. Reference [21], highlight that there is a growing increase in smartphone applications available to improve farmers' decision-making, where 95% of them already use a smartphone and out of these, 71% have specific applications that provide information about a specific culture, and detection and prediction of pests or diseases.

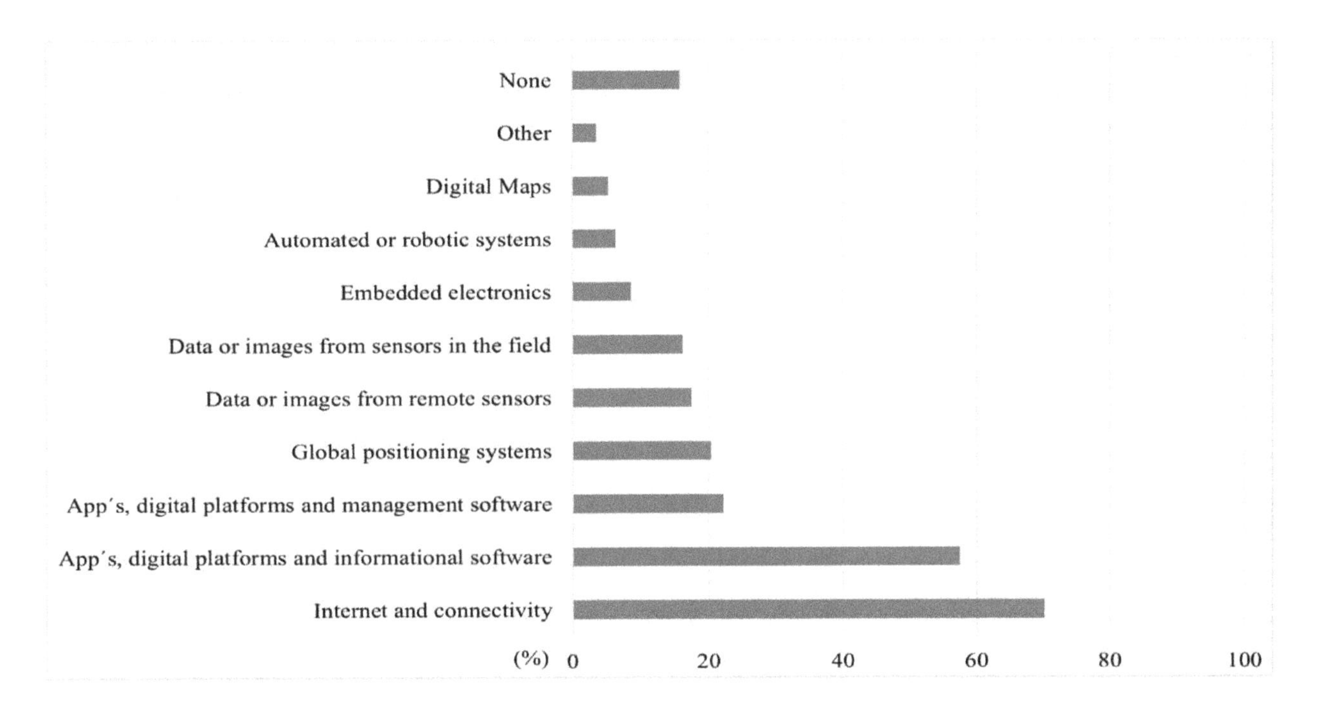

Figure 3. Precision and digital agriculture technologies used by farmers.

Medium complexity technologies, mobile applications, digital platforms, and management software (22.2%), global positioning systems (20.4%), data or images from remote sensors-satellite, airplane, UAV (17.5%), and field sensors-plant, animal, soil, water, climate, disease, or pests (16.3%) are used by a smaller portion of farmers. Here, the consolidated Global Positioning System (GPS) technology stands out, which since the beginning of the implementation of precision agriculture, has supported an increasing range of activities and generated countless benefits in rural areas. The applications of GPS in supporting soil sampling and data collection, enabling the analysis of spatial variation, improving the navigation process of machines and equipment for planting and handling crops, minimizing the use of agricultural inputs, in the possibility of working in low visibility field conditions, such as rain, dust, fog and darkness, and in supporting the generation of agricultural productivity maps and increasing the efficiency of aerial spraying [11].

The more complex technologies associated with machines or equipment with embedded electronics, e.g., autopilot, telemetry, and applications at varying rates (8.7%), automated or robotic systems (6.5%) and digital maps, e.g., spatial variability of soil, relief, harvest monitor, vegetation indexes, and productivity (5.4%), with high potential for economic return, present a relatively low percentage use by the interviewed farmers in relation to the other technologies. The greater the soil spatial variation in the property and its influence on the application of inputs, the greater the potential economic return of the application at a variable rate in comparison with the uniform applications of the traditional system [12]. That is, the variable application of inputs depends on many factors; however, the authors commit that the variability inherent in an agricultural soil and the relative responsiveness of the yield to fertilizer inputs at different levels of concentration in that soil are the most important factors to influence the economic gain.

When evaluating the use of precision agriculture technologies, specifically in the sugar cane sector in São Paulo, the main technologies pointed out were the use of satellite images (76%), autopilot systems (39%), and applications of inputs at a variable rate (29%) [29]. In the Brazilian states of Rio Grande do Sul, Paraná, Maranhão, Goiás, and Tocantins, 67% of soybean, corn, coffee, and cotton farmers used some technology in precision agriculture in the 2011/2012 harvest, with 56% using GPS mapping and 22% using remote sensors (satellite or airplane images), field sensors, and telemetry [30].

When analyzing functions of digital technologies used by farmers nowadays, a first group (between 40% and 65%) presented a broad and non-specific use of applications within production;

that is, applications focused on obtaining general information, property management, acquisition of inputs, and in trading a production or specific products (Figure 4). A second group of applications (between 20% and 40%) is focused on mapping and planning the use of land, predicting climatic risks, such as frost, hail, summer, and intense rain, depending on the Brazilian region, implementing animal welfare processes, and obtaining estimates of agricultural production and productivity. A third group of applications (below 20%), mostly related to processes with a higher technological complexity level, involves processes of detecting the control of nutritional deficit, diseases, pests, weeds, operational failures, and water deficit, in addition to applications associated with certifications and traceability of agricultural products.

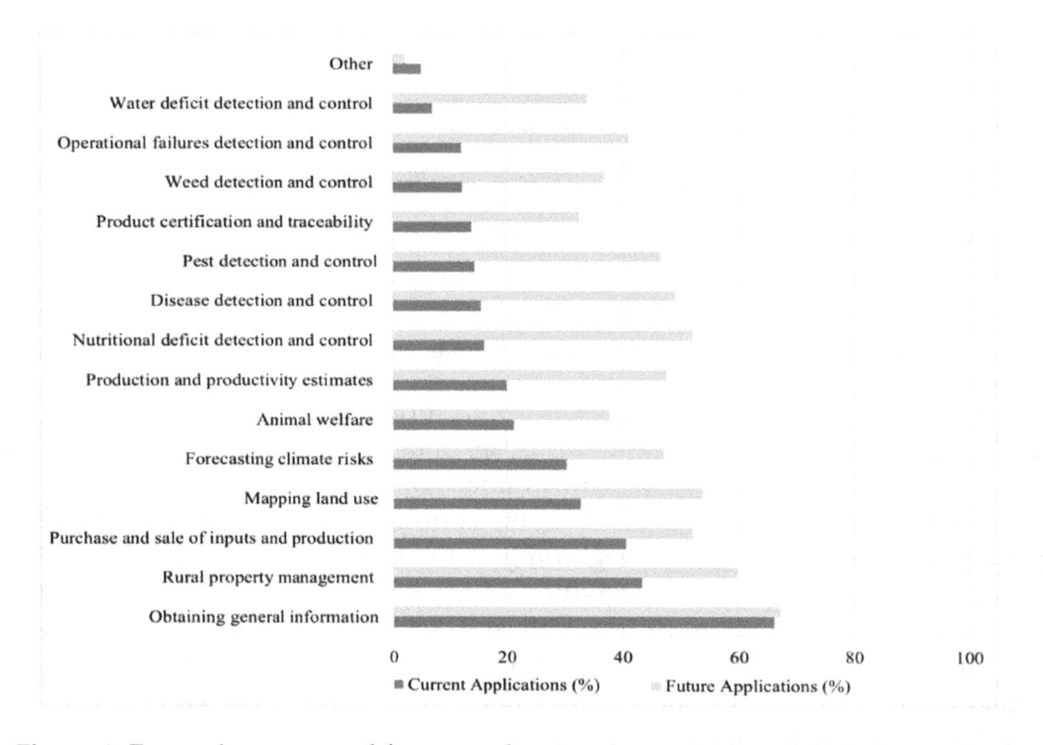

Figure 4. Farmer's current and future applications in precision and digital agriculture.

When analyzing future expectations, it was observed that farmers have an interest in initiating or strengthening the use of all the applications previously described (Figure 4). However, the most important topics in the percentage difference between current applications and future applications are in more complex activities involving the detection in the control of nutritional deficiencies (35.9%), diseases (33.5%), pests (32.2%), operational failures (29.0%), water deficit (27%), and weeds (24.8%). The applications for production and productivity estimates (27.6%) and in the mapping of land use (21.1%) are also noteworthy.

The prospect of increasing the applications of greater complexity observed is related to the growing evolution of remote sensors-orbital or field and data processing algorithms such as artificial neural networks. In remote sensing, the possibilities are highly diversified by increasing the spatial resolution, reaching the sub-metrical level, increasing the temporal resolution, with daily revisits, and better spectral resolution, with dozens of bands at different wavelengths of electromagnetic radiation. The sensors, initially focused on the visible or near infrared bands, today can range from ultraviolet to microwave, allowing even more advanced applications via hyperspectral sensors, light detection and ranging (LiDAR), fluorescence spectroscopy, and thermal spectroscopy. These conditions have driven the applications of greater complexity, allowing the evaluation of the properties of the soil and of an agricultural culture through analysis of specific compounds, molecular interactions and biophysical or biochemical characteristics [8]. There are also growing applications of spectral indexes of vegetation, such as the traditional NDVI (Normalized Difference Vegetation Index) and the EVI

(Enhanced Vegetation Index), adjusted version of NDVI with a focus on more dense vegetation, which allow management decision making of soil or agricultural crops in almost in real time.

In research with artificial neural networks to predict the harvest area, productivity, and production for soybean in Brazil by gathering data from a 1961–2016 time series, Reference [41] obtained reliable models for future predictions and support to farmers and the market in anticipating productive information. In the analysis carried [30], a large percentage of farmers in Brazil pointed to the expectation of implementing more complex and advanced technologies, such as sensor systems, to support the planting and application of fertilizers at different rates (67%), analysis, and integration of different databases in agriculture (78%).

The approach to access the technologies used by farmers was also addressed in a specific question, with farmers indicating that it occurs through direct acquisition and own use of machines, equipment and applications (68.8%); access through consultancy or services offered by associations, cooperatives, unions, NGOs (30.8%); access through consultancy or public services offered by city halls, state or federal government (20.8%); and via contracting services or consultancies specialized in digital agriculture (19.0%). Thinking about the future, it should be noted that 95% of the total respondents indicated that they were interested in receiving more information about digital agriculture and its applications.

Another aspect raised in the research was the perception of farmers about the positive impacts obtained in their production process considering the use of digital technologies aggregated in three groups: (i). images from remote sensing; (ii). field sensors, machinery, and equipment, and (iii). mobile applications, digital platforms, and software (Figure 5). All the impacts indicated were above 50% positive for farmers. This wide range of positive perceptions may be associated with the breadth of possibilities for the application of digital technologies throughout the farmer's production process. As highlights, there was a perception of increased agricultural productivity, pointed out by 64.7% of farmers, followed by perceptions of easier marketing and better planning of the daily activities of the property (62.7%); the reduction in production costs (62.3%) and the increase in the profit obtained (60.9%). The result indicates that the potential increase in agricultural productivity associated with commercialization is a key factor in managerial decision-making on the applications of digital technologies.

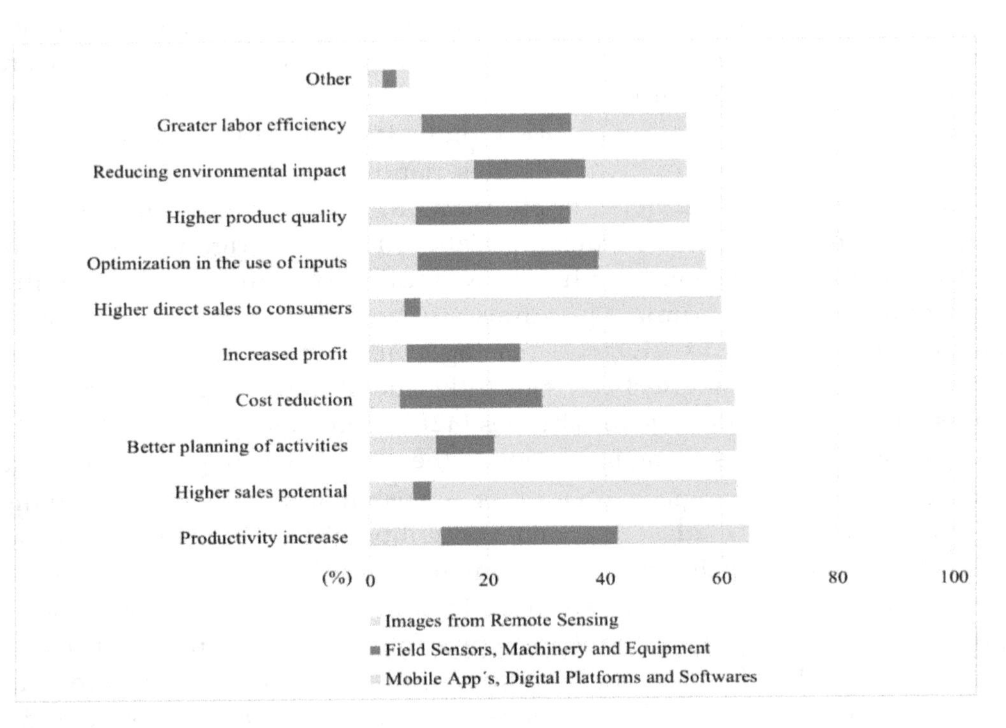

Figure 5. Farmer's perceptions about the advantages provided using precision and digital agriculture.

When assessing the perceived benefits of using four technologies in precision agriculture (variable rate fertilizer application, precision soil sampling, guidance and autosteer, and yield monitoring) with American farmers, also noted that the main perceived benefit is the increased productivity, followed by reduced production costs and greater convenience [26].

When analyzing individually the technologies adopted, in face of farmer's perception of benefits in the production system, it indicated that the use of mobile applications, digital platforms, and software is the group of technologies with the highest perception in gains, especially for increasing trade, greater direct sales with the consumer and to better plan the daily activities in the property. These results suggest that farmers are perceiving the marketing channels as an important aspect to leverage and implement a greater profit in the post-production process. An approach that is amplified by the greater smartphone dissemination in rural areas and "short marketing circuits" channels in the production based on the direct interaction between rural producer and urban consumer. In this sense, digital technologies in the management area that guide the integration of technical data with financial and management can collaborate in the decision-making of producers and in the planning of the rural business.

Field sensors, machinery, and equipment, on the other hand, was the group of technologies associated with a greater perception of positive impact on the optimization of the use of agricultural inputs, increased productivity and higher quality of products generated on the rural property. The results suggest that farmers seek to reduce production costs and/or increase yields for a more profitable system based on a better management of the agricultural input application process, probably associated with prior information from field sensors and variable rate applications, made possible by machines and equipment connected to global satellite positioning systems. Already emphasized that yield and production quality must be maximized, while the use of inputs, such as seeds, fertilizers, herbicides, and fungicides, must be optimized or minimized, pointing out the importance of using field sensors in these applications [5]. Thus, it is possible to maximize the economic return of agricultural production units through the measurement of traditional information, such as grain yield and moisture content, and new applications, such as grain protein content and straw yield.

The group that uses images from remote sensing had the lowest level of perception of positive impact in relation to the other groups, with a certain emphasis on the reduction of environmental impact. The greatest indication in the environmental issue, may be associated with the well-known and well-established use of images for mapping and monitoring land usage and coverage in Brazil. These issues are linked to the Brazilian Forest Code, which from 2012 made it mandatory to prove the maintenance of forest areas within the properties, ranging from 20% to 80%, depending on the region.

The developing new remote sensors in satellites, nanosatellites, and UAVs for agriculture have significantly enhance the potential of the applications in last year's [8–10], e.g., soil assessment, field mapping and monitoring-variable rate application, fertility, irrigation and drainage, harvest planning, and livestock monitoring. Further, new applications are developed with customized remote sensors with machine learning algorithms and computer vision to gather more accurate data and imagery. However, even with the increasing availability of images, the agricultural sector is yet to implement remote sensing technologies fully due to knowledge gaps on their sufficiency, appropriateness, and techno-economic feasibilities [42]. It may also be associated with the smaller number of farmers using remote sensing in relation to the other two groups of technologies evaluated.

Analyzing precision technologies with a focus on sustainability, Reference [43] highlights that, with its applications, the ability to identify spatial variability within the field increases and the use of this information for a more assertive crop management, operating resources with greater efficiency, making agriculture more productive, more sustainable, and more environmentally friendly. Attitudes like of confidence toward using the technologies, perceptions of net benefit, farm size, and farmer educational levels positively influenced the intention to adopt precision agriculture technologies [44].

In attempt to better understand the challenges and deterrents to the growth and expansion of digital technologies used by farmers, the survey also addressed to respondents what would be the main difficulties faced. These challenges were evaluated at the aggregate level of technologies in rural properties; that is, specific technologies were not assessed (Figure 6). Many of these challenges are related to the cost of digital technologies, be it purchase, upgrade or acquiring services. It was observed that even with the relative high use of digital technologies, indicated by the study (Figures 3 and 4), there are still important challenges or barriers, mainly associated with the investment value for the acquisition of machines, equipment or applications (67.1%); problems or lack of connection in rural areas (47.8%), the value for hiring specialized service providers (44%); and the lack of knowledge about which technologies are most appropriate to be used in their property (40.9%) (Figure 6). A similar analysis carried out in the United States indicated that the main barriers pointed out by farmers are also related to the costs associated with precision agriculture technologies and services [45].

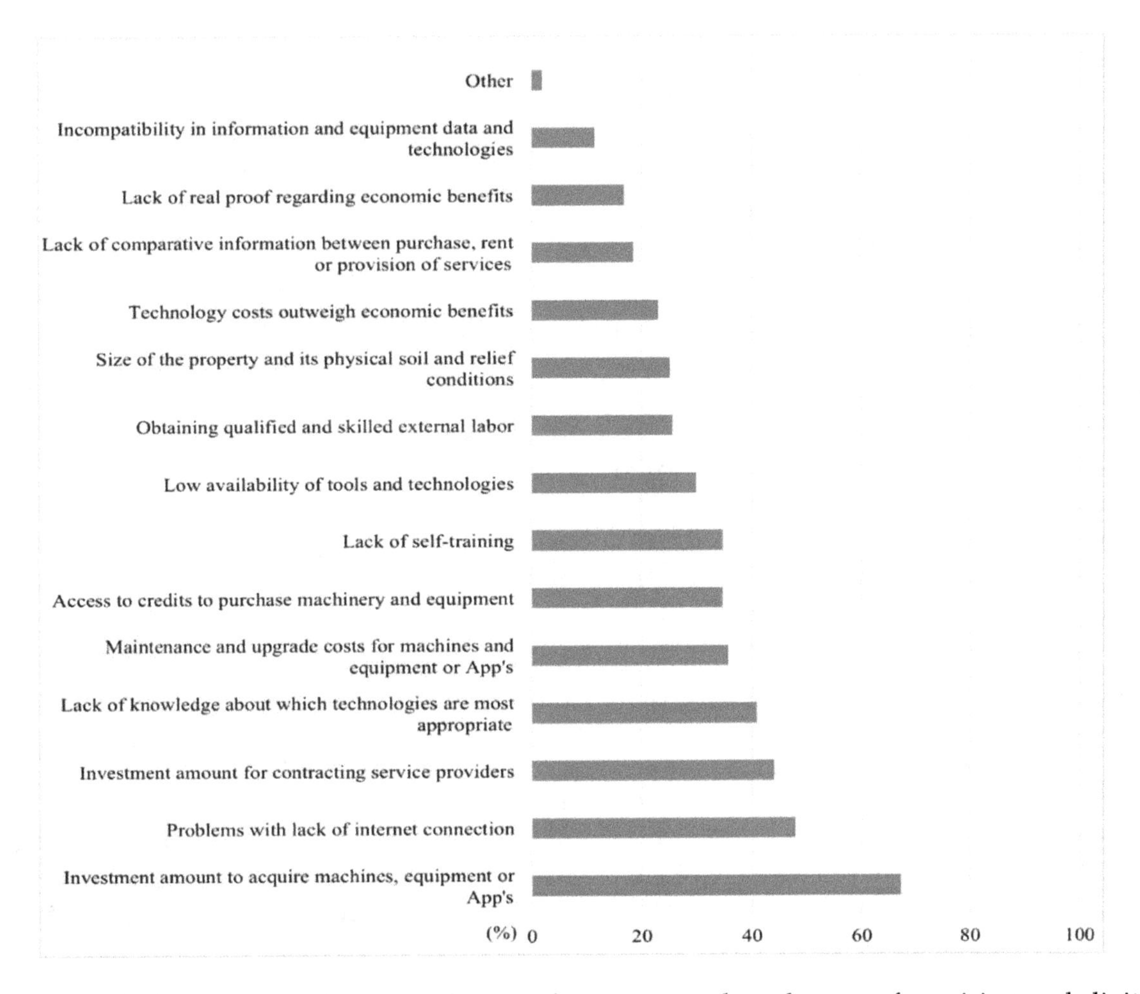

Figure 6. Difficulties and challenges farmers face to strengthen the use of precision and digital agriculture.

When analyzing the questions related to farmers investment considering only smaller farms (>20 ha), it was observed that 68.5% indicated the value for the acquisition of machines, equipment, or applications; and 45.3% indicated the value for hiring specialized service providers as the main difficulties and challenges. Thus, acquisition costs are considered a large difficulty/challenge across all farm sizes. However, it is interesting to note that for properties above 500 ha, despite large percentage indicating acquisition costs as a challenge (66%), connection issues were the main answer (69%). This difference is even more pronounced on larger properties (>1000 ha), where connection issues were indicated by 69% of the respondents and acquisition costs by 54%, still a large percentage.

The high percentage highlights difficulties in the acquisition, service value, and access to credits can be related, mainly, to the fact that currently the technologies available are mostly imported, thus indicating a significant space for investment in the national industry in digital technologies, making the acquisition process more accessible. Brazil has developed important regions in digital technologies; however, the scarcity of economic resources in many properties and the small number of public research organizations in the sector can inhibit decisions to adopt these technologies [37].

It should be noted that there are already several applications and free software on the market that can partially meet this demand. Strategic actions to make these public APPs solutions available to different production systems-animal or plant, can raise the interest and use of digital technologies even further [45,46]. In addition, the strategy of courses, training and availability of content that strengthens knowledge management can also contribute to the greater use of specific digital technologies.

Even though Brazil is among the ten main world markets for mobile telecommunications and fixed broadband, according to an analysis by the National Telecommunication Agency of Brazil [47], the last Agricultural Census pointed out that access to the Internet covers only about 30% of farmers [39]. The territorial dimensions, the low demographic density of a large part of the rural environment and the socioeconomic inequalities are some of the main obstacles to increase the access to Internet in the country. The National Bank for Economic and Social Development of Brazil [48] estimates that greater connectivity in agriculture through the Internet of Things (IoTs) could generate 50 to 200 billion dollars of annual economic impact in 2025. Thus, it is strategic that public and private actions are implemented to expand the availability of the Internet and connectivity in the field.

It is also noteworthy that only 16.9% of farmers pointed out that the lack of proof of economic benefits is a challenge to strengthen the use of digital technologies in Brazil. Result that reflects the high perception in potential impacts and future benefits of digital technologies that farmers have underscored in this research. In this sense, "The future of Smart Farming may unravel in a continuum of two extreme scenarios: (1) closed, proprietary systems in which the farmer is part of a highly integrated food supply chain or (2) open, collaborative systems in which the farmer and every other stakeholder in the chain network is flexible in choosing business partners as well for the technology as for the food production side. The further development of data and application infrastructures (platforms and standards) and their institutional embedment will play a crucial role in the battle between these scenarios" [3] (p. 69).

The main drivers of adopting digital agriculture technologies in Brazil were increased productivity, better process quality, reduced costs, and greater knowledge of cultivated areas [49]. The use of technologies such as GPS, field sensors, remote sensors and telemetry could grow rapidly in areas with higher value-added crops, such as citrus and sugar cane, in States with higher land value, and regions with a strong base of Brazilian agricultural research organizations [37]. Some indicators suggest that the availability of sensors, mapping technology, and tracking technologies have changed many farming systems and hat these technologies will lead to relevant analysis at every stage of the agricultural value chain-from producers to consumers [50].

The rapid development of the Internet services, and, consequently, the greater number of instruments and devices connected to the network, increased the amount of data produced and collected every day in rural areas. The land structure with many small, medium, and large-scale Brazilian properties, and the relief conditions, especially the Cerrado Biome and the news forms of market access directly with the consumer should further strengthen the use of digital technologies in the medium term and support Brazilian rural development.

In addition to the technologies already consolidated in precision agriculture in Brazil, such as field sensors, remote sensors, embedded electronics, telemetry and automation, digital agriculture also involves the use of mobile apps, digital platforms, social networks, big data, Internet of Things, artificial intelligence, cloud computing, blockchain and cryptography, allowing to support greater sustainability of production systems. The use of digital technologies, underscored by Brazilian farmers, are differentiated and currently cover a wide spectrum, mainly in obtaining information about property management, product sales, inputs acquisition, property mapping and climate forecasting, presenting future interests in a greater use in production estimation, detection and control in factors related to soil physic-biotics factors, and other stresses of the agricultural crops.

It is important that the public and private sectors evaluate their forms of aid to enhance the use of these digital technologies, since a large proportion of farmers, especially those with lower technical skills, still need support and technical monitoring to adapt and become more familiar with the use and applications of these technologies. Moreover, it is essential to structure a support process to enable the producer's decision making by presenting the costs and benefits in the acquisition and use of such technologies.

4. Conclusions

The digital agriculture can be understood as a set of technologies for communication, information, and analysis that allows farmers to plan, monitor, and manage the operational and strategic activities of the agriculture production systems, from pre-production, in production, and post-production. The results observed in the survey indicate that 84% of the Brazilian farmers interviewed use at least one digital technology in their production system, and this percentage decreases as the level of the application's technological complexity increases. Connectivity, mobile apps, digital platforms, software, global satellite positioning systems, remote sensing, and field sensors are the main technologies used.

Brazilian farmers' main perceptions in relation to benefits are linked to increased productivity, product higher selling potential, better planning, and management of production systems, 95% of farmers would like to learn more about new technologies and strengthen its applications. The use of such digital technologies has the potential to increase the sustainable management of natural resources (soil and water) and to reduce the use of agricultural inputs, making agricultural areas more productive and reducing their environmental impact. However, important difficulties to amplify its use were pointed out. The cost of purchasing machines, equipment and applications, problems with/or lack of connectivity in rural areas were the main issues presented. It is suggested that future studies focus on the new research questions from this work, more specifically on farmers' social and economic behavior towards adoption of new technologies and whether the adopters have a competitive advantage compared to non-adopters. Further research could also consider the way technology adoption will occur by the new generation, accordingly to the level of Internet access in the rural areas and hoping that they will continue to run the family business.

Author Contributions: Conceptualization, É.L.B., L.A.d.C.J., I.D.S. and A.L.J.; Methodology, É.L.B., L.A.d.C.J., I.D.S. A.L.J., C.C.d.C. and R.Y.I.; Software, C.C.d.C., D.d.C.V. and É.L.B.; Validation, É.L.B and A.L.J.; Formal Analysis, É.L.B., L.A.d.C.J., I.D.S., A.L.J., C.C.d.C., D.d.C.V., R.Y.I., C.R.G., V.R.F. and A.R.R.; Investigation, É.L.B., L.A.d.C.J., I.D.S. and A.L.J.; Resources, É.L.B., V.R.F. and A.R.R.; Data Curation, C.C.d.C. and D.d.C.V.; Writing—Original Draft Preparation, É.L.B., L.A.d.C.J., I.D.S. A.L.J. and D.d.C.V.; Writing—Review & Editing, É.L.B., C.C.d.C., D.d.C.V., R.Y.I., C.R.G., V.R.F. and A.R.R.; Visualization, É.L.B.; Supervision; Project Administration and Funding Acquisition, É.L.B. and V.R.F. All authors have read and agreed to the published version of the manuscript.

References

1. Food and Agriculture Organization (FAO). Digital Agriculture. Available online: http://www.fao.org/digital-agriculture/en/ (accessed on 10 July 2020).
2. Rose, D.C.; Chilvers, J. Agriculture 4.0: Broadening responsible innovation in an era of smart farming. *Front. Sustain. Food Syst.* **2018**, *2*, 87. [CrossRef]
3. Wolfert, S.; GE, L.; Verdouw, C.; Bogaardt, M.J. Big Data in Smart Farming—A review. *Agric. Syst.* **2017**, *153*, 69–80. [CrossRef]
4. Wolf, S.; Buttel, F. The Political Economy of Precision Farming. *Am. J. Agric. Econ.* **1996**, *78*, 1269–1274. [CrossRef]
5. Reyns, P.; Missotten, B.; Ramon, H.J.; De Baerdemaeker, J. A review of combine sensors for precision farming. *Precis. Agric.* **2002**, *3*, 169–182. [CrossRef]
6. Adamchuk, V.I.; Hummel, J.W.; Morgan, M.T.; Upadhyaya, S.K. On-the-go soil sensors for precision agriculture. *Comput. Electron. Agric.* **2004**, *44*, 71–91. [CrossRef]

7. Kayad, A.; Paraforos, D.S.; Marinello, F.; Fountas, S. Latest Advances in Sensor Applications in Agriculture. *Agriculture* **2020**, *10*, 362. [CrossRef]
8. Mulla, D.J. Twenty five years of remote sensing in precision agriculture: Key advances and remaining knowledge gaps. *Biosyst. Eng.* **2013**, *114*, 358–371. [CrossRef]
9. Mogili, U.; Deepak, B. Review on application of drone systems in precision agriculture. *Procedia Comput. Sci.* **2018**, *133*, 502–509. [CrossRef]
10. Oliveira, M.V.N.; Broadbent, E.N.; Oliveira, L.C.; Almeida, D.R.A.; Papa, D.A.; Ferreira, M.E.; Zambrano, A.M.A.; Silva, C.A.; Avino, F.S.; Prata, G.A.; et al. Aboveground Biomass Estimation in Amazonian Tropical Forests: A Comparison of Aircraft- and GatorEye UAV-borne LiDAR Data in the Chico Mendes Extractive Reserve in Acre, Brazil. *Remote Sens.* **2020**, *12*, 1754. [CrossRef]
11. Official U.S. Government Information about the Global Positioning System (GPS) and Related Topics. Available online: https://www.gps.gov/applications/agriculture/ (accessed on 8 September 2020).
12. Suciu, G.; Marcu, I.; Balaceanu, C.; Dobrea, M.; Botezat, E. Efficient IoT system for precision agriculture. *Int. Conf. Eng. Mod. Electr. Syst.* **2019**, 173–176. [CrossRef]
13. Havlin, J.L.; Heiniger, R.W. A variable-rate decision support tool. *Precis. Agric.* **2009**, *10*, 356–369. [CrossRef]
14. McKinion, J.M.; Turner, S.B.; Willers, J.L.; Read, J.J.; Jenkins, J.N.; McDade, J. Wireless technology and satellite internet access for high-speed whole farm connectivity in precision agriculture. *Agric. Syst.* **2004**, *81*, 201–212. [CrossRef]
15. Lin, Y.P.; Petway, J.R.; Anthony, J.; Mukhtar, H.; Liao, S.-W.; Chou, C.-F.; Ho, Y.-F. Blockchain: The Evolutionary Next Step for ICT E-Agriculture. *Environments* **2017**, *4*, 50. [CrossRef]
16. Chen, Y.; Li, Y.; Li, C. Electronic agriculture, blockchain and digital agricultural democratization: Origin, theory and application. *J. Clean. Prod.* **2020**, *268*, 122071. [CrossRef]
17. Liakos, K.G.; Busato, P.; Moshou, D.; Pearson, S.; Bochtis, D. Review machine learning in agriculture: A Review. *Sensors* **2018**, *18*, 2674. [CrossRef]
18. Kamilaris, A.; Prenafeta-Boldú, F.X. Deep learning in agriculture: A survey. *Comput. Electron. Agric.* **2018**, *147*, 70–90. [CrossRef]
19. Castro, W.; Junior, J.M.; Polidoro, C.; Osco, L.P.; Gonçalves, W.; Rodrigues, L.; Santos, M.F.; Jank, L.; Barrios, S.C.L.; Resende, R.M.S.; et al. Deep learning applied to phenotyping of biomass in forages with UAV-based RGB imagery. *Sensors* **2020**, *20*, 4802. [CrossRef]
20. Verdouw, C.; Sundmaeker, H.; Tekinerdogan, B.; Conzon, D.; Montanaro, T. Architecture framework of IoT-based food and farm systems: A multiple case study. *Comput. Electron. Agric.* **2019**, *165*, 104939. [CrossRef]
21. Michels, M.; Bonke, V.; Musshoff, O. Understanding the adoption of herd management smartphone apps. *J. Dairy Sci.* **2019**, *102*, 9422–9434. [CrossRef]
22. Michels, M.; Bonke, V.; Musshoff, O. Understanding the adoption of smartphone apps in crop protection. *Precis. Agric.* **2020**, *21*, 1209–1226. [CrossRef]
23. Ampatzidis, Y.; Partel, V.; Costa, L. Agroview: Cloud-based application to process, analyze and visualize UAV-collected data for precision agriculture applications utilizing artificial intelligence. *Comput. Electron. Agric.* **2020**, *174*, 105457. [CrossRef]
24. Bongiovanni, R.; Lowenberg-DeBoer, J. Precision agriculture and sustainability. *Precis. Agric.* **2004**, *5*, 359–387. [CrossRef]
25. Saiz-Rubio, V.; Rovira-Más, F. From smart farming towards agriculture 5.0: A review on crop data management. *Agronomy* **2020**, *10*, 207. [CrossRef]
26. Thompson, N.; Bir, C.; Widmar, D.A.; Mintert, J.R. Farmer perceptions of precision agriculture technology benefits. *J. Agric. Appl. Econ.* **2018**, *51*, 142–163. [CrossRef]
27. Cambra Baseca, C.; Sendra, S.; Lloret, J.; Tomas, J. A Smart Decision System for Digital Farming. *Agronomy* **2019**, *9*, 216. [CrossRef]
28. Ministério da Agricultura, Pecuária e Abastecimento. Brasil Projeções do Agronegócio 2020 a 2030. Available online: https://www.gov.br/agricultura/pt-br/assuntos/politica-agricola/todas-publicacoes-de-politica-agricola/projecoes-do-agronegocio/projecoes-do-agronegocio_2019_20-a-2029_30.pdf/view (accessed on 10 September 2020).
29. Silva, C.B.; De Moraes, M.A.F.D.; Molin, J.P. Adoption and use of precision agriculture technologies in the sugarcane industry of São Paulo state, Brazil. *Precis. Agric.* **2011**, *12*, 67–81. [CrossRef]

30. Borghi, E.; Avanzi, J.R.C.; Bortolon, L.; Luchiari, J.R.A.; Bortolon, E.S.O. Adoption and use of precision agriculture in Brazil: Perception of growers and service dealership. *J. Agric. Sci.* **2016**, *8*, 89–104. [CrossRef]
31. Bernardi, A.C.C.; Naime, J.M.; Resende, Á.V.; Bassoi, L.H.; Inamasu, R.Y. *Agricultura de Precisão: Resultados de um Novo Olhar*, 1st ed.; Embrapa: Brasília, Brazil, 2014; p. 596. Available online: http://ainfo.cnptia.embrapa.br/digital/bitstream/item/113993/1/Agricultura-de-precisao-2014.pdf (accessed on 5 September 2020).
32. Zhai, Z.; Martinez, J.F.; Beltran, V.B.; Martinez, N.L. Decision support systems for agriculture 4.0: Survey and challenges. *Comput. Electron. Agric.* **2020**, *170*, 105256. [CrossRef]
33. Klerkx, L.; Jakku, E.; Labarthe, P. A review of social science on digital agriculture, smart farming and agriculture 4.0: New contributions and a future research agenda. *Wagening. J. Life Sci.* **2019**, *90*, 100315. [CrossRef]
34. Larson, J.A.; Roberts, R.K.; English, B.C.; Larkin, S.L.; Marra, M.C.; Martin, S.W.; Paxton, K.W.; Reeves, J.M. Factors affecting farmer adoption of remotely sensed imagery for precision management in cotton production. *Precis. Agric.* **2008**, *9*, 195–208. [CrossRef]
35. Reichardt, M.; Jürgens, C. Adoption and future perspective of precision farming in Germany: Results of several surveys among different agricultural target groups. *Precis. Agric.* **2009**, *10*, 73–94. [CrossRef]
36. Zervopoulos, A.; Tsipis, A.; Alvanou, A.G.; Bezas, K.; Papamichail, A.; Vergis, S.; Stylidou, A.; Tsoumanis, G.; Komianos, V.; Koufoudakis, G.; et al. Wireless Sensor Network Synchronization for Precision Agriculture Applications. *Agriculture* **2020**, *10*, 89. [CrossRef]
37. Griffin, T.W.; Lowenberg-DeBoer, J. Worldwide adoption and profitability of precision agriculture implications for Brazil. *Rev. Política Agríc.* **2005**, *14*, 20–37. Available online: https://seer.sede.embrapa.br/index.php/RPA/article/view/549/498 (accessed on 15 September 2020).
38. LimeSurvey Manual. Available online: https://manual.limesurvey.org/LimeSurvey_Manual (accessed on 26 February 2020).
39. Instituto Brasileiro de Geografia e Estatística (IBGE). Censo Agropecuário: Resultados Definitivos-2017. Available online: https://biblioteca.ibge.gov.br/visualizacao/periodicos/3096/agro_2017_resultados_definitivos.pdf (accessed on 26 September 2020).
40. Instituto Brasileiro de Geografia e Estatística (IBGE). Valor Bruto da Produção Agrícola-2019. Available online: https://sidra.ibge.gov.br/pesquisa/lspa/referencias (accessed on 19 September 2020).
41. Abraham, E.R.; dos Reis, J.G.M.; Vendrametto, O.; de Costa Neto, P.O.; Toloi, R.C.; Souza, A.E.; De Oliveira Morais, M. Time Series Prediction with Artificial Neural Networks: An Analysis using Brazilian Soybean Production. *Agriculture* **2020**, *10*, 475. [CrossRef]
42. Khanal, S.; KC, K.; Fulton, J.P.; Shearer, S.; Ozkan, E. Remote Sensing in Agriculture—Accomplishments, Limitations, and Opportunities. *Remote Sens.* **2020**, *12*, 3783. [CrossRef]
43. Lowenberg-DeBoer, J. The economics of precision agriculture. In *Precision Agriculture for Sustainability*, 1st ed.; Sttaford, J., Ed.; Burleigh Dodds Science Publishing Limited: Cambridge, UK, 2019; pp. 461–482. [CrossRef]
44. Adrian, A.M.; Norwood, S.H.; Mask, P.L. Producers' perceptions and attitudes toward precision agriculture technologies. *Comput. Electron. Agric.* **2005**, *48*, 256–271. [CrossRef]
45. Erickson, B.; Deboer, J.L. *Precision Agriculture Dealership Survey-2020*; Purdue University: West Lafayette, IN, USA, 2020; p. 30. Available online: https://ag.purdue.edu/digital-ag-resources/wp-content/uploads/2020/03/2019-CropLife-Purdue-Precision-Survey-Report-4-Mar-2020-1.pdf (accessed on 16 September 2020).
46. Brazilian Agricultural Research Corporation (EMBRAPA). Mobile APP. Available online: https://play.google.com/store/apps/developer?id=Embrapa (accessed on 26 September 2020).
47. National Telecommunication Agency. Relatório e Comparação Internacional. Available online: https://www.anatel.gov.br/dados/relatorios-de-acompanhamento (accessed on 10 September 2020).
48. National Bank for Economic and Social Development. Internet das Coisas. Available online: https://www.bndes.gov.br/wps/portal/site/home/conhecimento/pesquisaedados/estudos/estudo-internet-das-coisas-iot/estudo-internet-das-coisas-um-plano-de-acao-para-o-brasil (accessed on 10 September 2020).
49. Pivoto, D.; Barham, B.; Dabdab, P.; Zhang, D.; Talamini, E. Factors influencing the adoption of smart farming by Brazilian grain farmers. *Int. Food Agribus. Manag. Rev.* **2019**, *22*, 571–588. [CrossRef]
50. Coble, K.H.; Mishra, A.K.; Ferrell, S.; Griffin, T. Big data in agriculture: A challenge for the future. *Appl. Econ. Perspect. Policy* **2018**, *40*, 79–96. [CrossRef]

2

Sourdough Technology as a Tool for the Development of Healthier Grain-Based Products: An Update

Juan Fernández-Peláez [1], Candela Paesani [1,2] and Manuel Gómez [1,*]

1 Food Technology Area, College of Agricultural Engineering, University of Valladolid, 34004 Palencia, Spain; juan.fernandez.pelaez@uva.es (J.F.-P.); candepaesani@agro.unc.edu.ar (C.P.)
2 Área de Mejora Nutricional y Alimentos Nutricionales, Instituto de Ciencia y Tecnología de Alimentos Córdoba (ICYTAC-CONICET-UNC), X5016BMB Córdoba, Argentina
* Correspondence: pallares@iaf.uva.es

Abstract: There has been growing demand by consumers for grain-based products with well-balanced nutritional profiles and health-promoting properties. The components of the flours obtained from different grains can be modified or improved at a nutritional level by using sourdough technology, which has gained increasing interest in recent years. Sourdough hydrolyse dietary fibre, reduces fat rancidity, and enables an increase in starch and protein digestibility, as well as vitamin levels and mineral bioavailability. In addition, bioactive compounds are synthesized during fermentation, while components that interfere with the digestion of grain-based products or digestion-linked pathologies, such as gluten sensitivity or gastrointestinal syndromes, are reduced. Finally, it has been observed that sourdough fermented products can play a role in gut microbiota regulation. Thanks to this health-promoting potential, sourdough can stand out among other fermentation processes and opens up a new range of healthier commercial products to be developed. The current review discusses the extensive research carried out in the last 15 years and aims at updating and deepening understanding on how sourdough can enhance the nutritional and health-related characteristics of the different components present in the grains.

Keywords: sourdough; fermentation; nutrition; bread making; microbiota

1. Introduction

Sourdough is the result of the fermentation of flour from cereals and pseudocereals or legumes, among others, by the action of the microorganisms present in the preparation [1]. Some sourdoughs can also incorporate added microorganisms. Therefore, sourdoughs can be defined as stable ecosystems composed of the lactic acid bacteria (LAB) and yeasts used in the production of bakery products [2]. Traditionally, sourdough was used as a leavening agent, but today, it is increasingly being used to improve organoleptic characteristics and to reduce the need for additives [3].

Sourdough provides multiple benefits to the quality of the products obtained and it allows extending of the life of the bakery products [4]. This increased shelf life is mainly due to the lowering of pH, which induces the inhibition of microbial development, and to the decomposition of starch during lactic acid fermentation, resulting in staling delay. The use of sourdough also makes it possible to obtain products with greater aroma and sweetness, due to the hydrolysis processes and the compounds generated in the Maillard reaction during the baking process [5]. However, the application of sourdough can also lead to complications, such as the increased workload required or the difficulty in achieving homogeneous productions [6].

The modification of cereal and grain nutrients and flours by the microorganisms contained in the sourdough to improve bakery products has gained great interest from researchers in recent years.

Consequently, the number of publications addressing the interrelationship between sourdough and the different components of bread has increased significantly in recent years (Figure 1a). Similarly, recently published reviews, such as Sakandar et al. (2019) [7], highlight the processing benefits provided by this technology, focusing on bread making.

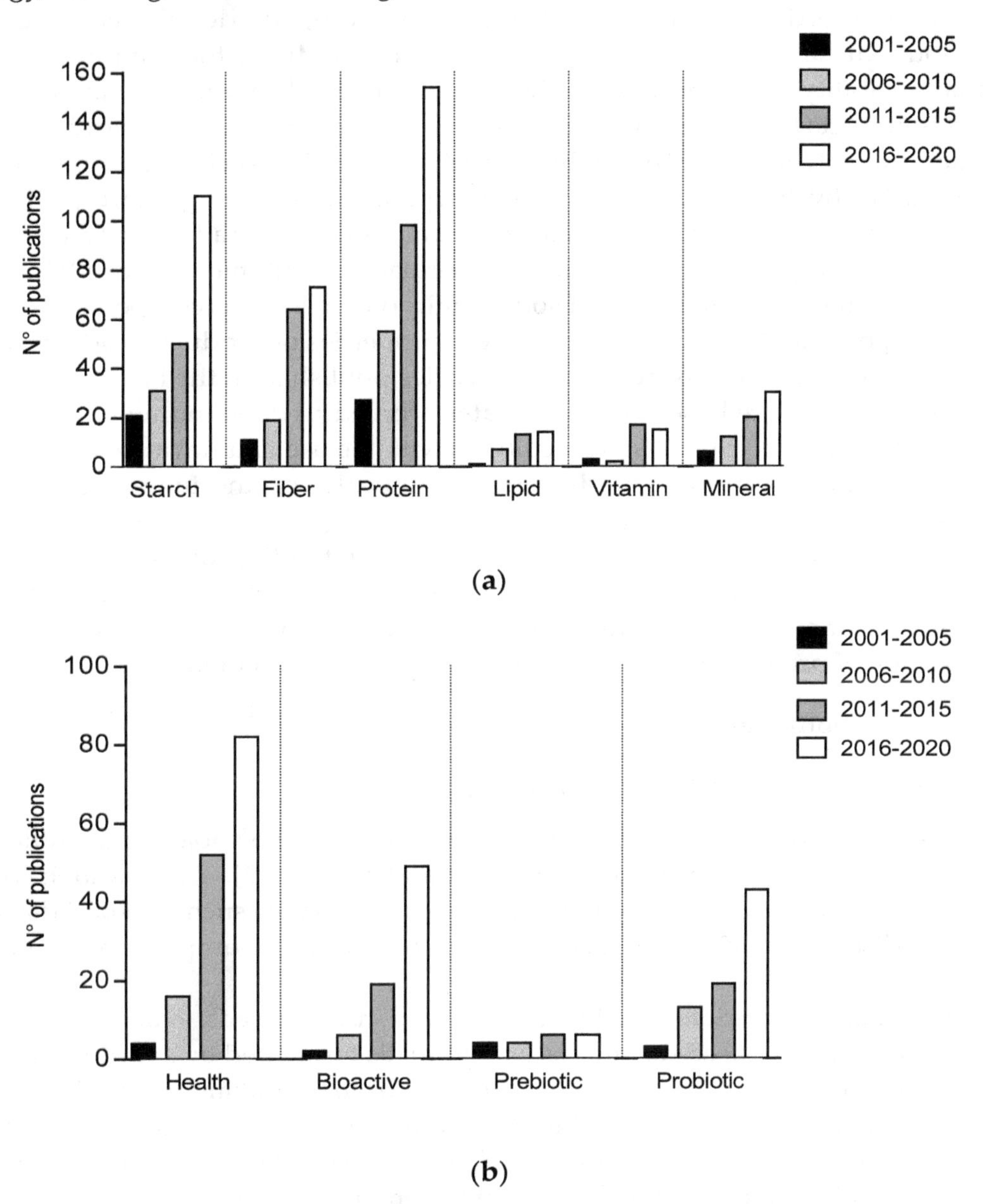

Figure 1. Number of publications about sourdough over the period 2001–2020, separated in 5-year periods, according to the Scopus database (last accessed 30 June 2020). (**a**) Publications on sourdough according to the relation with sourdough components. Search: "sourdough AND starch/fibre/protein/lipid/vitamin/mineral". (**b**) Publications on the sourdough-health relationship, and functioning components. Search: "sourdough AND health/bioactive/prebiotic/probiotic".

It has been observed that sourdoughs allow a reduction in gluten content, which may be particularly interesting for lowering the risks of gluten contamination in the case of gluten-free products [8,9]. Even if this does not involve a nutritional improvement, sourdough can contribute as well to improve the organoleptic quality of the gluten-free products, expanding the offer of products suitable for celiac patients or those allergic to wheat [10–12]. However, it is necessary to be cautious of this possibility, since the complete elimination of toxicity for people with celiac disease implies important modifications in the gluten network, affecting gluten functionality and bread quality.

Some studies show that the use of sourdough enables an increase in the sensation of satiety that the breads generate when they are consumed as well as in the postprandial insulinemic response [13]. In addition, other research studies point to a greater digestibility of sourdough-fermented cereal products, compared to yeast-fermented ones [14,15].

Although most sourdoughs are made from wheat flour, for doughs based on legume or pseudocereal flours, it has been shown that this technique reduces the antinutrient content and increases the acceptability of the products. This has also been observed with the incorporation of bran or other by-products [16,17]. Compounds such as fermentable oligosaccharides, disaccharides, monosaccharides, and polyols (FODMAP), which lie behind disorders such as irritable bowel syndrome (IBS), can be diminished by the action of the fermentation process of sourdough [18,19]. Other innovative approaches include the use of sourdough to reduce the amount of sugar added to bakery products [20] and the ability of sourdough-fermented products to regulate the gut microbiome [21].

Despite reviews concerning the effect of sourdoughs on the various components of flour [22,23], there is still a great potential for improving the health properties of products made from sourdough. This is evidenced by the high number of research articles published in the last 15 years on the topic (Figure 1b). Many of them are based on this greater accrued understanding of microbial ecology, which has enabled a better selection of starter cultures with the use of innovative techniques [1,24,25]. Other articles focus on the nutritional modifications and health-promoting potential of sourdough fermentation, as will be discussed hereafter.

The aim of this review is to update the current understanding of the relationship between fermentation with sourdough, changes in the chemical composition of the products obtained, and the nutritional improvement of cereal and grain products. In addition, we will address the health benefits that the consumption of products with this type of fermentation can entail.

2. Sourdough and Carbohydrates

2.1. *Influence of Sourdough Fermentation on Starch Digestibility*

Starch digestibility after sourdough fermentation has been the subject of significant research, probably because starch is the main fraction in grains and cereals [22]. This research covers aspects such as glycaemic response or its effect on chronic metabolic diseases, such as type 2 diabetes, obesity, or cardiovascular disease, which have been linked to the high consumption of rapidly digestible starch [26].

Sourdough fermentation reduces the insulinemic response of breads. Nordlund et al. (2016) [27] explain this in terms of the reduced disintegration of breads fermented by sourdough. Furthermore, different studies have shown that the addition of sourdoughs to the baking process helps diminish the glycaemic index (GI) of wheat bread, whether it is made with white, wholemeal, or fibre-enriched flour [28–34]. This is related to the lower pH generated by fermentation before gelatinisation of the starch, as will be explained later. While the consumption of wholemeal bread has been associated with a decrease in glycaemic index as compared with white flour breads, this index is reduced even further when sourdough is used [35,36]. The glycaemic response to bread, however, is specific to each person [37]. The reduced digestibility of starch has been observed not only in bread [38], but also in products such as pasta made from sourdough, due to the higher level of retrograded starch that is inaccessible to enzymatic degradation [39,40]. However, in general, the beneficial effect of sourdough on the GI is attributed to the production of organic acids, especially lactic and acetic acid. These acids represent major fermentation end-products during sourdough fermentation [41]. Lactic acid production by LAB in sourdough-fermented bread reduces postprandial glycaemic and insulinemic responses in healthy adults [14,33,42,43]. The addition of lactic acid is effective to lower the postprandial glucose and insulin responses in humans, when done before heat application, and therefore, before the gelatinisation of the starch, which reduces its bioavailability [44]. In turn, acetic acid produces a delayed gastric emptying rate, but is not required to be present before the gelatinisation of the starch [45].

This suggests that the action of the sourdoughs may depend on the lactic/acetic ratio generated by them. De Angelis et al. (2007) [28] explain that lactic acid is more effective than acetic acid in reducing starch hydrolysis.

In this regard, to reduce starch digestibility, acidification techniques are currently being applied to some products, which promote the interaction between starch and proteins, gluten in particular, and reduce the bioavailability [46]. In the literature, it is stated that the reduction of compounds such as amylase and trypsin inhibitors also facilitates the digestion of products made from wheat flour fermented with sourdough [47]. Another aspect to consider is the increased formation of resistant starch, and the consequent lower digestibility due to the increment of organic acid in the fermentation of the dough [48]. It has been shown that digestibility also decreases if sourdough fermentation is combined with subsequent application of freezing, with reductions in the GI of the products by over 40% [49].

Finally, considering the effect of sourdough over the GI of breads, the type of flour has to be taken into account. Despite GI reduction in wheat breads when sourdough is used, sourdough fermentation showed an increase in the estimated GI of gluten-free breads made with buckwheat flour, quinoa, and teff [34,50]. According to Giuberti and Gallo (2018) [51], this could be because the use of pseudocereal flours, with or without sourdough, does not always guarantee an increase in slowly digestible starch. Moreover, Wolter et al. (2013) [52] suggest that it can be explained in terms of the higher GI of the flours with smaller starch granule size. Nevertheless, studies on gluten-free breads are still limited and should be interpreted considering the processing parameters.

2.2. *Interaction between Sourdough Fermentation and Dietary Fibre*

Currently, there is a tendency to look for new sources of dietary fibre (DF) as a functional component for the food industry, because of its potential to promote numerous beneficial physiological effects. Dietary fibre contributes to lower blood cholesterol and glucose levels, plays an important role in the health-promoting effect, and there is evidence of a strong association between chronic disease and obesity and a poor dietary fibre intake [53]. However, these effects depend on the type of fibre. The whole grain dietary fibre complex is an important source of beneficial molecules for the host, such as β-glucans, fructans, resistant starch, and arabinoxylans. The bioavailability of these compounds in whole grain products generally depends on different technological issues [54]. It is known that the enrichment of breads with DF requires certain adjustments of several process parameters to obtain a high quality that is acceptable to most consumers; one option for improving it is the use of sourdough biotechnology [55]. When a sourdough process is adopted, the fibres change their chemical and physical properties according to the degree of fermentation [54].

Enzymatic action can change the ratio of insoluble fibres to soluble ones during bread making [56]. In sourdough breads, fibre can undergo two types of enzymatic hydrolysis. When flour is hydrated, several hydrolytic enzymes intrinsic to cereals are activated [57], such as hemicellulase enzymes that degrade hemicelluloses. On the other hand, lactic acid bacteria release enzymes with glycolytic activity [58], which are also capable of acting on the fibre present in the dough. Recently, the effects of *Lactobacillus plantarum* fermentation were reported on β-glucans of barley [59]. Fermentation also decreased the molecular weight of β-glucans, which could have a negative impact on the physiological activities of barley β-glucan, especially in relation to glucose and lipid metabolism.

Different studies have been carried out on bread made using the sourdough technique for its enrichment with dietary fibre. Mihhalevski et al. (2013) [60] evaluated dietary fibre in a rye sourdough bread and concluded that the proportion of both soluble and insoluble dietary fibre increased during rye sourdough processing. They attributed this difference to biochemical and microbiological processes that occur during sourdough bread making. The increase in total dietary fibre content was caused by the formation of resistant starch, and the increase in soluble dietary fibre content can be explained by the conversion of insoluble to soluble fibre during fermentation of rye flour. Such a redistribution of dietary fibre formed by the activities of the intrinsic enzymes in rye flour (amylase, xylosidase, arabinofuranosidase, b-glucanase, endo-xylanase, and cinnamoyl esterase) was also suggested by Boskov Hansen et al. (2002) [57].

De Angelis et al. (2009) [29] used sourdough to improve the organoleptic quality of breads enriched with fibres and to increase the reduction in the glycaemic index achieved with this enrichment. A sourdough elaboration process, with the appropriate selection of lactic acid bacteria according to the added fibre [29], can be considered one of the tools to make low GI bread. Moreover, the breads' specific volume was higher and the sensory analysis indicated that they were preferred for their acid smell, taste, and aroma.

The possibility of using naked barley as a food product is gaining popularity because of its dietary fibre, especially β-glucans. Pejcz et al. (2017) [61] reported that the fermentation of sourdough breads with barley improved the volume, colour, and sensory properties of breads with this cereal. The fermentation process resulted in a higher concentration of dietary fibre, arabinoxylans, and β-glucans. In another study, sourdoughs made from barley flour without sole husks, or from a mixture of 50 g/100 g barley and wheat flour, were characterized from a microbiological and technological point of view, compared to a single wheat flour from sourdough [62]. Results showed that the use of sourdough can be a strategy to improve the quality of barley bread with higher nutritional value and high DF. Furthermore, despite the lower specific volume and higher density of barley breadcrumbs compared to wheat bread, no significant differences were observed after baking or during the shelf life, thus confirming the possibility of successful exploitation of barley flour in the baking industry.

It has also been shown that the fermentation of sourdough leads to an increased activity of certain enzymes, leading to solubility of arabinoxylans expansion [63], just as sourdough was shown to improve dough and bread structure quality of breads containing whole grains of barley, but little effect on breads prepared with refined barley flour [64]. This positive effect on bread quality could be due to a softening effect on bran particles during fermentation, which would result in less impediment of the gluten network and in the formation of gas cells in the dough [64]. On the other hand, wheat bran fermentation by *Lactobacillus rhamnosus* tripled water extractable arabinoxylans in wheat bran due to endoxylanases activity on high molecular weight arabinoxylans [65,66], which makes these more easily usable by the intestinal microbiota [67]. The soluble dietary fibre (SDF) content was also increased in the fermented wheat bran, which is consistent with the data for SDF in rye sourdough production [60].

Other carbohydrates that have gained interest in recent years are the fermentable oligosaccharides, disaccharides, monosaccharides, and polyols (FODMAPs). The components of the raffinose family oligosaccharides (RFOs) in legume flours, and the fructans in cereal flours, are particularly noteworthy [68]. In general, they are not easily digested carbohydrates that can lead to the development of pathologies, such as irritable bowel syndrome (IBS) or non-celiac gluten sensitivity (NCGS) [69]. They are equally related to osmotic diarrhoea, inflammation, and abdominal distension, and it is known that a reduction in FODMAP content improves the health status of people suffering from any of these pathologies [18].

However, the use of sourdough can contribute to a significant reduction in FODMAPs content without affecting the content of slowly fermented dietary fibre [18]. Therefore, fermented sourdough products are suitable for consumption in low-FODMAP diets and are safe for IBS patients [19,69].

In the last five years, the use of sourdough has gained popularity as a tool to reduce the content of FODMAPs, as well as antinutrients, in grain and cereal products. In this line, Ziegler et al. (2016) [70] indicate that long fermentations of more than 4 h allow for a reduction in FODMAPs, especially fructans, due to their degradation, by up to 90%, in doughs of flours from *Triticum monococcum* and *T. dicoccum*.

3. Influence of Sourdough Fermentation over Proteins

The hydrolysis of the proteins contained in the flours helps, from a health perspective, create products intended to reduce the adverse reactions caused by gluten, or to reduce the risk of contamination by gluten [8]. In recent years, there has been significant progress in the research on the role of sourdoughs in enabling individuals affected by gluten sensitivity or various gastrointestinal disorders to consume cereal-based products [47].

This is due, among other factors, to the fact that the LAB and yeast of the sourdough synthesize proteases that promote the hydrolysis of gluten [10,71]. Similarly, the reduction in pH facilitates the activation of endogenous enzymes in cereals, as well as the solubilisation and depolymerisation of gluten proteins [72–74]. After fermentation, there is an increase in the presence of protein fragments from the degradation of gliadins and of the high molecular weight subunits of glutenins [75]. It is noteworthy that proteases, both those contributed by microorganisms and those endogenous to the raw materials, influence the volume and texture of products, especially bread [76].

In this sense, there are several strategies to increase protein hydrolysis. Phenomena such as grain germination prior to flour production are positioned as a tool to boost gluten degradation by increasing protease content [77,78]. It may also be beneficial to use flours with high proteolytic activity [79], such as those from brewers' spent grains [80].

It should be noted that if complete degradation of gluten is achieved, the products obtained could be safe for celiac patients [81,82]. With respect to bread, Rizzello et al. (2016) [83] managed to elaborate gluten-free bread, with a high protein digestibility, after the elimination of gluten by the action of sourdough. Some studies, such as Curiel et al. (2014) [84], show that it is possible to reduce gluten to a residual concentration of less than 10 ppm also in products other than bread, such as pasta. However, a total lack of gluten affects the commercial quality of the products [12], as it is difficult to produce breads with adequate volume and texture, and this may have an effect on the consumer acceptability. Therefore, total degradation of the gluten is not always convenient.

Although protein degradation can be an alternative to obtain products suitable for the celiac collective, this technique presents serious disadvantages. On the one hand, most research studies on sourdough do not manage to reach the maximum doses of gluten allowed by the different legislations, or they do not measure it. On the other hand, in the case of managing to reach such maximum, the degradation of gluten would be so extensive that it would be difficult to make bread, where gluten plays an essential role. Then, when taking protein degradation to the commercial practice, an exhaustive control of gluten levels in all batches would be necessary.

Finally, when sourdough is used, the concentration of free amino acids is higher than when only baker's yeast is used [15], although it is possible to combine both. Inoculation with certain LAB strains succeeds in doubling the concentration of certain amino acids, such as leucine, isoleucine, histidine, and lysine, thanks to proteolysis processes [85]. Protein degradation during fermentation that takes place in the sourdough processing also gives rise to potentially health-promoting peptides, such as short branched-chain amino acids and small-sized peptides [86]. These compounds have been shown to contribute to the regulation of insulinemic response, providing protection against type 2 diabetes mellitus and cardiovascular disease [86].

4. Influence of Sourdough Fermentation over Lipids

The presence of lipids in the sourdough is very limited compared to other compounds, such as starch or protein, due to the low lipid content of the flours. This content will be somewhat higher in the case of sourdoughs obtained with whole flours, due to the higher fat content of the germ [87]. Particularly in these cases, it is known that sourdough fermentation reduces fat rancidity by limiting the lipase activity due to the lowering of the pH [88]. This reduction in lipid rancidity in the sourdough contributes positively to the final bread aroma by minimizing unpleasant aromas [89]. The action of the enzymes present in the dough modifies the lipid profile of the flours by partial hydrolysis of the triglycerides and diglycerides. This increases the percentage of monoglycerides and maintains a stable sterol ester content, although the influence on the nutritional characteristics of the bread is limited due to the low level of lipids in the product [90,91]. Considering the reduction of the fat rancidity phenomenon due to the low pH generated, it should be assumed that the hydrolysis of the triglycerides occurs in the first phases of the development of the sourdough. As already known, monoglycerides are products that help in retarding phenomena as bread staling, as they reduce starch retrogradation, allowing the extending of bread shelf life [92].

As a remarkable aspect, the action of certain microorganisms on some lipids may have other positive aspects. Thus, it has been proven that certain strains of *Lactobacillus hammesii* can convert linoleic acid into monohydroxy octadecenoic acid, which has shown antifungal activity in bread [93]. Therefore, its incorporation in sourdough could increase the shelf life of breads, beyond its beneficial effect by simply lowering the pH.

5. Sourdough Fermentation and Vitamins

The most relevant vitamins in cereal products are vitamin E, thiamine, and folates, which are concentrated in the germ and bran [22,23]. Pseudocereals, in turn, contain vitamin A, as well as vitamin E and folate [94]. To increase the vitamin content, fermentation with sourdough can be applied. Certain strains of LAB can synthesise vitamins such as riboflavin, thiamine, and folate [17,24,95–97]. Mihhalevski et al. (2013) [60] also showed that the content of nicotinamide increased during processing by tenfold, presumably due to microbial activity during sourdough fermentation.

Therefore, the use of LAB in sourdough is well positioned as an alternative to help in the prevention of clinical and subclinical vitamin deficiencies [98]. Strategies such as the selection of vitamin-overproducing strains to increase bioavailability should be applied [99]. In addition, it is presumed that *S. cerevisiae* may stimulate the growth of LAB by producing vitamins not contained in wheat flour such as B12, C, or D [100]. In this regard, Chawla and Nagal (2015) [101] point out that the presence of yeast promotes the formation of folates and thiamine.

Additionally, the use of flour from sprouted grains can improve the availability of the vitamins present in them, since they are generated together with other bioactive compounds when they leave the dormant state [102].

Despite the techniques mentioned above for increasing the concentration and bioavailability of vitamins, the baking process leads to a reduction in the vitamin content. Mihhalevski et al. (2013) [60] establish that the concentration of thiamine, nicotinic acid, pyridoxal, riboflavin, and pyridoxine is reduced by temperature increases. Vitamin B12 is also reduced in the process, both in its added form, cyanocobalamin, in its natural form, hydroxocobalamin, and in its synthesized in situ form [103]. The search for techniques to maintain accessibility and vitamin levels in the products is therefore still necessary. However, the higher the amount of vitamins before baking, the higher the final amount of vitamins after the process.

6. Sourdough Fermentation and Mineral Bioavailability

The most relevant minerals in cereals are K, P, Mg, and Zn [22], but they are deficient in others such as Fe [104]. Certain compounds contained in cereals and grains, especially phytic acid, can reduce the bioavailability of minerals as explained above. It is an acid present in grains and cereals in a variable quantity, and an antinutritional factor that prevents the absorption of Ca, Fe, K, Mg, Mn, and Zn, with which it forms phytates, making them insoluble [10]. Phytates are concentrated in the outer layers of the grain and can be hydrolysed by the phytase present within it [105]. Therefore, the efficiency of wheat bran sourdough fermentation in the hydrolysis of phytates and the solubility of minerals has been studied in comparison with whole wheat flour [106]. It was concluded that the pre-fermentation process of whole grains or bran, under adequate conditions of hydration, allows the degradation of most of the phytic acid and an optimal bioavailability of the minerals. In the fermentation stage, the breakdown of polysaccharides is generally greater with sourdough fermentation than with yeast fermentation [107]. However, this action needs the right pH, time, and temperature conditions, which usually occur during fermentation. The simple fermentation and the phytase activity itself contribute to improve the bioavailability of the minerals, as well as the use of sourdough. Therefore, fermentation with sourdough is well positioned as an effective tool to increase the bioavailability of minerals. Lopez et al. (2003) [107] point out that it improves bioavailability, mainly of Mg, Fe, and Zn. Other studies note that sourdough fermentation can increase the release of Fe up to eight times [108]. It may even help in the protection against oxidative stress through the biotransformation of Se, thanks to its conversion to a bioaccessible form [109]. The action of strains of the genus *Lactobacillus* spp. manages to increase the bioavailability of Ca, Zn, and Mg [110]. Yeast strains isolated from sourdoughs, such as *Kluyveromyces marxianus*, can also reduce the phytate content [111]. Combination strains of LAB and yeasts lead to reductions in phytic acid content of more than 40% [112]. Regarding the combination of baker's yeast and sourdough, it can increase the concentration of Ca and Mn in the bread, and, in addition, prolonged fermentations lead to an improvement in the solubility of Mg and P due to acidification [113].

In addition, the pH reduction caused by acidification as a consequence of sourdough fermentation increases the activity of endogenous phytase in the grains, making them even more effective than those of microbial origin [114,115]. The resulting action of a combination of endogenous phytases from different cereals can even completely hydrolyse phytic acid [116]. During the fermentation of legume flour, reductions in phytates due to degradation have also been recorded [117,118].

It has been possible to successfully isolate different LAB strains of the genera *Pediococcus* and *Bifidobacterium*, with high phytase activity [119–121] or to increase their phytase expression through their modification [122].

7. The Role of Sourdough Fermentation on the Levels and the Stability of Bioactive Compounds

Fermentation enhances the presence of bioactive compounds which allow the prevention of various pathologies and diseases related to metabolic syndrome or to cancer [123]. As for cereals, the phytochemicals are divided into flavonoids and non-flavonoid phenols, the latter being more abundant [16]. Since they are mainly found in the outer layers of grains, wholegrain flours are rich in phytochemicals. However, these compounds are sensitive to air contact and their level is reduced in baking processes [23].

The presence of phytochemicals, with antioxidant activity potentially beneficial to health, such as free phenolic acids, total phenolic compounds, or alkylresorcinols, increases with the reduction in pH caused by fermentation processes [124]. Sourdough fermentation contributes to a significant increase in bioactive compound levels and antioxidant activity [123,125–127].

A number of LAB strains are capable of synthesising peptides with an ex vivo antioxidant activity, and with anti-inflammatory and free radical scavenging activities during the fermentation of cereal flours [128,129]. These peptides also maintain their activity in the final product [130]. Colosimo et al. (2020) [131] estimate that fermentations of 72 to 96 h achieve optimal antioxidant activity by increasing the presence of amino acids, organic acids, and aromatic compounds with antioxidant potential. The amount of these compounds is also increased by the protein hydrolysis caused by microbial activity. It should be noted that wheat flour harvested at an advanced stage of maturation presents a greater quantity of bioactive compounds [132], which can reduce the previously mentioned fermentation times.

Meanwhile, other studies have pointed out that the antioxidant activity depends on the type of inoculum and the substrate used in the fermentation [133,134], with numerous examples found in the literature. In the case of *Pediococcus acidilactici* strains incorporated to barley sourdough, they can increase the content of phenolic compounds by 34.6% and the activity of free radical scavengers by 79.7% [135]. The fermentation of durum wheat grain and khamut flour with *L. sanfranciscensis* and *L. brevis* strains slightly increases the release of flavonoids [54]. Lactic acid fermentation also acts in the delivery of bioactive peptides from legume proteins and with the biotransformation of phenols into compounds with higher activity [118]. Strains of *L. acidophilus* that can produce exopolysaccharides (EPS) with antioxidant potential [136] have also been identified. The use of probiotic microorganisms, such as *Enterococcus faecium* and *Kluyveromyces aestuarii*, is currently being studied to boost the antioxidant capacity of bread by increasing the concentration of phenolic compounds [111]. Brewer's spent grain might also be an option to consider since the enzymatic activity it has been subjected to increases its antioxidant potential [137].

In the sourdough, due to the hydrolysis of the proteins, low molecular weight peptides with angiotensin-converting enzyme (ACE) inhibitory activity are also released [138,139]. These peptides with antihypertensive effects are mainly found in wholemeal flours [140]. Research suggests that, despite the reduction in the content of these peptides during heat treatment, their levels in bread are at a concentration required for their activity in vivo [141]. Fermentation with certain strains of the genus *Lactobacillus* spp. can contribute to increase the concentration of this peptide up to seven times [142].

Moreover, γ-aminobutyric acid (GABA) is a non-protein amino acid, acts as a neurotransmitter, has hypotensive, diuretic, and tranquilizing effects, derived from the decarboxylation of glutamate mediated by enzymes, and it is present in the grain or is generated by some LAB [88]. A remarkable fact is that breads fermented with sourdough have higher levels of GABA than other commercial breads [139,143]. Certain *L. plantarum* and *L. lactis* strains can increase the concentration of GABA in cereal, pseudocereal, and legume flours [144]. Curiel et al. (2015) [117] also detected significant increases in legume flours after sourdough fermentation.

Among the bioactive compounds that see their concentration incremented with fermentation, lunasin stands out. *L. curvatus* and *L. brevis* strains detected in sourdough fermentation can synthesize this compound, a cancer-preventing peptide, increasing its concentration two to four times during fermentation [145]. Studies on legume flour have identified that lunasin-like polypeptides' concentration rose during fermentation due to proteolysis processes [146]. *L. plantarum* and *L. rossiae* strains have also shown a potential for the release of biologically active benzoquinones, which have exhibited antitumour activity in ex vivo trials [147].

Lastly, during the fermentation of buckwheat flour, it has been observed that compounds with antioxidant activity, such as quercetin, remain both in the dough and in the final product [148]. By fermenting rye bran, the availability of phytochemicals is enhanced by increasing the presence of free phenolic acids, released by the degradation of benzoxazinoids and alkylresorcinols [149].

8. Sourdough and Regulation of Gut Microbiota

The term microbiome refers to the total number of microbes and their genetic material, while the microbiota is the microbial population present in different parts of the human body [150]. These microbial populations explain critical features of human biology and play an important role in the regulation of human health and disease. This microbiota is able to degrade and ferment complex carbohydrates in the large intestine to produce basic energy recovery and regulate the health of the intestine through the metabolites produced [151]. The diet and with it, the intake of specific polysaccharides, can change the composition and metabolic activities of the intestinal microbiota, promoting the growth of healthy microbes (*Bifidobacterium*, *Lactobacillus*, or butyrate producing bacteria), the production of short chain fatty acids (SCFA), and the reduction in pH, inhibiting effects towards pathogens [67]. They thus affect the health of the host and the gut by modulating inflammation, and glucose and lipid metabolisms [151]. Fermented products in general and cereal-based foods in particular are characterized by an abundance of ingredients that reach the gastrointestinal tract and are accessible by the gut microorganisms of the host [152]. It is known that sourdough fermentation has the ability to actively delay the digestibility of starch, which can increase the production of indigestible polysaccharides that escape from the small intestine, along with the grain fibres, and can then be fermented by the colonic microbiota [33]. The fermentation of the sourdough can influence intestinal health by different mechanisms, such as modulating the dietary fibre complex and its subsequent fermentation pattern, producing exopolysaccharides (EPS) with prebiotic properties, and/or providing metabolites from LAB fermentation that influence the intestinal microbiota [23]. Certain LAB produce exopolysaccharides, such as glucan, fructans, and gluco- and fructo-oligosaccharides that have potential gut health-promoting properties. Intestinal microbes can metabolize these compounds that have been shown to possess prebiotic properties [153–155]. In this way, the metabolism of these compounds can generate propionic acid, which has several beneficial effects [156], such as reducing cholesterol and triglyceride levels, and increasing insulin sensitivity.

Although research suggests that the consumption of sourdough breads may be especially beneficial for the modulation and maintenance of the human intestinal microbiome profile, studies on how the consumption of sourdough breads may have effects on both the intestinal microbiota and the release of health-promoting metabolites, are scarce. In this regard, Abbondio et al. (2019) [157] evaluated how these effects influence the composition and function of intestinal microbiota by feeding rats a diet supplemented with sourdough breads. As a result, supplementation of the diet with sourdough bread led to a reduction in specific members of the intestinal microbiota associated with low-protein diets or known intestinal pathogens, and *Bacteroides* ssp. and *Clostridium* spp. were detected in larger quantities. In addition, recent research interestingly suggests that the cell wall components of *Lactobacillus plantarum*, a strain present in the sourdough, have the ability to stimulate the immune response in the intestine, even when the bacteria are not alive [158].

Another important contribution of sourdough to the health of the human intestine and the regulation of its microbiome is that the bacterial strains isolated from it have shown to be potential probiotics [159–161]. However, alternatives are still being studied that may respond to the fact that probiotic bacteria present in the sourdough die during the baking process. Although it is undeniable that the process of sourdough fermentation results in a compound-rich product shown to promote human health benefits, it is important to study in more detail the specific effect of sourdough on the intestinal microbiome.

9. Future Trends

Advances in understanding the capacity of improving health-related properties with sourdough fermentation are explained by the extensive research carried out in the last 15 years. Figure 2 provides an overview of the potential of this technology for grain processing.

More than 10 years ago, the reviews by Katina et al. (2005) [22] and Poutanen et al. (2009) [23] ventured a series of future trends, most notably the following: (1) regulation of microbiota, (2) development of products with probiotic properties, (3) optimization of processing conditions, (4) improvement of products other than bread, and (5) increased concentration of bioactive compounds and microbial metabolites.

The last three points are those where most progress has been made [3,10,29,84,102]. Points 3 and 4 are not specific to this review and, although much research has been conducted in recent years, there are still aspects to be studied about point 5, such as the incorporation of certain ingredients rich in fibre, highly digestible starches, or proteins in these microbial metabolites. In addition, there is a greater need for an in-depth regulation of the microbiota and the way to obtain products with probiotic potential, especially because of their close relation with a potential improvement in health. On the other hand, the optimization of the process must focus on a balance between healthy properties and sensory quality.

Many of the trends mentioned are still under study, especially the increase in the concentration of bioactive compounds and microbial metabolites. However, there are other future trends that are currently emerging and have not been addressed by previous reviews. Sourdough fermentation can also help fight inflammatory and high oxidative stress processes among consumers in developing countries [162]. Sourdough fermentation is therefore positioned as a useful tool for health promotion in societies where access to healthcare is not widely spread.

Another trend for the future would be the use of sourdough to reduce the content of harmful components in grain and cereal products, although the literature on this phenomenon is not yet extensive. Nevertheless, it has already been shown that sourdoughs may help reduce the content of sugar added to products by the presence of polyol-producing LAB [163]. They can also prevent the development of pathogens, such as *Clostridium*, by generating conditions that induce states of metabolic latency in them [157]. In addition, some LAB isolated from sourdoughs are able to prevent the growth of mould [164] and have an anti-aphlatoxigenic effect [165]. Finally, the reduction in acrylamide, another compound of major concern in grain and cereal products, by sourdough fermentation, has been addressed, for instance, by Bartkiene et al., 2017 [166]. For future research, it would be interesting to expand knowledge about the microbiota, the temperature, or the raw materials employed for obtaining sourdough.

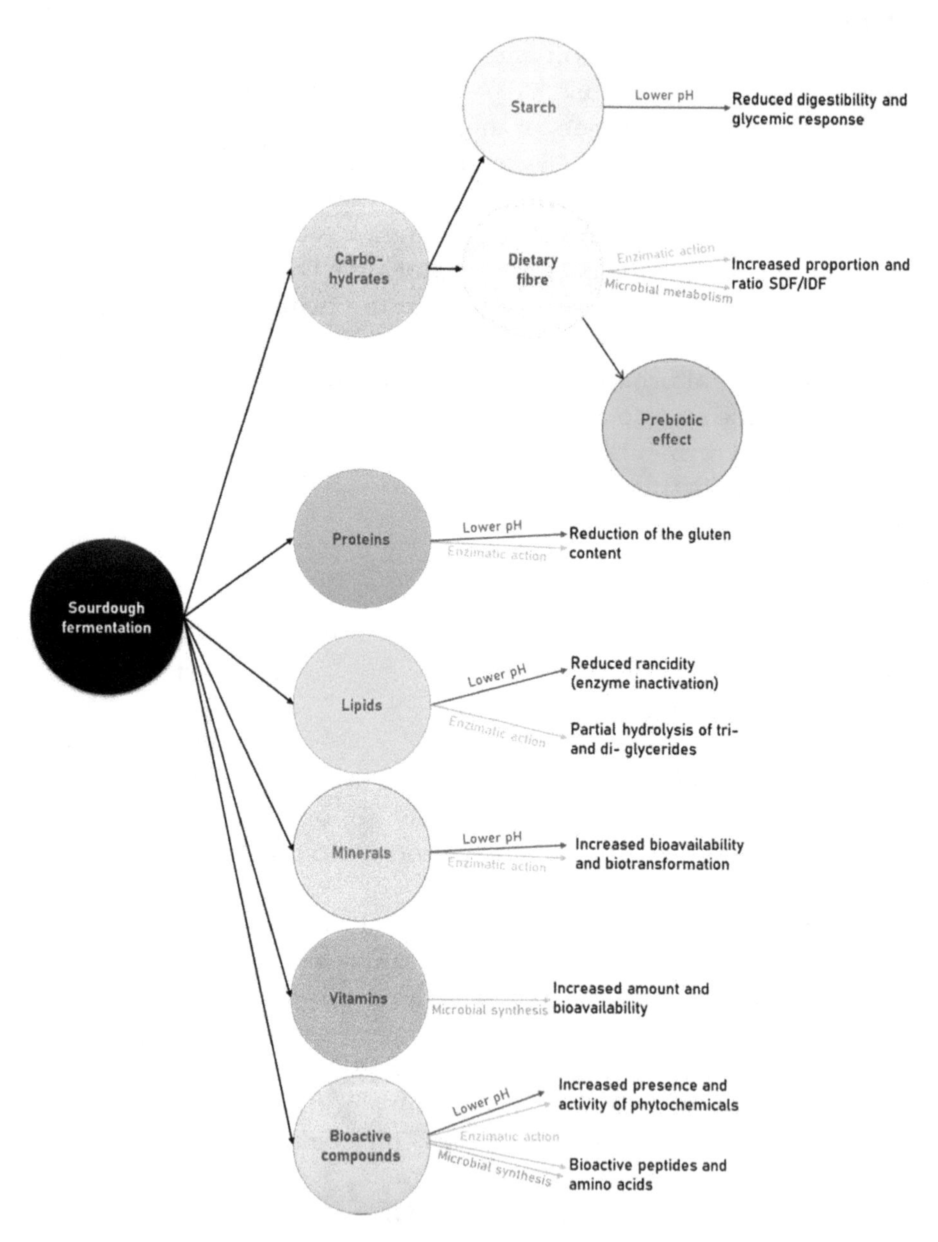

Figure 2. Potential of sourdough fermentation to modify the nutritional quality of cereal and grain-based products.

10. Conclusions

Research on sourdough conducted before the period under study here focused on technological improvements and on the reduction in the gluten content and the glycaemic index of breads. In the last 15 years, although these aspects have remained under study, new research has been initiated on the use of sourdough to improve the potential of grain as well as cereal products and thus, enhance health through various modifications. Improvements cover the modification of nutrients quantity and bioavailability, the generation of bioactive compounds, as well as the reduction of compound contents which could be harmful to certain segments of the population.

Author Contributions: J.F.-P. and C.P. were responsible for the bibliographic search and writing original draft preparation. M.G. was responsible for supervision, writing—review and editing. All authors have read and agreed to the published version of the manuscript.

References

1. Weckx, S.; Van Kerrebroeck, S.; De Vuyst, L. Omics approaches to understand sourdough fermentation processes. *Int. J. Food Microbiol.* **2019**, *302*, 90–102. [CrossRef] [PubMed]
2. Zhang, G.; Tu, J.; Sadiq, F.A.; Zhang, W.; Wang, W. Prevalence, Genetic Diversity, and Technological Functions of the Lactobacillus sanfranciscensis in Sourdough: A Review. *Compr. Rev. Food Sci. Food Saf.* **2019**, *18*, 1209–1226. [CrossRef]
3. Gänzle, M.; Ripari, V. Composition and function of sourdough microbiota: From ecological theory to bread quality. *Int. J. Food Microbiol.* **2016**, *239*, 19–25. [CrossRef] [PubMed]
4. Torrieri, E.; Pepe, O.; Ventorino, V.; Masi, P.; Cavella, S. Effect of sourdough at different concentrations on quality and shelf life of bread. *LWT* **2014**, *56*, 508–516. [CrossRef]
5. Cavallo, N.; De Angelis, M.; Calasso, M.; Quinto, M.; Mentana, A.; Minervini, F.; Cappelle, S.; Gobbetti, M. Microbial cell-free extracts affect the biochemical characteristics and sensorial quality of sourdough bread. *Food Chem.* **2017**, *237*, 159–168. [CrossRef]
6. Wehrle, K.; Arendt, E.K. Rheological Changes in Wheat Sourdough During Controlled and Spontaneous Fermentation. *Cereal Chem. J.* **1998**, *75*, 882–886. [CrossRef]
7. Sakandar, H.A.; Hussain, R.; Kubow, S.; Sadiq, F.A.; Huang, W.; Imran, M. Sourdough bread: A contemporary cereal fermented product. *J. Food Process. Preserv.* **2019**, *43*, e13883. [CrossRef]
8. Nionelli, L.; Rizzello, C.G. Sourdough-Based Biotechnologies for the Production of Gluten-Free Foods. *Foods* **2016**, *5*, 65. [CrossRef]
9. Scherf, K.A.; Wieser, H.; Koehler, P. Novel approaches for enzymatic gluten degradation to create high-quality gluten-free products. *Food Res. Int.* **2018**, *110*, 62–72. [CrossRef]
10. Arendt, E.K.; Moroni, A.; Zannini, E. Medical nutrition therapy: Use of sourdough lactic acid bacteria as a cell factory for delivering functional biomolecules and food ingredients in gluten free bread. *Microb. Cell Fact.* **2011**, *10*, S15. [CrossRef]
11. Capriles, V.D.; Santos, F.G.; Arêas, J.A.G. Gluten-free breadmaking: Improving nutritional and bioactive compounds. *J. Cereal Sci.* **2016**, *67*, 83–91. [CrossRef]
12. Foschia, M.; Horstmann, S.; Arendt, E.K.; Zannini, E. Nutritional therapy—Facing the gap between coeliac disease and gluten-free food. *Int. J. Food Microbiol.* **2016**, *239*, 113–124. [CrossRef] [PubMed]
13. Zamaratskaia, G.; Johansson, D.P.; Junqueira, M.A.; Deissler, L.; Langton, M.; Hellström, P.M.; Landberg, R. Impact of sourdough fermentation on appetite and postprandial metabolic responses—A randomised cross-over trial with whole grain rye crispbread. *Br. J. Nutr.* **2017**, *118*, 686–697. [CrossRef] [PubMed]
14. Polese, B.; Nicolai, E.; Genovese, D.; Verlezza, V.; La Sala, C.N.; Aiello, M.; Inglese, M.; Incoronato, M.; Sarnelli, G.; De Rosa, T.; et al. Postprandial Gastrointestinal Function Differs after Acute Administration of Sourdough Compared with Brewer's Yeast Bakery Products in Healthy Adults. *J. Nutr.* **2018**, *148*, 202–208. [CrossRef] [PubMed]
15. Rizzello, C.G.; Portincasa, P.; Montemurro, M.; Di Palo, D.M.; Lorusso, M.P.; De Angelis, M.; Bonfrate, L.; Genot, B.; Gobbetti, M. Sourdough Fermented Breads are More Digestible than Those Started with Baker's Yeast Alone: An In Vivo Challenge Dissecting Distinct Gastrointestinal Responses. *Nutrients* **2019**, *11*, 2954. [CrossRef] [PubMed]
16. Gobbetti, M.; De Angelis, M.; Di Cagno, R.; Polo, A.; Rizzello, C.G. The sourdough fermentation is the powerful process to exploit the potential of legumes, pseudo-cereals and milling by-products in baking industry. *Crit. Rev. Food Sci. Nutr.* **2020**, *60*, 2158–2173. [CrossRef]
17. Rollán, G.; Gerez, C.L.; Leblanc, J.G. Lactic Fermentation as a Strategy to Improve the Nutritional and Functional Values of Pseudocereals. *Front. Nutr.* **2019**, *6*, 98. [CrossRef]
18. Loponen, J.; Gänzle, M. Use of Sourdough in Low FODMAP Baking. *Foods* **2018**, *7*, 96. [CrossRef]
19. Menezes, L.A.A.; Minervini, F.; Filannino, P.; Sardaro, M.L.S.; Gatti, M.; Lindner, J.D.D. Effects of Sourdough on FODMAPs in Bread and Potential Outcomes on Irritable Bowel Syndrome Patients and Healthy Subjects. *Front. Microbiol.* **2018**, *9*, 1972. [CrossRef]

20. Sahin, A.W.; Zannini, E.; Coffey, A.; Arendt, E.K. Sugar reduction in bakery products: Current strategies and sourdough technology as a potential novel approach. *Food Res. Int.* **2019**, *126*, 108583. [CrossRef]
21. Longoria-García, S.; Cruz-Hernández, M.A.; Flores-Verástegui, M.I.M.; Contreras-Esquivel, J.C.; Montañez-Sáenz, J.C.; Belmares-Cerda, R.E. Potential functional bakery products as delivery systems for prebiotics and probiotics health enhancers. *J. Food Sci. Technol.* **2018**, *55*, 833–845. [CrossRef] [PubMed]
22. Katina, K.; Arendt, E.K.; Liukkonen, K.-H.; Autio, K.; Flander, L.; Poutanen, K. Potential of sourdough for healthier cereal products. *Trends Food Sci. Technol.* **2005**, *16*, 104–112. [CrossRef]
23. Poutanen, K.; Flander, L.; Katina, K. Sourdough and cereal fermentation in a nutritional perspective. *Food Microbiol.* **2009**, *26*, 693–699. [CrossRef] [PubMed]
24. Arena, M.P.; Russo, P.; Spano, G.; Capozzi, V. From Microbial Ecology to Innovative Applications in Food Quality Improvements: The Case of Sourdough as a Model Matrix. *J Multidiscip. Sci. J.* **2020**, *3*, 9–19. [CrossRef]
25. Minervini, F.; De Angelis, M.; Di Cagno, R.; Gobbetti, M. Ecological parameters influencing microbial diversity and stability of traditional sourdough. *Int. J. Food Microbiol.* **2014**, *171*, 136–146. [CrossRef]
26. Sanna, M.; Fois, S.; Falchi, G.; Marco, C.; Roggio, T.; Catzeddu, P. Effect of liquid sourdough technology on the pre-biotic, texture, and sensory properties of a crispy flatbread. *Food Sci. Biotechnol.* **2019**, *28*, 721–730. [CrossRef]
27. Nordlund, E.; Katina, K.; Mykkänen, H.; Poutanen, K. Distinct Characteristics of Rye and Wheat Breads Impact on Their in Vitro Gastric Disintegration and in Vivo Glucose and Insulin Responses. *Foods* **2016**, *5*, 24. [CrossRef]
28. De Angelis, M.; Rizzello, C.G.; Alfonsi, G.; Arnault, P.; Cappelle, S.; Di Cagno, R.; Gobbetti, M. Use of sourdough lactobacilli and oat fibre to decrease the glycaemic index of white wheat bread. *Br. J. Nutr.* **2007**, *98*, 1196–1205. [CrossRef]
29. De Angelis, M.; Damiano, N.; Rizzello, C.G.; Cassone, A.; Di Cagno, R.; Gobbetti, M. Sourdough fermentation as a tool for the manufacture of low-glycemic index white wheat bread enriched in dietary fibre. *Eur. Food Res. Technol.* **2009**, *229*, 593–601. [CrossRef]
30. Lappi, J.; Selinheimo, E.; Schwab, U.; Katina, K.; Lehtinen, P.; Mykkänen, H.; Kolehmainen, M.; Poutanen, K. Sourdough fermentation of wholemeal wheat bread increases solubility of arabinoxylan and protein and decreases postprandial glucose and insulin responses. *J. Cereal Sci.* **2010**, *51*, 152–158. [CrossRef]
31. Mackay, K.; Tucker, A.; Duncan, A.; Graham, T.; Robinson, L. Whole grain wheat sourdough bread does not affect plasminogen activator inhibitor-1 in adults with normal or impaired carbohydrate metabolism. *Nutr. Metab. Cardiovasc. Dis.* **2012**, *22*, 704–711. [CrossRef] [PubMed]
32. Novotni, D.; Curić, D.; Bituh, M.; Barić, I.C.; Skevin, D.; Cukelj, N. Glycemic index and phenolics of partially-baked frozen bread with sourdough. *Int. J. Food Sci. Nutr.* **2010**, *62*, 26–33. [CrossRef] [PubMed]
33. Scazzina, F.; Del Rio, D.; Pellegrini, N.; Brighenti, F. Sourdough bread: Starch digestibility and postprandial glycemic response. *J. Cereal Sci.* **2009**, *49*, 419–421. [CrossRef]
34. Wolter, A.; Hager, A.-S.; Zannini, E.; Arendt, E.K. Influence of sourdough on in vitro starch digestibility and predicted glycemic indices of gluten-free breads. *Food Funct.* **2014**, *5*, 564–572. [CrossRef]
35. Maioli, M.; Pes, G.M.; Sanna, M.; Cherchi, S.; Dettori, M.; Manca, E.; Farris, G.A. Sourdough-leavened bread improves postprandial glucose and insulin plasma levels in subjects with impaired glucose tolerance. *Acta Diabetol.* **2008**, *45*, 91–96. [CrossRef]
36. Najjar, A.M.; Parsons, P.M.; Duncan, A.M.; Robinson, L.E.; Yada, R.Y.; Graham, T.E. The acute impact of ingestion of breads of varying composition on blood glucose, insulin and incretins following first and second meals. *Br. J. Nutr.* **2008**, *101*, 391–398. [CrossRef]
37. Korem, T.; Zeevi, D.; Zmora, N.; Weissbrod, O.; Bar, N.; Lotan-Pompan, M.; Avnit-Sagi, T.; Kosower, N.; Malka, G.; Rein, M.; et al. Bread Affects Clinical Parameters and Induces Gut Microbiome-Associated Personal Glycemic Responses. *Cell Metab.* **2017**, *25*, 1243–1253.e5. [CrossRef]
38. Johansson, D.; Gutiérrez, J.L.V.; Landberg, R.; Alminger, M.; Langton, M. Impact of food processing on rye product properties and their in vitro digestion. *Eur. J. Nutr.* **2018**, *57*, 1651–1666. [CrossRef]
39. Montemurro, M.; Coda, R.; Rizzello, C.G. Recent Advances in the Use of Sourdough Biotechnology in Pasta Making. *Foods* **2019**, *8*, 129. [CrossRef]
40. Fois, S.; Piu, P.P.; Sanna, M.; Roggio, T.; Catzeddu, P. Starch digestibility and properties of fresh pasta made with semolina-based liquid sourdough. *LWT* **2018**, *89*, 496–502. [CrossRef]

41. Gobbetti, M.; De Angelis, M.; Corsetti, A.; Di Cagno, R. Biochemistry and physiology of sourdough lactic acid bacteria. *Trends Food Sci. Technol.* **2005**, *16*, 57–69. [CrossRef]
42. Bondia-Pons, I.; Nordlund, E.; Mattila, I.; Katina, K.; Aura, A.-M.; Kolehmainen, M.; Orešič, M.; Mykkänen, H.; Poutanen, K. Postprandial differences in the plasma metabolome of healthy Finnish subjects after intake of a sourdough fermented endosperm rye bread versus white wheat bread. *Nutr. J.* **2011**, *10*, 116. [CrossRef] [PubMed]
43. Liljeberg, H.; Björck, I.M. Delayed gastric emptying rate as a potential mechanism for lowered glycemia after eating sourdough bread: Studies in humans and rats using test products with added organic acids or an organic salt. *Am. J. Clin. Nutr.* **1996**, *64*, 886–893. [CrossRef] [PubMed]
44. Östman, E.M.; Nilsson, M.; Elmståhl, H.L.; Molin, G.; Björck, I. On the Effect of Lactic Acid on Blood Glucose and Insulin Responses to Cereal Products: Mechanistic Studies in Healthy Subjects and In Vitro. *J. Cereal Sci.* **2002**, *36*, 339–346. [CrossRef]
45. Liljeberg, H.; Björck, I. Delayed gastric emptying rate may explain improved glycaemia in healthy subjects to a starchy meal with added vinegar. *Eur. J. Clin. Nutr.* **1998**, *52*, 368–371. [CrossRef] [PubMed]
46. Pellegrini, N.; Vittadini, E.; Fogliano, V. Designing food structure to slow down digestion in starch-rich products. *Curr. Opin. Food Sci.* **2020**, *32*, 50–57. [CrossRef]
47. Laatikainen, R.; Koskenpato, J.; Hongisto, S.-M.; Loponen, J.; Poussa, T.; Huang, X.; Sontag-Strohm, T.; Salmenkari, H.; Ič, R.K. Pilot Study: Comparison of Sourdough Wheat Bread and Yeast-Fermented Wheat Bread in Individuals with Wheat Sensitivity and Irritable Bowel Syndrome. *Nutrients* **2017**, *9*, 1215. [CrossRef]
48. Buddrick, O.; Jones, O.A.H.; Hughes, J.; Kong, I.; Small, D.M. The effect of fermentation and addition of vegetable oil on resistant starch formation in wholegrain breads. *Food Chem.* **2015**, *180*, 181–185. [CrossRef]
49. Borczak, B.; Sikora, E.; Sikora, M.; Van Haesendonck, I. The impact of sourdough addition to frozen stored wheat-flour rolls on glycemic response in human volunteers. *Starch Stärke* **2011**, *63*, 801–807. [CrossRef]
50. Shumoy, H.; Van Bockstaele, F.; Devecioglu, D.; Raes, K. Effect of sourdough addition and storage time on in vitro starch digestibility and estimated glycemic index of tef bread. *Food Chem.* **2018**, *264*, 34–40. [CrossRef]
51. Giuberti, G.; Gallo, A. Reducing the glycaemic index and increasing the slowly digestible starch content in gluten-free cereal-based foods: A review. *Int. J. Food Sci. Technol.* **2017**, *53*, 50–60. [CrossRef]
52. Wolter, A.; Hager, A.-S.; Zannini, E.; Arendt, E.K. In vitro starch digestibility and predicted glycaemic indexes of buckwheat, oat, quinoa, sorghum, teff and commercial gluten-free bread. *J. Cereal Sci.* **2013**, *58*, 431–436. [CrossRef]
53. Izydorczyk, M.; Dexter, J. Barley β-glucans and arabinoxylans: Molecular structure, physicochemical properties, and uses in food products—A Review. *Food Res. Int.* **2008**, *41*, 850–868. [CrossRef]
54. Saa, D.L.T.; Di Silvestro, R.; Dinelli, G.; Gianotti, A. Effect of sourdough fermentation and baking process severity on dietary fibre and phenolic compounds of immature wheat flour bread. *LWT* **2017**, *83*, 26–32. [CrossRef]
55. Katina, K. *Sourdough: A Tool for the Improved Flavour, Texture and Shelf-Life of Wheat Bread*; ESPOO 2005; Technical Research Centre of Finland; VTT Publications: Espoo, Finland, 2005; pp. 13–41, 53–75.
56. Elleuch, M.; Bedigian, D.; Roiseux, O.; Besbes, S.; Blecker, C.; Attia, H. Dietary fibre and fibre-rich by-products of food processing: Characterisation, technological functionality and commercial applications: A review. *Food Chem.* **2011**, *124*, 411–421. [CrossRef]
57. Hansen, H.B.; Andreasen, M.F.; Nielsen, M.; Larsen, L.; Knudsen, B.K.; Meyer, A.S.; Christensen, L.P.; Hansen, Å.S. Changes in dietary fibre, phenolic acids and activity of endogenous enzymes during rye bread-making. *Eur. Food Res. Technol.* **2002**, *214*, 33–42. [CrossRef]
58. Axelsson, L. *Lactic Acid Bacteria Classification and Physiology*, 2nd ed.; Marcel Dekker: New York, NY, USA, 1998; pp. 1–66.
59. Xiao, X.; Tan, C.; Sun, X.; Zhao, Y.; Zhang, J.; Zhu, Y.; Bai, J.; Dong, Y.; Zhou, X. Effects of fermentation on structural characteristics and in vitro physiological activities of barley β-glucan. *Carbohydr. Polym.* **2020**, *231*, 115685. [CrossRef]
60. Mihhalevski, A.; Nisamedtinov, I.; Hälvin, K.; Ošeka, A.; Paalme, T. Stability of B-complex vitamins and dietary fiber during rye sourdough bread production. *J. Cereal Sci.* **2013**, *57*, 30–38. [CrossRef]
61. Pejcz, E.; Czaja, A.; Wojciechowicz-Budzisz, A.; Gil, Z.; Spychaj, R. The potential of naked barley sourdough to improve the quality and dietary fibre content of barley enriched wheat bread. *J. Cereal Sci.* **2017**, *77*, 97–101. [CrossRef]

62. Mariotti, M.; Garofalo, C.; Aquilanti, L.; Osimani, A.; Fongaro, L.; Tavoletti, S.; Hager, A.-S.; Clementi, F. Barley flour exploitation in sourdough bread-making: A technological, nutritional and sensory evaluation. *LWT Food Sci. Technol.* **2014**, *59*, 973–980. [CrossRef]
63. Gänzle, M. Enzymatic and bacterial conversions during sourdough fermentation. *Food Microbiol.* **2014**, *37*, 2–10. [CrossRef] [PubMed]
64. Rieder, A.; Holtekjølen, A.K.; Sahlstrøm, S.; Moldestad, A. Effect of barley and oat flour types and sourdoughs on dough rheology and bread quality of composite wheat bread. *J. Cereal Sci.* **2012**, *55*, 44–52. [CrossRef]
65. Spaggiari, M.; Ricci, A.; Calani, L.; Bresciani, L.; Neviani, E.; Dall'Asta, C.; Lazzi, C.; Galaverna, G. Solid state lactic acid fermentation: A strategy to improve wheat bran functionality. *LWT* **2020**, *118*, 108668. [CrossRef]
66. Zhao, H.-M.; Guo, X.-N.; Zhu, K.-X. Impact of solid state fermentation on nutritional, physical and flavor properties of wheat bran. *Food Chem.* **2017**, *217*, 28–36. [CrossRef] [PubMed]
67. Paesani, C.; Degano, A.L.; Salvucci, E.; Zalosnik, M.I.; Fabi, J.P.; Sciarini, L.S.; Perez, G.T. Soluble arabinoxylans extracted from soft and hard wheat show a differential prebiotic effect in vitro and in vivo. *J. Cereal Sci.* **2020**, *93*, 102956. [CrossRef]
68. Gänzle, M.G. Food fermentations for improved digestibility of plant foods—An essential ex situ digestion step in agricultural societies? *Curr. Opin. Food Sci.* **2020**, *32*, 124–132. [CrossRef]
69. Menezes, L.A.A.; Molognoni, L.; Ploêncio, L.A.D.S.; Costa, F.B.M.; Daguer, H.; Lindner, J.D.D. Use of sourdough fermentation to reducing FODMAPs in breads. *Eur. Food Res. Technol.* **2019**, *245*, 1183–1195. [CrossRef]
70. Ziegler, J.U.; Steiner, D.; Longin, C.F.H.; Würschum, T.; Schweiggert, R.; Carle, R. Wheat and the irritable bowel síndrome—FODMAP levels of modern and ancient species and their retention during bread making. *J. Funct. Foods* **2016**, *25*, 257–266. [CrossRef]
71. Rizzello, C.G.; De Angelis, M.; Di Cagno, R.; Camarca, A.; Silano, M.; Losito, I.; De Vincenzi, M.; De Bari, M.D.; Palmisano, F.; Maurano, F.; et al. Highly Efficient Gluten Degradation by Lactobacilli and Fungal Proteases during Food Processing: New Perspectives for Celiac Disease. *Appl. Environ. Microbiol.* **2007**, *73*, 4499–4507. [CrossRef]
72. Wu, J.; Loponen, J.; Gobbetti, M. Proteolysis in sourdough fermentations: Mechanisms and potential for improved bread quality. *Trends Food Sci. Technol.* **2008**, *19*, 513–521. [CrossRef]
73. Thiele, C.; Grassl, S.; Gänzle, M. Gluten Hydrolysis and Depolymerization during Sourdough Fermentation. *J. Agric. Food Chem.* **2004**, *52*, 1307–1314. [CrossRef] [PubMed]
74. Yin, Y.; Wang, J.; Yang, S.; Feng, J.; Jia, F.; Zhang, C. Protein Degradation in Wheat Sourdough Fermentation with Lactobacillus plantarum M616. *Interdiscip. Sci. Comput. Life Sci.* **2015**, *7*, 205–210. [CrossRef] [PubMed]
75. Zotta, T.; Piraino, P.; Ricciardi, A.; McSweeney, P.L.H.; Parente, E. Proteolysis in Model Sourdough Fermentations. *J. Agric. Food Chem.* **2006**, *54*, 2567–2574. [CrossRef] [PubMed]
76. Arendt, E.K.; Ryan, L.A.; Bello, F.D. Impact of sourdough on the texture of bread. *Food Microbiol.* **2007**, *24*, 165–174. [CrossRef]
77. Freitag, T.L.; Loponen, J.; Messing, M.; Zevallos, V.; Andersson, L.C.; Sontag-Strohm, T.; Saavalainen, P.; Schuppan, D.; Salovaara, H.; Meri, S. Testing safety of germinated rye sourdough in a celiac disease model based on the adoptive transfer of prolamin-primed memory T cells into lymphopenic mice. *Am. J. Physiol. Liver Physiol.* **2014**, *306*, G526–G534. [CrossRef]
78. Loponen, J.; Sontag-Strohm, T.; Venäläinen, J.; Salovaara, H. Prolamin Hydrolysis in Wheat Sourdoughs with Differing Proteolytic Activities. *J. Agric. Food Chem.* **2007**, *55*, 978–984. [CrossRef]
79. Tuukkanen, K.; Loponen, J.; Mikola, M.; Sontag-Strohm, T.; Salovaara, H. Degradation of Secalins During Rye Sourdough Fermentation. *Cereal Chem. J.* **2005**, *82*, 677–682. [CrossRef]
80. Waters, D.M.; Jacob, F.; Titze, J.; Arendt, E.K.; Zannini, E. Fibre, protein and mineral fortification of wheat bread through milled and fermented brewer's spent grain enrichment. *Eur. Food Res. Technol.* **2012**, *235*, 767–778. [CrossRef]
81. Di Cagno, R.; Barbato, M.; Di Camillo, C.; Rizzello, C.G.; De Angelis, M.; Giuliani, G.; De Vincenzi, M.; Gobbetti, M.; Cucchiara, S. Gluten-free Sourdough Wheat Baked Goods Appear Safe for Young Celiac Patients: A Pilot Study. *J. Pediatr. Gastroenterol. Nutr.* **2010**, *51*, 777–783. [CrossRef]
82. Greco, L.; Gobbetti, M.; Auricchio, R.; Di Mase, R.; Landolfo, F.; Paparo, F.; Di Cagno, R.; De Angelis, M.; Rizzello, C.G.; Cassone, A.; et al. Safety for Patients With Celiac Disease of Baked Goods Made of Wheat Flour Hydrolyzed during Food Processing. *Clin. Gastroenterol. Hepatol.* **2011**, *9*, 24–29. [CrossRef]

83. Rizzello, C.G.; Montemurro, M.; Gobbetti, M. Characterization of the Bread Made with Durum Wheat Semolina Rendered Gluten Free by Sourdough Biotechnology in Comparison with Commercial Gluten-Free Products. *J. Food Sci.* **2016**, *81*, H2263–H2272. [CrossRef]
84. Curiel, J.A.; Coda, R.; Limitone, A.; Katina, K.; Raulio, M.; Giuliani, G.; Rizzello, C.G.; Gobbetti, M. Manufacture and characterization of pasta made with wheat flour rendered gluten-free using fungal proteases and selected sourdough lactic acid bacteria. *J. Cereal Sci.* **2014**, *59*, 79–87. [CrossRef]
85. Chiș, M.S.; Păucean, A.; Stan, L.; Suharoschi, R.; Socaci, S.A.; Man, S.M.; Pop, C.R.; Muste, S. Impact of protein metabolic conversion and volatile derivatives on gluten-free muffins made with quinoa sourdough. *CyTA J. Food* **2019**, *17*, 744–753. [CrossRef]
86. Koistinen, V.M.; Mattila, O.; Katina, K.; Poutanen, K.; Aura, A.-M.; Hanhineva, K. Metabolic profiling of sourdough fermented wheat and rye bread. *Sci. Rep.* **2018**, *8*, 5684. [CrossRef] [PubMed]
87. Rosa-Sibakov, N.; Poutanen, K.; Micard, V. How does wheat grain, bran and aleurone structure impact their nutritional and technological properties? *Trends Food Sci. Technol.* **2015**, *41*, 118–134. [CrossRef]
88. Rizzello, C.G.; Nionelli, L.; Coda, R.; De Angelis, M.; Gobbetti, M. Effect of sourdough fermentation on stabilisation, and chemical and nutritional characteristics of wheat germ. *Food Chem.* **2010**, *119*, 1079–1089. [CrossRef]
89. Pétel, C.; Onno, B.; Prost, C. Sourdough volatile compounds and their contribution to bread: A review. *Trends Food Sci. Technol.* **2017**, *59*, 105–123. [CrossRef]
90. Rocha, J.M.; Kalo, P.J.; Malcata, F.X. Composition of neutral lipid classes and content of fatty acids throughout sourdough breadmaking. *Eur. J. Lipid Sci. Technol.* **2011**, *114*, 294–305. [CrossRef]
91. Rocha, J.M.; Kalo, P.J.; Malcata, F.X. Fatty Acid Composition of Non-Starch and Starch Neutral Lipid Extracts of Portuguese Sourdough Bread. *J. Am. Oil Chem. Soc.* **2012**, *89*, 2025–2045. [CrossRef]
92. Gómez, M.; Del Real, S.; Rosell, C.M.; Ronda, F.; Blanco, C.A.; Caballero, P.A. Functionality of different emulsifiers on the performance of breadmaking and wheat bread quality. *Eur. Food Res. Technol.* **2004**, *219*, 145–150. [CrossRef]
93. Black, B.A.; Zannini, E.; Curtis, J.M.; Gänzle, M.G. Antifungal Hydroxy Fatty Acids Produced during Sourdough Fermentation: Microbial and Enzymatic Pathways, and Antifungal Activity in Bread. *Appl. Environ. Microbiol.* **2013**, *79*, 1866–1873. [CrossRef] [PubMed]
94. Axel, C.; Röcker, B.; Brosnan, B.; Zannini, E.; Furey, A.; Coffey, A.; Arendt, E.K. Application of Lactobacillus amylovorus DSM19280 in gluten-free sourdough bread to improve the microbial shelf life. *Food Microbiol.* **2015**, *47*, 36–44. [CrossRef] [PubMed]
95. Capozzi, V.; Menga, V.; Digesù, A.M.; De Vita, P.; Van Sinderen, D.; Cattivelli, L.; Fares, C.; Spano, G. Biotechnological Production of Vitamin B2-Enriched Bread and Pasta. *J. Agric. Food Chem.* **2011**, *59*, 8013–8020. [CrossRef] [PubMed]
96. Gujska, E.; Majewska, K. Effect of Baking Process on Added Folic Acid and Endogenous Folates Stability in Wheat and Rye Breads. *Plant Foods Hum. Nutr.* **2005**, *60*, 37–42. [CrossRef]
97. Okoroafor, I.; Banwo, K.; Olanbiwoninu, A.A.; Odunfa, S.A. Folate Enrichment of Ogi (a Fermented Cereal Gruel) Using Folate Producing Starter Cultures. *Adv. Microbiol.* **2019**, *9*, 177–193. [CrossRef]
98. Leblanc, J.G.; Laiño, J.; Del Valle, M.J.; Vannini, V.; Van Sinderen, D.; Taranto, M.; De Valdez, G.F.; De Giori, G.S.; Sesma, F. B-Group vitamin production by lactic acid bacteria—Current knowledge and potential applications. *J. Appl. Microbiol.* **2011**, *111*, 1297–1309. [CrossRef]
99. Capozzi, V.; Russo, P.; Dueñas, M.T.; López, P.; Spano, G. Lactic acid bacteria producing B-group vitamins: A great potential for functional cereals products. *Appl. Microbiol. Biotechnol.* **2012**, *96*, 1383–1394. [CrossRef]
100. Sieuwerts, S.; Bron, P.A.; Smid, E.J. Mutually stimulating interactions between lactic acid bacteria and Saccharomyces cerevisiae in sourdough fermentation. *LWT* **2018**, *90*, 201–206. [CrossRef]
101. Chawla, S.; Nagal, S. Sourdough in bread-making: An ancient technology to solve modern issues. *Int. J. Ind. Biotechnol. Biomater.* **2015**, *1*. [CrossRef]
102. Montemurro, M.; Pontonio, E.; Gobbetti, M.; Rizzello, C.G. Investigation of the nutritional, functional and technological effects of the sourdough fermentation of sprouted flours. *Int. J. Food Microbiol.* **2019**, *302*, 47–58. [CrossRef]
103. Edelmann, M.; Chamlagain, B.; Santin, M.; Kariluoto, S.; Piironen, V. Stability of added and in situ-produced vitamin B12 in breadmaking. *Food Chem.* **2016**, *204*, 21–28. [CrossRef] [PubMed]
104. Balk, J.; Connorton, J.M.; Wan, Y.; Lovegrove, A.; Moore, K.L.; Uauy, C.; Sharp, P.A.; Shewry, P.R. Improving wheat as a source of iron and zinc for global nutrition. *Nutr. Bull.* **2019**, *44*, 53–59. [CrossRef] [PubMed]

105. Rizzello, C.G.; Coda, R.; Mazzacane, F.; Minervini, D.; Gobbetti, M. Micronized by-products from debranned durum wheat and sourdough fermentation enhanced the nutritional, textural and sensory features of bread. *Food Res. Int.* **2012**, *46*, 304–313. [CrossRef]
106. Lioger, D.; Leenhardt, F.; Demigné, C.; Rémésy, C. Sourdough fermentation of wheat fractions rich in fibres before their use in processed food. *J. Sci. Food Agric.* **2007**, *87*, 1368–1373. [CrossRef]
107. Lopez, W.; Duclos, V.; Coudray, C.; Krespine, V.; Feillet-Coudray, C.; Messager, A.; Demigné, C.; Rémésy, C. Making bread with sourdough improves mineral bioavailability from reconstituted whole wheat flour in rats. *Nutrition* **2003**, *19*, 524–530. [CrossRef]
108. Rodríguez-Ramiro, I.; Brearley, C.A.; Bruggraber, S.; Perfecto, A.; Shewry, P.; Fairweather-Tait, S. Assessment of iron bioavailability from different bread making processes using an in vitro intestinal cell model. *Food Chem.* **2017**, *228*, 91–98. [CrossRef] [PubMed]
109. Di Nunzio, M.; Bordoni, A.; Aureli, F.; Cubadda, F.; Gianotti, A. Sourdough Fermentation Favorably Influences Selenium Biotransformation and the Biological Effects of Flatbread. *Nutrients* **2018**, *10*, 1898. [CrossRef]
110. Di Cagno, R.; Rizzello, C.G.; De Angelis, M.; Cassone, A.; Giuliani, G.; Benedusi, A.; Limitone, A.; Surico, R.F.; Gobbetti, M. Use of Selected Sourdough Strains of Lactobacillus for Removing Gluten and Enhancing the Nutritional Properties of Gluten-Free Bread. *J. Food Prot.* **2008**, *71*, 1491–1495. [CrossRef]
111. Fekri, A.; Torbati, M.; Khosrowshahi, A.Y.; Shamloo, H.B.; Azadmard-Damirchi, S. Functional effects of phytate-degrading, probiotic lactic acid bacteria and yeast strains isolated from Iranian traditional sourdough on the technological and nutritional properties of whole wheat bread. *Food Chem.* **2020**, *306*, 125620. [CrossRef]
112. Karaman, K.; Sagdic, O.; Durak, M.Z. Use of phytase active yeasts and lactic acid bacteria isolated from sourdough in the production of whole wheat bread. *LWT* **2018**, *91*, 557–567. [CrossRef]
113. Helou, C.; Gadonna-Widehem, P.; Robert, N.; Branlard, G.; Thebault, J.; Librere, S.; Jacquot, S.; Mardon, J.; Piquet-Pissaloux, A.; Chapron, S.; et al. The impact of raw materials and baking conditions on Maillard reaction products, thiamine, folate, phytic acid and minerals in white bread. *Food Funct.* **2016**, *7*, 2498–2507. [CrossRef] [PubMed]
114. García-Mantrana, I.; Yebra, M.J.; Haros, M.; Monedero, V. Expression of bifidobacterial phytases in Lactobacillus casei and their application in a food model of whole-grain sourdough bread. *Int. J. Food Microbiol.* **2016**, *216*, 18–24. [CrossRef] [PubMed]
115. Leenhardt, F.; Levrat-Verny, M.-A.; Chanliaud, E.; Rémésy, C. Moderate Decrease of pH by Sourdough Fermentation Is Sufficient To Reduce Phytate Content of Whole Wheat Flour through Endogenous Phytase Activity. *J. Agric. Food Chem.* **2005**, *53*, 98–102. [CrossRef] [PubMed]
116. Baye, K.; Mouquet-Rivier, C.; Icard-Vernière, C.; Rochette, I.; Guyot, J.-P. Influence of flour blend composition on fermentation kinetics and phytate hydrolysis of sourdough used to make injera. *Food Chem.* **2013**, *138*, 430–436. [CrossRef]
117. Curiel, J.A.; Coda, R.; Centomani, I.; Summo, C.; Gobbetti, M.; Rizzello, C.G. Exploitation of the nutritional and functional characteristics of traditional Italian legumes: The potential of sourdough fermentation. *Int. J. Food Microbiol.* **2015**, *196*, 51–61. [CrossRef]
118. Gabriele, M.; Sparvoli, F.; Bollini, R.; Lubrano, V.; Longo, V.; Pucci, L. The Impact of Sourdough Fermentation on Non-Nutritive Compounds and Antioxidant Activities of Flours from Different Phaseolus Vulgaris L. Genotypes. *J. Food Sci.* **2019**, *84*, 1929–1936. [CrossRef]
119. Doğan, M.; Tekiner, I.H. Extracellular phytase activites of lactic acid bacteria in sour-dough mix prepared from traditionally produced boza as starter culture. *Food Health* **2020**, *6*, 117–127. [CrossRef]
120. Moroni, A.V.; Bello, F.D.; Arendt, E.K. Sourdough in gluten-free bread-making: An ancient technology to solve a novel issue? *Food Microbiol.* **2009**, *26*, 676–684. [CrossRef]
121. Palacios, M.C.; Haros, M.; Rosell, C.M.; Sanz, Y. Selection of phytate-degrading human bifidobacteria and application in whole wheat dough fermentation. *Food Microbiol.* **2008**, *25*, 169–176. [CrossRef]
122. Garcia-Mantrana, I.; Monedero, V.; Haros, M. Myo-inositol hexakisphosphate degradation by Bifidobacterium pseudocatenulatum ATCC 27919 improves mineral availability of high fibre rye-wheat sour bread. *Food Chem.* **2015**, *178*, 267–275. [CrossRef]
123. Rizzello, C.G.; Tagliazucchi, D.; Babini, E.; Rutella, G.S.; Saa, D.L.T.; Gianotti, A. Bioactive peptides from vegetable food matrices: Research trends and novel biotechnologies for synthesis and recovery. *J. Funct. Foods* **2016**, *27*, 549–569. [CrossRef]

124. Katina, K.; Liukkonen, K.-H.; Kaukovirta-Norja, A.; Adlercreutz, H.; Heinonen, S.-M.; Lampi, A.-M.; Pihlava, J.-M.; Poutanen, K. Fermentation-induced changes in the nutritional value of native or germinated rye. *J. Cereal Sci.* **2007**, *46*, 348–355. [CrossRef]
125. Đorđević, T.M.; Šiler-Marinković, S.S.; Dimitrijević-Branković, S.I. Effect of fermentation on antioxidant properties of some cereals and pseudo cereals. *Food Chem.* **2010**, *119*, 957–963. [CrossRef]
126. Gandhi, A.; Dey, G. Fermentation responses andin vitroradical scavenging activities ofFagopyrum esculentum. *Int. J. Food Sci. Nutr.* **2012**, *64*, 53–57. [CrossRef]
127. Hayta, M.; Ertop, M.H. Optimisation of sourdough bread incorporation into wheat bread by response surface methodology: Bioactive and nutritional properties. *Int. J. Food Sci. Technol.* **2017**, *52*, 1828–1835. [CrossRef]
128. Coda, R.; Rizzello, C.G.; Pinto, D.; Gobbetti, M. Selected Lactic Acid Bacteria Synthesize Antioxidant Peptides during Sourdough Fermentation of Cereal Flours. *Appl. Environ. Microbiol.* **2011**, *78*, 1087–1096. [CrossRef]
129. Galli, V.; Mazzoli, L.; Luti, S.; Venturi, M.; Guerrini, S.; Paoli, P.; Vincenzini, M.; Granchi, L.; Pazzagli, L. Effect of selected strains of lactobacilli on the antioxidant and anti-inflammatory properties of sourdough. *Int. J. Food Microbiol.* **2018**, *286*, 55–65. [CrossRef]
130. Luti, S.; Mazzoli, L.; Ramazzotti, M.; Galli, V.; Venturi, M.; Marino, G.; Lehmann, M.; Guerrini, S.; Granchi, L.; Paoli, P.; et al. Antioxidant and anti-inflammatory properties of sourdoughs containing selected Lactobacilli strains are retained in breads. *Food Chem.* **2020**, *322*, 126710. [CrossRef]
131. Colosimo, R.; Gabriele, M.; Cifelli, M.; Longo, V.; Domenici, V.; Pucci, L. The effect of sourdough fermentation on Triticum dicoccum from Garfagnana: 1H NMR characterization and analysis of the antioxidant activity. *Food Chem.* **2020**, *305*, 125510. [CrossRef]
132. Saa, D.L.T.; Di Silvestro, R.; Nissen, L.; Dinelli, G.; Gianotti, A. Effect of sourdough fermentation and baking process severity on bioactive fiber compounds in immature and ripe wheat flour bread. *LWT* **2018**, *89*, 322–328. [CrossRef]
133. Banu, I.; Vasilean, I.; Aprodu, I. Effect of Lactic Fermentation on Antioxidant Capacity of Rye Sourdough and Bread. *Food Sci. Technol. Res.* **2010**, *16*, 571–576. [CrossRef]
134. Pallin, A.; Agback, P.; Jonsson, H.; Roos, S. Evaluation of growth, metabolism and production of potentially bioactive components during fermentation of barley with Lactobacillus reuteri. *Food Microbiol.* **2016**, *57*, 159–171. [CrossRef] [PubMed]
135. Bartkiene, E.; Vizbickiene, D.; Bartkevics, V.; Pugajeva, I.; Krungleviciute, V.; Zadeike, D.; Zavistanaviciute, P.; Juodeikiene, G. Application of Pediococcus acidilactici LUHS29 immobilized in apple pomace matrix for high value wheat-barley sourdough bread. *LWT* **2017**, *83*, 157–164. [CrossRef]
136. Abedfar, A.; Abbaszadeh, S.; Hosseininezhad, M.; Taghdir, M. Physicochemical and biological characterization of the EPS produced by L. acidophilus isolated from rice bran sourdough. *LWT* **2020**, *127*, 109373. [CrossRef]
137. Ktenioudaki, A.; Alvarez-Jubete, L.; Smyth, T.J.; Kilcawley, K.; Rai, D.K.; Gallagher, E. Application of bioprocessing techniques (sourdough fermentation and technological aids) for brewer's spent grain breads. *Food Res. Int.* **2015**, *73*, 107–116. [CrossRef]
138. Nakamura, T.; Yoshida, A.; Komatsuzaki, N.; Kawasumi, T.; Shima, J. Isolation and Characterization of a Low Molecular Weight Peptide Contained in Sourdough. *J. Agric. Food Chem.* **2007**, *55*, 4871–4876. [CrossRef]
139. Peñas, E.; Diana, M.; Frias, J.; Quílez, J.; Emartinez-Villaluenga, C. A Multistrategic Approach in the Development of Sourdough Bread Targeted Towards Blood Pressure Reduction. *Plant Foods Hum. Nutr.* **2015**, *70*, 97–103. [CrossRef]
140. Rizzello, C.G.; Cassone, A.; Di Cagno, R.; Gobbetti, M. Synthesis of Angiotensin I-Converting Enzyme (ACE)-Inhibitory Peptides and γ-Aminobutyric Acid (GABA) during Sourdough Fermentation by Selected Lactic Acid Bacteria. *J. Agric. Food Chem.* **2008**, *56*, 6936–6943. [CrossRef]
141. Zhao, C.J.; Hu, Y.; Schieber, A.; Gänzle, M. Fate of ACE-inhibitory peptides during the bread-making process: Quantification of peptides in sourdough, bread crumb, steamed bread and soda crackers. *J. Cereal Sci.* **2013**, *57*, 514–519. [CrossRef]
142. Hu, Y.; Stromeck, A.; Loponen, J.; Lopes-Lutz, D.; Schieber, A.; Gänzle, M.G. LC-MS/MS Quantification of Bioactive Angiotensin I-Converting Enzyme Inhibitory Peptides in Rye Malt Sourdoughs. *J. Agric. Food Chem.* **2011**, *59*, 11983–11989. [CrossRef]
143. Diana, M.; Rafecas, M.; Quilez, J. Free amino acids, acrylamide and biogenic amines in gamma-aminobutyric acid enriched sourdough and commercial breads. *J. Cereal Sci.* **2014**, *60*, 639–644. [CrossRef]

144. Coda, R.; Rizzello, C.G.; Gobbetti, M. Use of sourdough fermentation and pseudo-cereals and leguminous flours for the making of a functional bread enriched of γ-aminobutyric acid (GABA). *Int. J. Food Microbiol.* **2010**, *137*, 236–245. [CrossRef] [PubMed]

145. Rizzello, C.G.; Nionelli, L.; Coda, R.; Gobbetti, M. Synthesis of the Cancer Preventive Peptide Lunasin by Lactic Acid Bacteria During Sourdough Fermentation. *Nutr. Cancer* **2012**, *64*, 111–120. [CrossRef] [PubMed]

146. Rizzello, C.G.; Hernández-Ledesma, B.; Fernández-Tomé, S.; Curiel, J.A.; Pinto, D.; Marzani, B.; Coda, R.; Gobbetti, M. Italian legumes: Effect of sourdough fermentation on lunasin-like polypeptides. *Microb. Cell Fact.* **2015**, *14*, 1–20. [CrossRef]

147. Rizzello, C.; Mueller, T.; Coda, R.; Reipsch, F.; Nionelli, L.; Curiel, J.A.; Gobbetti, M. Synthesis of 2-methoxy benzoquinone and 2,6-dimethoxybenzoquinone by selected lactic acid bacteria during sourdough fermentation of wheat germ. *Microb. Cell Fact.* **2013**, *12*, 105. [CrossRef]

148. Lukšič, L.; Bonafaccia, G.; Timoracka, M.; Vollmannova, A.; Trček, J.; Nyambe, T.K.; Melini, V.; Acquistucci, R.; Germ, M.; Kreft, I. Rutin and quercetin transformation during preparation of buckwheat sourdough bread. *J. Cereal Sci.* **2016**, *69*, 71–76. [CrossRef]

149. Koistinen, V.M.; Katina, K.; Nordlund, E.; Poutanen, K.; Hanhineva, K. Changes in the phytochemical profile of rye bran induced by enzymatic bioprocessing and sourdough fermentation. *Food Res. Int.* **2016**, *89*, 1106–1115. [CrossRef]

150. Bakhtiar, S.M.; Leblanc, A.D.M.D.; Salvucci, E.; Yoon, B.-J.; Martín, R.; Langella, P.; Chatel, J.-M.; Miyoshi, A.; Bermúdez-Humarán, L.G.; Azevedo, V. Implications of the human microbiome in inflammatory bowel diseases. *FEMS Microbiol. Lett.* **2013**, *342*, 10–17. [CrossRef]

151. Chassard, C.; Lacroix, C. Carbohydrates and the human gut microbiota. *Curr. Opin. Clin. Nutr. Metab. Care* **2013**, *16*, 453–460. [CrossRef]

152. Tsafrakidou, P.; Michaelidou, A.-M.; Biliaderis, C.G. Fermented Cereal-based Products: Nutritional Aspects, Possible Impact on Gut Microbiota and Health Implications. *Foods* **2020**, *9*, 734. [CrossRef]

153. Korakli, M.; Ganzle, M.; Vogel, R. Metabolism by bifidobacteria and lactic acid bacteria of polysaccharides from wheat and rye, and exopolysaccharides produced by Lactobacillus sanfranciscensis. *J. Appl. Microbiol.* **2002**, *92*, 958–965. [CrossRef] [PubMed]

154. Nam, S.-H.; Ko, E.-A.; Jin, X.-J.; Breton, V.; Abada, E.; Kim, Y.-M.; Kimura, A.; Kim, D. Synthesis of Thermo- and Acid-stable Novel Oligosaccharides by Using Dextransucrase with High Concentration of Sucrose. *J. Appl. Glycosci.* **2007**, *54*, 147–155. [CrossRef]

155. Tieking, M.; Wu, J. Exopolysaccharides from cereal-associated lactobacilli. *Trends Food Sci. Technol.* **2005**, *16*, 79–84. [CrossRef]

156. Jann, A.; Arragoni, E.; Florence, R.; Schmid, D.; Bauche, A. Method for Increasing the Production of Propionate in the Gastrointestinal Tract. U.S. Patent 7,091,194, 15 August 2006.

157. Abbondio, M.; Palomba, A.; Tanca, A.; Fraumene, C.; Pagnozzi, D.; Serra, M.; Marongiu, F.; Laconi, E.; Uzzau, S. Fecal Metaproteomic Analysis Reveals Unique Changes of the Gut Microbiome Functions after Consumption of Sourdough Carasau Bread. *Front. Microbiol.* **2019**, *10*, 1733. [CrossRef]

158. Van Baarlen, P.; Troost, F.J.; Van Hemert, S.; Van Der Meer, C.; De Vos, W.M.; De Groot, P.J.; Hooiveld, G.J.E.J.; Brummer, R.-J.; Kleerebezem, M. Differential NF-κB pathways induction by Lactobacillus plantarum in the duodenum of healthy humans correlating with immune tolerance. *Proc. Natl. Acad. Sci. USA* **2009**, *106*, 2371–2376. [CrossRef]

159. Denkova, R.; Georgieva, L.; Denkova, Z.; Urshev, Z. Biochemical and technological properties of Lactobacillus plantarum X2 from naturally fermented sourdough Biochemical and technological properties of Lactobacillus plantarum X2 from naturally fermented sourdough. *J. Food Packag. Sci. Tech. Technol.* **2012**, *1*, 2–8.

160. Ilha, E.C.; Da Silva, T.; Lorenz, J.G.; Rocha, G.D.O.; Sant'Anna, E.S. Lactobacillus paracasei isolated from grape sourdough: Acid, bile, salt, and heat tolerance after spray drying with skim milk and cheese whey. *Eur. Food Res. Technol.* **2014**, *240*, 977–984. [CrossRef]

161. Penaloza-Vazquez, A.; Ma, L.M.; Rayas-Duarte, P. Isolation and characterization of Bacillus spp. strains as potential probiotics for poultry. *Can. J. Microbiol.* **2019**, *65*, 762–774. [CrossRef]

162. Laurent-Babot, C.; Guyot, J.-P. Should Research on the Nutritional Potential and Health Benefits of Fermented Cereals Focus More on the General Health Status of Populations in Developing Countries? *Microorganisms* **2017**, *5*, 40. [CrossRef]

163. Jeske, S.; Zannini, E.; Lynch, K.M.; Coffey, A.; Arendt, E.K. Polyol-producing lactic acid bacteria isolated from sourdough and their application to reduce sugar in a quinoa-based milk substitute. *Int. J. Food Microbiol.* **2018**, *286*, 31–36. [CrossRef]
164. Bartkiene, E.; Lele, V.; Ruzauskas, M.; Domig, K.J.; Starkute, V.; Zavistanaviciute, P.; Bartkevics, V.; Pugajeva, I.; Klupsaite, D.; Juodeikiene, G.; et al. Lactic Acid Bacteria Isolation from Spontaneous Sourdough and Their Characterization Including Antimicrobial and Antifungal Properties Evaluation. *Microorganisms* **2019**, *8*, 64. [CrossRef] [PubMed]
165. Sadeghi, A.; Ebrahimi, M.; Raeisi, M.; Nematollahi, Z. Biological control of foodborne pathogens and aflatoxins by selected probiotic LAB isolated from rice bran sourdough. *Biol. Control* **2019**, *130*, 70–79. [CrossRef]
166. Zadeike, D.; Bartkevics, V.; Krungleviciute, V.; Pugajeva, I.; Zadeike, D.; Juodeikiene, G. Lactic Acid Bacteria Combinations for Wheat Sourdough Preparation and Their Influence on Wheat Bread Quality and Acrylamide Formation. *J. Food Sci.* **2017**, *82*, 2371–2378. [CrossRef]

3

Modified Fish Gelatin as an Alternative to Mammalian Gelatin in Modern Food Technologies

Svetlana R. Derkach, Nikolay G. Voron'ko *, Yuliya A. Kuchina and Daria S. Kolotova

Department of Chemistry, Murmansk State Technical University, 183010 Murmansk, Russia;
derkachsr@mstu.edu.ru (S.R.D.); kuchinayua@mstu.edu.ru (Y.A.K.); kolotovads@mstu.edu.ru (D.S.K.)
* Correspondence: voronkong@mstu.edu.ru

Abstract: This review considers the main properties of fish gelatin that determine its use in food technologies. A comparative analysis of the amino acid composition of gelatin from cold-water and warm-water fish species, in comparison with gelatin from mammals, which is traditionally used in the food industry, is presented. Fish gelatin is characterized by a reduced content of proline and hydroxyproline which are responsible for the formation of collagen-like triple helices. For this reason, fish gelatin gels are less durable and have lower gelation and melting temperatures than mammalian gelatin. These properties impose significant restrictions on the use of fish gelatin in the technology of gelled food as an alternative to porcine and bovine gelatin. This problem can be solved by modifying the functional characteristics of fish gelatin by adding natural ionic polysaccharides, which, under certain conditions, are capable of forming polyelectrolyte complexes with gelatin, creating additional nodes in the spatial network of the gel.

Keywords: fish gelatin; amino-acid composition; sole-gel transition; rheology

---

1. Introduction

Gelatin from bone and connective tissue of pigs and cattle is traditionally used in the food industry as a gelling agent [1–4]. However, the consumption of gelatin from these mammalian species contradicts ethnocultural and religious norms of a number of religions, and is also associated with the risk of contracting prion diseases (in particular, spongiform encephalopathy) [2,5–8]. In this regard, it seems relevant to look for alternative sources of food gelatin [5,9]. Such a source may be the connective tissue of fish [2,10–13], the industrial processing of which partially solves the problem of disposal and integrated use of waste from the fish processing industry [8,14–17]. The total world volume of fishing in 2018 is estimated at 179 million tons, and this figure is increasing every year [18]. According to various estimates, waste from fish processing can be up to85% of the total catch [14,18,19]. A significant percentage of waste (about 30%) is skin, bones, and scales with a high collagen content [8,14,16,18]. Therefore, the production and use of fish gelatin as a food structure-forming agent seems to be very promising. Currently, only 1.5% of gelatin is produced from fish collagen-containing raw materials, while 41% is produced from pig skin, 28.5% from bovine hides, and 29.5% from bovine bones [20].

Fish gelatin is characterized by a lower content of proline and hydroxyproline compared to gelatin from mammals [1,2,21,22]. This is especially true for gelatin obtained from cold-water species [1,23]. This leads to a deterioration in the gelling ability of fish gelatin, a decrease in the gelation and melting temperatures [21,24], a decrease in the gel strength [24,25], and increased consumption of gelatin as a food component for hydrogel formation [7,26]. Many studies have been devoted to finding ways to eliminate these serious disadvantages by treating fish gelatin with various physical and chemical cross-linkers: irradiation in various frequency ranges [7,27–29], high pressure [28,30], enzymatic modification [7,31], additions of mono- and disaccharides [7,32] and ferulic and caffeic acids [33,34]. However, the most effective and common way to improve gelling ability and rheological characteristics are the modification of fish gelatin with natural polysaccharides, for example, κ-carrageenan [6,35,36], sodium alginate [37–39], chitosan [29,40], gellan [6,36], gum arabic [41,42], and pectin [20,43]. This method of improving functional properties can increase the nutritional value of gelatin gels [29,39,43].

This review analyses the composition, structure, and functional properties of fish gelatin as an alternative to mammalian gelatin when used as a food gelling agent. The mechanism of modifying the gel-forming properties of fish gelatin and the rheological properties of gelatin gel by adding natural ionic polysaccharides is considered.

2. Amino-Acid Composition of Fish Gelatin

The amino acid sequence of a gelatin macromolecule can be described by the general formula Gly-X-Y, where the X position usually occupied by proline and Y position by hydroxyproline [1,7,9,19]. The amino acid triads Gly-Pro-Y, Gly-X-Hyp, and Gly-Pro-Hyp in macromolecular chains play a major role in the formation of triple collagen-like helices [19,44,45]. The comparative amino acid composition of gelatin obtained from various sources is presented in Table 1. Compared with gelatin from mammals, fish gelatin is characterized by a lower content of proline and hydroxyproline. At the same time, in terms of the content of these amino acids, gelatin from the skin of warm-water fish species (tilapia, tuna, black carp) is similar to gelatin from pork and calf skin.

Table 1. Amino-acid composition of some fish gelatins compared to pork and calfskin gelatine.

	Cold Water Fish Skin			Warm Water Fish Skin				
Source	Cod [10]	Hake [10]	Alaska Pollock [2]	Tilapia [1,2]	Tuna [46]	Black Carp [19,47]	Pork Skin [2]	Calf Skin [21]
Amino Acid Composition (Residues Per 1000 Total Amino Acid Residues)								
Glycine (Gly)	344	331	358	347	336	314	330	313
Basic groups	99	97	91	86	90	88	86	101
Lysine (Lys)	29	28	26	25	25	29	27	34
Hydroxylysine(Hyl)	6	5	6	8	6	2	6	11
Histidine (His)	8	10	8	6	7	4	4	5
Arginine (Arg)	56	54	51	47	52	53	49	51
Carboxylic groups	130	123	125	117	115	126	118	116
Aspartic acid (Asp)	52	49	51	48	44	48	46	45
Glutamic acid (Glu)	78	74	74	69	71	78	72	71
Hydroxylic groups	142	134	146	140	150	131	147	144
Serine (Ser)	64	49	63	35	48	37	35	37
Threonine (Thr)	25	22	25	24	21	25	18	18
Hydroxyproline (Hyp)	50	59	55	79	78	69	91	86
Tyrosine (Tyr)	3	4	3	2	3	0	3	3
Hydrophobic groups	286	314	280	309	321	336	322	326
Alanine (Ala)	96	119	108	122	119	119	112	114
Valine (Val)	18	19	18	15	28	22	26	22
Leucine (Leu)	22	23	20	23	21	22	24	25
Isoleucine (Ile)	11	9	11	8	7	12	10	11
Proline (Pro)	106	114	95	119	117	133	132	135
Phenylalanine (Phe)	16	15	12	13	13	14	14	13
Methionine (Met)	17	15	16	9	16	14	4	6

The low content of proline and hydroxyproline, which are involved in the stabilization of collagen-like triple helices, leads to the fact that the secondary structure of fish gelatin is represented more by β-turn/β-shift structures than triple helices, as was found by FTIR spectroscopy [23].

This distinguishes it from mammalian gelatin [5,48,49]. In addition, [2] reported a negative impact of a large number of β-structures on the functional properties of fish gelatin.

3. Sole-Gel Transition and Rheological Properties of Fish Gelatin Gel

The amino acid composition, and hence the structural features, of fish gelatin is the reason that fish gelatin as a food gelling agent is inferior to mammalian gelatin in some functional characteristics (gelation and melting temperatures, rheological characteristics of solutions and gels) [1,2,42].

It was established by optical rotation that the conformational transition random coil → helix ends when the solution is cooled to a temperature of about 35 °C for gelatin from mammals [50], and to a temperature of 15 to 20 °C for gelatin from various cold-water fish species [1,21]. Below the conformational transition temperature, the triple collagen-like helices of gelatin are combined into a spatial network, forming a thermo-reversible viscoelastic hydrogel [44,45,51].

Table 2 shows the gelation and melting temperatures of 10% (*w/v*) gelatin gels from cold and warm water fish. For comparison, data for mammalian gelatin (porcine and bovine) are shown. Data is obtained from various literature sources.

Table 2. Gelling and melting temperatures of cold and warm water fish gelatin gels compared to mammalian gelatin gels for 6.67–10% (*w/v*) gel.

Gelatin	Gelling Temperature, °C	Melting Temperature, °C	References
Cold water fish gelatin	4–8	16–18	[1]
	7–11	11–19	[2]
	4–5	12–13	[5]
	4–12	<17	[9]
	4–10	13–16	[21]
		16–21	[34]
	7–9	18–20	[52]
	12	14–21	[53]
Warm water fish gelatin	21–22	28–29	[1]
	15–20	20–27	[2]
	18–19	24–29	[9]
		22	[24]
		22–29	[34]
	19–22	24–25	[53]
Mammalian gelatin	26–27	33–34	[1]
	20–25	28–31	[2,34]
		29	[24]

As seen in Table 2, fish gelatin gels generally form and melt at a lower temperature than mammalian gelatin gels. At the same time, the values of the temperatures of gelation and the melting of gelatin gels from warm-water fish approximately correspond to the temperatures of gelatin from mammals. Thus, the melting point of gelatin gel from Black Tilapia leather is 28.9 °C [34]. There is a correlation between the content of proline and hydroxyproline (see Table 1) in gelatin and the temperatures of gelation and the melting of gels (see Table 2), which is also noted by Karim [2] and Wasswa [14].

Another significant limitation to the use of fish gelatin as a gelling agent in food technologies is the low rheological parameters (strength, elastic moduli) of gels as compared to mammalian gelatin [1,42,53]. This is also due to the reduced content of the amino acids, proline and hydroxyproline. Traditionally, the standard Bloom method is used to assess the strength of 6.67% (*w/v*) gelatin gels held at 10 °C for 17 h [1,42]. The Bloom strength of gelatin gels for cold-water species does not usually exceed 100 g; for warm-water species this is—200 g, while for gelatin from mammals it reaches 320 g and higher [25,42,53]. At the same time, the Bloom strength of gelatin gels for some species of warm-water fish (Nile tilapia, catfish, grass carp) is comparable and may even exceed the strength of mammalian gelatin gels [2,25].

Solution viscosity is the second most important commercial property of gelatin [25,45]. The viscosity of a gelatin solution is measured by a standard method at a concentration of 6.67% (*w/v*) and a temperature of 60 °C [1,34,53]. Commercial gelatin solutions have a viscosity of 2.0 to 7.0 MPas for most types, and over 13.0 Mpas for specialized types [34,45]. The viscosity of gelatin obtained from different fish species has different values, which depend on the specific characteristics of the fish raw materials, production conditions, molecular weight, and degree of polydispersity [14,34,53]. Thus, the viscosity of gelatin from the skin and bones of African catfish was 1.13 and 0.66 MPas, respectively; the viscosity of gelatin from the rainbow trout leather was 3.53 MPas; the viscosity of gelatin from the skin and bones of tiger-toothed croaker was 10.5 and 8.3 MPa, while the viscosity of farmed giant catfish skin gelatin is 112.5 MPas [34,53].

It should be noted that gelatin, which forms the least stable gel at 10 °C, is, at the same time, characterized by the highest values of solution viscosity at 60 °C in the region of about pI of gelatin [25]. For example, the viscosity of Alaska Pollock skin gelatin (cold water type) is 120 MPa (with a Bloom gel strength of 98 g), and the viscosity of tilapia skin gelatin (warm water type) is 38 MPas (Bloom 273 g) [14,34]. For comparison, the viscosity of pork skin gelatin is 47 MPas (Bloom 240 g) [14], while the viscosity of bovine gelatin is only 5.5 MPas (Bloom 323 g) [25].

Bloom gel strength and gelatin solution viscosity are spot measurements [1,45]. More complete information on the kinetics of changes in the viscoelastic characteristics of gelatin during gelation can be obtained by periodic oscillations [52]. The gelation process is kinetic because the storage modulus G′ can increase almost infinitely over time, although the most dramatic increase is observed in the first two hours [1,54,55]. It was shown by Yoshimura [24] that during cooling of a gelatin solution of 3.0% (*w/v*), the elastic modulus G′ began to increase sharply (the beginning of gelation) upon reaching a temperature of ~30 °C for pig gelatin, and only upon reaching ~21 °C for shark gelatin. Accordingly, at the same temperature values, a sharp drop in the tangent of the angle of mechanical losses, tanδ, was observed. Besides, when the gels were kept for 4 h at 4 °C, G′ reached ~2000 Pa for pig gelatin and was almost twice that for shark gelatin. When comparing the G′ values of gelatin gels obtained from cold-water fish and mammals, a much more significant difference is observed. Thus, the G′ of a bovine gelatin gel practically does not differ from the G′ of tilapia gelatin (warm-water species) [56], but exceeds the G′ of cod gelatin by about 10 times in a wide concentration range [26]. It was also noted by Haug [21] that gelatin gels from cold-water fish species (cod, haddock, pollack) have significantly lower G′ values than gelatin gels from cattle hide.

It is well known that the physical and functional properties of gelatin depend not only on their amino acid composition but also on their Bloom index [57–59], molecular weight distribution, on the relative contents of α-, β-, γ-chains and on the presence of protein fragments of low molecular weight [1,44,53]. So the viscosity of solutions [25] and strength of gels [1] are partially controlled by molecular weight and polydispersity of gelatin regardless of their source of origin. The fish gelatin extracted under mild conditions (pH5) showed higher content of α-chains, lower content of low molecular components and narrower molecular weight distribution then gelatin obtained under more aggressive acidic and alkaline conditions. This was found for gelatin from the skin of cold water species: Atlantic cod [23] and Atlantic salmon [60–62]. At the same time, salmon gelatin showed lower molecular weight compared with bovine gelatin [60]. Accordingly, fish gelatin samples extracted under pH5 with higher content of α-chains formed gels with higher values of gel strength, viscoelastic properties, gelling and melting temperatures [23,61,62].

4. Improving the Functional Properties of Fish Gelatin Food Gel

The noted serious disadvantages of fish gelatin in comparison with gelatin from mammals (see paragraph 3) can be eliminated by using natural polysaccharides that have the status of food additives [20,63]. Macromolecules of gelatin carry charged basic and acidic groups, as well as hydroxyl groups and hydrophobic hydrocarbon radicals (see Table 1). Therefore, gelatin can form polyelectrolyte complexes with ionic polysaccharides, entering with them into intermolecular electrostatic interactions [63–65]. Hydrogen bonds [66,67] and hydrophobic interactions [63,67] also play an important role in the stabilization of the polysaccharide–gelatin complexes.

The complexation of fish gelatin with polysaccharides depends on pH, as shown by the example of gelatin from cold-water fish with the anionic polysaccharide gum arabic in a wide pH range [41]. Since gelatin was extracted under alkaline conditions, its isoelectric point pI was 4.8. Concentrated emulsions, stabilized by gum arabic–fish gelatin complexes, with high values of the storage modulus G′ and low values of the loss modulus G″, are formed in the acidic pH range (3.6) in the region below the pI of gelatin. Obviously, under these conditions (provided that the dissociation of the carboxyl groups of gelatins is suppressed), the electrostatic interactions of negatively charged gum arabic with basic gelatin residues are most effective. Measurement of the turbidity of the solution shows that, with a further decrease in pH, large insoluble aggregates of gum arabic with fish gelatin are formed [42]. A similar effect of an increase in the G′ of emulsions stabilized with fish gelatin, upon the addition of another anionic polysaccharide, pectin, has been considered by [43].

The anionic marine polysaccharide proposed to improve the functional properties of fish gelatin, in order to replace mammalian gelatin in food technologies; is sodium alginate [37–39]. At low pH values of 3.5 (below pI gelatin), mixtures of sodium alginate and gelatin from the skin of cold-water fish species (cod, pollock, haddock) exhibit a high foaming ability and form stable food foams [39]. This is due to the favourable conditions for electrostatic interactions of gelatin with alginate in this pH range. In the pH region below pI, fish gelatin forms large insoluble aggregates with sodium alginate [37]. The addition of the anionic polysaccharide sodium alginate leads to an increase in the elastic moduli (storage modulus G′ and loss modulus G″) of fish gelatin gel [68] (Figure 1).

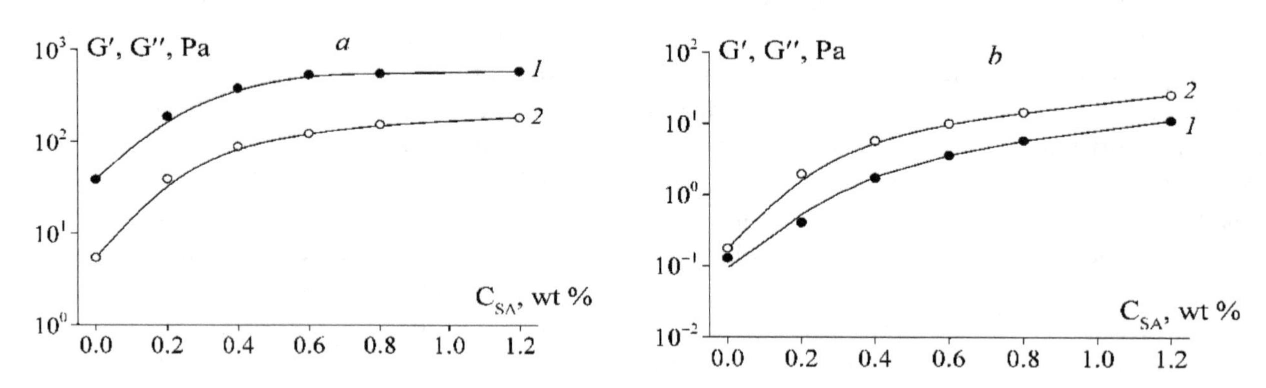

Figure 1. Dependencies of storage G′ (1) and loss G″ (2) moduli on the concentration of sodium alginate C_{SA} in fish gelatin gels at 4 °C (**a**) and 14 °C (**b**). Gelatin concentration is 10 wt%.

This is in agreement with the results of studying the interactions of anionic gum arabic with fish gelatin described by Anvari [41], and Chung [42].

Sow [38] proposed a schematic model of the mechanism for the formation of the spatial network of the complex gel of sodium alginate with fish gelatin from tilapia, due to electrostatic interactions between biopolymers. The area of ratios between polysaccharide and gelatin, in which the most durable gels were formed, was established.

The melting point and elastic modulus of gelatin gel from the skin of cold-water fish species increases with the addition of κ-carrageenan [35]. This is due to the formation of a complex κ-carrageenan–fish gelatin network, in which electrostatic interactions between negative κ-carrageenan groups and positive gelatin groups create additional nodes. A similar effect of κ-carrageenan additives on the melting point and strength characteristics of gelatin gel from tilapia fish skin has been described by Pranoto [36]. Another anionic polysaccharide, gellan, has a similar effect [36]. The strengthening effect of charged polysaccharides is primarily due to the presence of electrostatic interactions. In this case, hydrogen bonds play a significant role in the complexation of κ-carrageenan and gellan with fish gelatin during the modification of the gel nanostructure [6]. This is in agreement with the conclusions obtained when describing the mechanism of the complexation of κ-carrageenan with mammalian gelatin [64,67].

The schematic model of the formation of a complex κ-carrageenan–gelatin gel network is presented in Figure 2. Using high resolution ^{1}H NMR spectroscopy it was shown that, upon cooling, electrostatic interactions and hydrogen bonds between κ-carrageenan and gelatin start to dominate, while the role of hydrophobic interactions minimizes [67]. The contribution of electrostatic interactions and hydrogen bonds in κ-carrageenan–gelatin complexation was also indicated by FTIR spectroscopy [64]. Complexes aggregation occurs due to the formation of intermolecular triple collagen-like helices of gelatin α-chains [64,67]. In this case, as is well known, hydrogen-bonded water network makes a major contribution to the stabilization of the collagen-like helices, increasing their crystallinity [44,69,70]. In particular, such a stabilizing role of water has been noted for gelatin from the skin of Atlantic salmon [60,61].

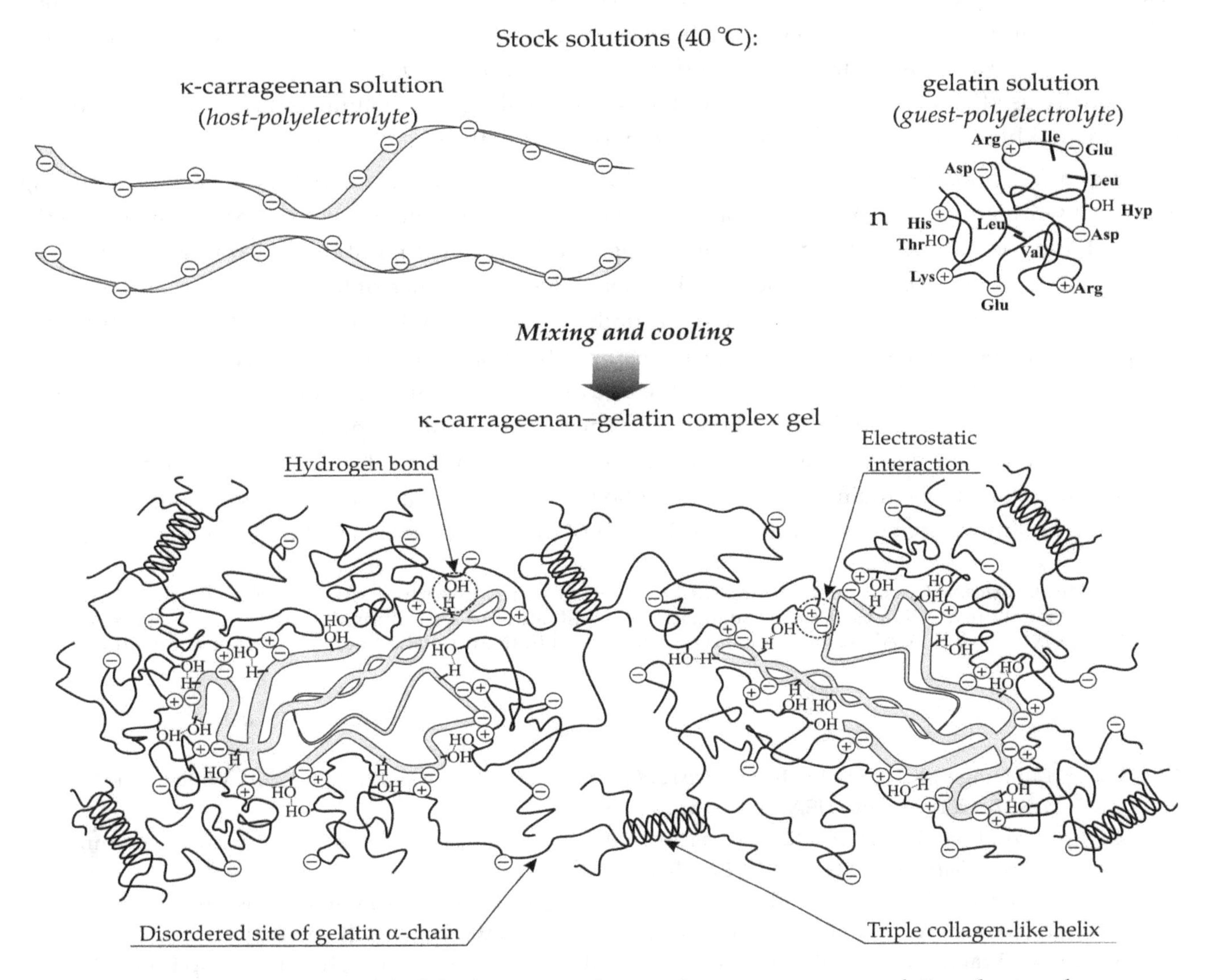

Figure 2. Schematic model of the formation of a complex κ-carrageenan–gelatin gel network.

In general, it should be noted [6,35,36] that fish gelatin modified with anionic polysaccharides from seaweed κ-carrageenan and gellan can be a good alternative to mammalian gelatin (porcine or bovine) in the food industry.

Unlike the anionic polysaccharides discussed above, the cationic polysaccharide chitosan can be effectively used to improve the functional properties of fish gelatin in the pH range above pI, where the electrostatic interactions of charged carboxyl groups of gelatin and positive groups of chitosan will be most pronounced [29]. However, it was found using the FTIR method that, in the acidic pH region (5) below pI of gelatin extracted under acidic conditions (8.9), where both biopolymers carry a positive charge, fish gelatin interacts with chitosan, mainly due to hydrogen bonds, but electrostatic interactions also make a small contribution [40]. A similar feature of the interaction of chitosan with gelatin from bovine skin at pH range of 3.2–3.9 below pI of gelatin extracted under alkaline conditions (4.7) was noted by Voron'ko [71].

5. Conclusions

Fish gelatins, in contrast to gelatin obtained from mammals, has a lower content of proline and hydroxyproline, amino acids responsible for the stabilization of collagen-like triple helices, and also of lower molecular weight. This is especially true for cold-water fish species and, respectively, for fish gelatin extracted under more aggressive conditions. In this regard, the secondary structure of fish gelatin, in contrast to mammalian gelatin, is represented more by β-turn/β-shift structures than triple helices. As a result, fish gelatin as a food gelling agent has several limitations compared to the traditionally used porcine or bovine gelatin. It has lower gelling and melting temperatures, reduced gel strength, and higher consumption in the production of structured food products.

Replacing porcine and bovine gelatin in the food industry with fish gelatin is very tempting because it will significantly expand the sales market by attracting potential consumers, for whom, at the moment, the consumption of products containing gelatin from the mentioned mammalian species is unacceptable for religious and ethnocultural reasons. Besides, the use of fish gelatin will eliminate the risk of infection by prion diseases and will partially solve the problem of disposal of waste from the fish processing industry. The disadvantages of fish gelatin as a food gelling agent can be eliminated by modifying its functional characteristics by the addition of natural ionic polysaccharides capable of forming polyelectrolyte complexes with gelatin. The formation of complexes in a pH range below pI of gelatin (for anionic polysaccharides) and above pI of gelatin (for cationic polysaccharides), due to electrostatic interactions (mainly) and hydrogen bonds, will lead to an increase in the temperatures for formation and melting of gels, and a strengthening of the spatial network of the gelatin hydrogel.

Author Contributions: Conceptualization, S.R.D. and N.G.V.; project administration, S.R.D.; formal analysis, N.G.V.; investigation, Y.A.K. and D.S.K.; visualization—Y.A.K. and D.S.K.; writing of the original draft preparation, N.G.V.; supervision, S.R.D. All authors have read and agreed to the published version of the manuscript.

References

1. Haug, I.J.; Draget, K.I. Gelatin. In *Handbook of Hydrocolloids*, 2nd ed.; Phillips, G.O., Williams, P.A., Eds.; CRC Press: Boca Raton, FL, USA, 2009; pp. 142–163.
2. Karim, A.A.; Bhat, R. Fish gelatin: Properties, challenges, and prospects as an alternative to mammalian gelatins. *Food Hydrocoll.* **2009**, *23*, 563–576. [CrossRef]
3. Sultana, S.; Ali, M.M.; Ahamad, M.N.U. Gelatine, collagen, and single cell proteins as a natural and newly emerging food ingredients. *Prep. Process. Relig. Cult. Foods* **2018**, 215–239. [CrossRef]
4. Kouhi, M.; Prabhakaran, M.P.; Ramakrishna, S. Edible polymers: An insight into its application in food, biomedicine and cosmetics. *Trends Food Sci. Technol.* **2020**, *103*, 248–263. [CrossRef]
5. Karim, A.A.; Bhat, R. Gelatin alternatives for the food industry: Recent developments, challenges and prospects. *Trends Food Sci. Technol.* **2008**, *19*, 644–656. [CrossRef]
6. Sow, L.C.; Kong, K.; Yang, H. Structural Modification of Fish Gelatin by the Addition of Gellan, κ-Carrageenan, and Salts Mimics the Critical Physicochemical Properties of Pork Gelatin. *J. Food Sci.* **2018**, *83*, 1280–1291. [CrossRef] [PubMed]
7. Huang, T.; Tu, Z.; Shangguan, X.; Sha, X.; Wang, H.; Zhang, L.; Bansal, N. Fish gelatin modifications: A comprehensive review. *Trends Food Sci. Technol.* **2019**, *86*, 260–269. [CrossRef]

8. Ahmed, M.; Verma, A.K.; Patel, R. Collagen extraction and recent biological activities of collagen peptides derived from sea-food waste: A review. *Sustain. Chem. Pharm.* **2020**, *18*, 100315. [CrossRef]
9. Gomez-Guillen, M.C.; Gimenez, B.; Lopez-Caballero, M.E.; Montero, M.P. Functional and bioactive properties of collagen and gelatin from alternative sources: A review. *Food Hydrocoll.* **2011**, *25*, 1813–1827. [CrossRef]
10. Gomez-Guillen, M.C.; Turnay, J.; Fernandez-Diaz, M.D.; Ulmo, N.; Lizarbe, M.A.; Montero, P. Structural and physical properties of gelatin extracted from different marine species: A comparative study. *Food Hydrocoll.* **2002**, *16*, 25–34. [CrossRef]
11. Akbar, I.; Jaswir, I.; Jamal, P.; Octavianti, F. Fish gelatin nanoparticles and their food applications: A review. *Int. Food Res. J.* **2017**, *24*, 255–264.
12. Bhagwat, P.K.; Dandge, P. Collagen and collagenolytic proteases: A review. *Biocatal. Agric. Biotechnol.* **2018**, *15*, 43–55. [CrossRef]
13. Lv, L.; Huang, Q.; Ding, W.; Xiao, X.; Zhang, H.; Xiong, L. Fish gelatin: The novel potential application. *J. Funct. Foods* **2019**, *63*, 103581. [CrossRef]
14. Wasswa, J.B.; Tang, J.; Gu, X. Utilization of Fish processing By-Products in the Gelatin Industry. *Food Rev. Int.* **2007**, *23*, 159–174. [CrossRef]
15. Arvanitoyannis, I.S.; Kassaveti, A. Fish industry waste: Treatments, environmental impacts, current and potential uses. *Int. J. Food Sci. Technol.* **2008**, *43*, 726–745. [CrossRef]
16. Bruno, S.F.; Ekorong, F.J.A.A.; Karkal, S.S.; Cathrine, M.S.B.; Kudre, T.G. Green and innovative techniques for recovery of valuable compounds from seafood by-products and discards: A review. *Trends Food Sci. Technol.* **2019**, *85*, 10–22. [CrossRef]
17. Uranga, J.; Etxabide, A.; Cabezudo, S.; de la Caba, K.; Guerrero, P. Valorization of marine-derived biowaste to develop chitin/fish gelatin products as bioactive carries and moisture scavengers. *Sci. Total Environ.* **2020**, *706*, 135747. [CrossRef] [PubMed]
18. Food and Agriculture Organization (FAO). *The State of World Fisheries and Aquaculture 2020. Sustainability in Action*; FAO: Rome, Italy, 2020; p. 244. [CrossRef]
19. Salvatore, L.; Gallo, N.; Natali, M.L.; Campa, L.; Lunetti, P.; Madaghiele, M.; Blasi, F.S.; Corallo, A.; Capobianco, L.; Sannino, A. Marine collagen and its derivatives: Versatile and sustainable bio-resources for healthcare. *Mater. Sci. Eng. C* **2020**, *113*, 110963. [CrossRef]
20. Milovanovic, I. Marine Gelatin from Rest Raw Materials. *Appl. Sci.* **2018**, *8*, 2407. [CrossRef]
21. Haug, I.J.; Draget, K.I.; Smidsrod, O. Physical and rheological properties of fish gelatin compared to mammalian gelatin. *Food Hydrocoll.* **2004**, *18*, 203–213. [CrossRef]
22. Oliveira, V.D.M.; Assis, C.R.D.; Costa, B.D.A.M.; Neri, R.C.D.A.; Monte, F.T.; Freitas, H.M.S.D.C.V.; França, R.C.P.; Santos, J.; Bezerra, R.D.S.; Porto, A.L.F. Physical, biochemical and spectroscopic techniques for characterization collagen from alternative sources: A review based on the sustainable valorization of aquatic by-products. *J. Mol. Struct.* **2021**, *1224*, 129023. [CrossRef]
23. Derkach, S.R.; Kuchina, Y.A.; Baryshnikov, A.V.; Kolotova, D.S.; Voron'ko, N.G. Tailoring Cod Gelatin Structure and Physical Properties with Acid and Alkaline Extraction. *Polymers* **2019**, *11*, 1724. [CrossRef] [PubMed]
24. Yoshimura, K.; Terashima, M.; Hozan, D.; Ebato, T.; Nomura, Y.; Ishii, Y.; Shirai, K. Physical Properties of Shark Gelatin Compared with Pig Gelatin. *J. Agric. Food Chem.* **2000**, *48*, 2023–2027. [CrossRef] [PubMed]
25. See, S.F.; Hong, P.K.; Ng, K.L.; Wan Aida, W.M.; Babji, A.S. Physicochemical properties of gelatins extracted from skins of different freshwater fish species. *Int. Food Res. J.* **2010**, *17*, 809–816.
26. Gilsenan, P.M.; Ross-Murphy, S.B. Rheological characterization of gelatin gels from mammalian and marine sources. *Food Hydrocoll.* **2000**, *14*, 191–195. [CrossRef]
27. Bhat, R.; Karim, A.A. Ultraviolet irradiation improves gel strength of fish gelatin. *Food Chem.* **2009**, *113*, 1160–1164. [CrossRef]
28. Da Silva, R.S.G.; Pinto, L.A.A. Physical Cross-linkers: Alternatives to Improve the Mechanical Properties of Fish Gelatin. *Food Eng. Rev.* **2012**, *4*, 165–170. [CrossRef]
29. Hosseini, S.F.; Gomez-Guillen, M.C. A state-of-the-art review on the elaboration of fish gelatin as bioactive packaging: Special emphasis on nanotechnology-based approaches. *Trends Food Sci. Technol.* **2018**, *79*, 125–135. [CrossRef]
30. Gomez-Guillen, M.C.; Gimenez, B.; Montero, P. Extraction of gelatin from fish skins by high pressure treatment. *Food Hydrocoll.* **2005**, *19*, 923–928. [CrossRef]

31. Kołodziejska, I.; Kaczorowski, K.; Piotrowska, B.; Sadowska, M. Modification of the properties of gelatin from skins of Baltic cod (*Gadus morhua*) with transglutaminase. *Food Chem.* **2004**, *86*, 203–209. [CrossRef]
32. Li, X.; Liu, X.; Lai, K.; Fan, Y.; Liu, Y.; Huang, Y. Effects of sucrose, glucose and fructose on the large deformation behaviors of fish skin gelatin gels. *Food Hydrocoll.* **2020**, *101*, 105537. [CrossRef]
33. Araghi, M.; Moslehi, Z.; Nafchi, A.M.; Mostahsan, A.; Salamat, N.; Garmakhany, A.D. Cold water fish gelatin modification by a natural phenolic cross-linker (ferulic acid and caffeic acid). *Food Sci. Nutr.* **2015**, *3*, 370–375. [CrossRef] [PubMed]
34. Lin, L.; Regenstein, J.M.; Lv, S.; Lu, J.; Jiang, S. An overview of gelatin derived from aquatic animals: Properties and modification. *Trends Food Sci. Technol.* **2017**, *68*, 102–112. [CrossRef]
35. Haug, I.J.; Draget, K.I.; Smidsrod, O. Physical behavior of fish gelatin—κ-carrageenan mixtures. *Carbohyd. Polym.* **2004**, *56*, 11–19. [CrossRef]
36. Pranoto, Y.; Lee, C.M.; Park, N.J. Characterization of fish gelatin films added with gellan and κ-carrageenan. *LWT Food Sci. Technol.* **2007**, *40*, 766–774. [CrossRef]
37. Razzak, M.A.; Kim, M.; Chung, D. Elucidation of aqueous interactions between fish gelatin and sodium alginate. *Carbohydr. Polym.* **2016**, *148*, 181–188. [CrossRef] [PubMed]
38. Sow, L.C.; Toh, N.Z.Y.; Wong, C.W.; Yang, H. Combination of sodium alginate with tilapia fish gelatin for improved texture properties and nanostructure modification. *Food Hydrocoll.* **2019**, *94*, 459–467. [CrossRef]
39. Phawaphuthanon, N.; Yu, D.; Ngamnikom, P.; Shin, I.-S.; Chung, D. Effect of fish gelatin-sodium alginate interactions on foam formation and stability. *Food Hydrocoll.* **2019**, *88*, 119–126. [CrossRef]
40. Staroszczyk, H.; Sztuka, K.; Wolska, J.; Wojtasz-Pajak, A.; Kolodziejska, I. Interactions of fish gelatin and chitosan in uncrosslinked and crosslinked with EDC films: FT-IR study. *Spectrochim. Acta A* **2014**, *117*, 707–712. [CrossRef]
41. Anvari, M.; Joyner, H.S. Effect of fish gelatin-gum arabic interactions on structural and functional properties of concentrated emulsions. *Food Res. Int.* **2017**, *102*, 1–7. [CrossRef]
42. Chung, D. Fish gelatin: Molecular interactions and applications. In *Biopolymer-Based Formulations. Biomedical and Food Applications*; Pal, K., Banerjee, I., Eds.; Elsevier: Amsterdam, The Netherlands, 2020; pp. 67–85. [CrossRef]
43. Cheng, L.H.; Lim, B.L.; Chow, K.H.; Chong, S.M.; Chang, Y.C. Using fish gelatin and pectin to make a low-fat spread. *Food Hydrocoll.* **2008**, *22*, 1637–1640. [CrossRef]
44. Veis, A. *The Macromolecular Chemistry of Gelatin*; Academic Press: New York, NY, USA, 1964; p. 478.
45. Johnston-Banks, F.A. Gelatin. In *Food Gels*; Harris, P., Ed.; Springer: Dordrecht, The Netherlands, 1990; pp. 233–289.
46. Gomez-Guillen, M.C.; Lopez-Caballero, M.E.; Aleman, A.; Lopez de Lacey, A.; Gimenez, B.; Montero, P. Antioxidant and antimicrobial peptide fractions from squid and tuna skin gelatin. In *Sea By-Products as Real Material: New Ways of Application*; Le Bihan, E., Ed.; Transworld Research Network: Trivandrum, Kerala, India, 2010; pp. 89–115.
47. Liu, D.; Zhou, P.; Li, T.; Regenstein, J.M. Comparison of acid-soluble collagens from the skins and scales of four carp species. *Food Hydrocoll.* **2014**, *41*, 290–297. [CrossRef]
48. Al-Saidi, G.S.; Al-Alawi, A.; Rahman, M.S.; Guizani, N. Fourier transform infrared (FTIR) spectroscopic study of extracted gelatin from shaari (*Lithrinus microdon*) skin: Effects of extraction conditions. *Int. Food Res. J.* **2012**, *19*, 1167–1173.
49. Zhang, T.; Sun, R.; Ding, M.; Li, L.; Tao, N.; Wang, X.; Zhong, J. Commercial cold-water fish skin gelatin and bovine bone gelatin: Structural, functional, and emulsion stability differences. *LWT Food Sci. Technol.* **2020**, *125*, 109207. [CrossRef]
50. Djabourov, M.; Papon, P. Influence of thermal treatment on the structure and stability of gelatin gels. *Polymer* **1983**, *24*, 537–541. [CrossRef]
51. Ledward, D.A. Gelation of gelatin. In *Functional Properties of Food Macromolecules*; Mitchell, J.R., Ledward, D.A., Eds.; Elsevier Applied Science Publishers: London, UK, 1986; pp. 171–201.
52. Gimenez, B.; Gomez-Guillen, M.C.; Montero, P. Storage of dried fish skins on quality characteristics of extracted gelatin. *Food Hydrocoll.* **2005**, *19*, 958–963. [CrossRef]
53. Da Trindade Alfaro, A.; Balbinot, E.; Weber, C.I.; Tonial, I.B. Fish Gelatin: Characteristics, Functional Properties, Applications and Future Potentials. *Food Eng. Rev.* **2015**, *7*, 33–44. [CrossRef]

54. Ross-Murphy, S.B. Structure and rheology of gelatin gels—Recent progress. *Polymer* **1992**, *33*, 2622–2627. [CrossRef]
55. Ross-Murphy, S.B. Structure and Rheology of Gelatin Gels. *Imaging Sci. J.* **1997**, *45*, 205–209. [CrossRef]
56. Gilsenan, P.M.; Ross-Murphy, S.B. Shear creep of gelatin gels from mammalian and piscine collagens. *Int. J. Biol. Macromol.* **2001**, *29*, 53–61. [CrossRef]
57. Lai, J.Y.; Lin, P.K.; Hsiue, G.H.; Cheng, H.Y.; Huang, S.J.; Li, Y.T. Low Bloom strength gelatin as a carrier for potential use in retinal sheet encapsulation and transplantation. *Biomacromolecules* **2009**, *10*, 310–319. [CrossRef]
58. Lai, J.Y. The Role of Bloom Index of Gelatin on the Interactions with Retinal Pigment Epithelial Cells. *Int. J. Mol. Sci.* **2009**, *10*, 3442–3456. [CrossRef] [PubMed]
59. Chou, S.F.; Luo, L.J.; Lai, J.Y.; Ma, D.H.K. On the Importance of Bloom Number of Gelatin to the Development of Biodegradable in Situ Gelling Copolymers for Intracameral Drug Delivery. *Int. J. Pharm.* **2016**, *511*, 30–43. [CrossRef] [PubMed]
60. Diaz-Calderon, P.; Lopez, D.; Matiacevich, D.L.S.; Osorio, F.; Enrione, J. State diagram of salmon (*Salmo salar*) gelatin films. *J. Sci. Food Agric.* **2011**, *91*, 2558–2565. [CrossRef] [PubMed]
61. Diaz-Calderon, P.; Flores, E.; Gonzales-Munoz, A.; Perczynska, M.; Quero, F.; Enrione, J. Influence of extraction variables on the structure and physical properties of salmon gelatin. *Food Hydrocoll.* **2017**, *71*, 118–128. [CrossRef]
62. Enrione, J.; Char, C.; Perczynska, M.; Padilla, C.; Gonzales-Munoz, A.; Olguin, Y.; Quinzio, C.; Iturriaga, L.; Diaz-Calderon, P. Rheological and Structural Study of Salmon Gelatin with Controlled Molecular Weight. *Polymers* **2020**, *12*, 1587. [CrossRef]
63. Turgeon, S.L.; Laneuville, S.I. Protein + Polysaccharide Coacervates and Complexes: From Scientific Background to their Application as Functional Ingredients in Food Products. In *Modern Biopolymer Science*; Kasapis, S., Norton, I.T., Eds.; Elsevier: London, UK, 2009; pp. 327–363.
64. Derkach, S.R.; Voron'ko, N.G.; Kuchina, Y.A.; Kolotova, D.S.; Gordeeva, A.M.; Faizullin, D.A.; Gusev, Y.A.; Zuev, Y.F.; Makshakova, O.N. Molecular structure and properties of κ-carrageenan-gelatin gels. *Carbohydr. Polym.* **2018**, *197*, 66–74. [CrossRef]
65. Derkach, S.R.; Kuchina, Y.A.; Kolotova, D.S.; Voron'ko, N.G. Polyelectrolyte Polysaccharide–Gelatin Complexes: Rheology and Structure. *Polymers* **2020**, *12*, 266. [CrossRef]
66. Cao, Y.; Wang, L.; Zhang, K.; Fang, Y.; Nishinari, K.; Phillips, G.O. Mapping the Complex Phase Behaviors of Aqueous Mixtures of κ-Carrageenan and Type B Gelatin. *J. Phys. Chem. B* **2015**, *119*, 9982–9992. [CrossRef]
67. Voron'ko, N.G.; Derkach, S.R.; Vovk, M.A.; Tolstoy, P.M. Complexation of κ-carrageenan with gelatin in the aqueous phase analysed by ^{1}H NMR kinetics and relaxation. *Carbohyd. Polym.* **2017**, *169*, 117–126. [CrossRef]
68. Derkach, S.R.; Kolotova, A.V.; Voron'ko, N.G.; Malkin, A.Y. Modification of the rheological properties of fish gelatin with sodium alginate. *Polymers* **2020**, in press.
69. Yannas, I.V. Collagen and Gelatin in the Solid State. *Polym. Rev.* **1972**, *7*, 40–106.
70. Brodsky, B.; Shah, N.K. The triple-helix motif in proteins. *FASEB J.* **1995**, *9*, 1537–1546. [CrossRef] [PubMed]
71. Voron'ko, N.G.; Derkach, S.R.; Kuchina, Y.A.; Sokolan, N.I. The chitosan–gelatin (bio)polyelectrolyte complexes formation in an acidic medium. *Carbohyd. Polym.* **2016**, *138*, 265–272. [CrossRef] [PubMed]

4

The Effect of Biochar used as Soil Amendment on Morphological Diversity of Collembola

Iwona Gruss [1,*], Jacek P. Twardowski [1], Agnieszka Latawiec [2,3], Jolanta Królczyk [4] and Agnieszka Medyńska-Juraszek [5]

1 Department of Plant Protection, Wroclaw University of Environmental and Life Sciences, 50-363 Wrocław, Poland
2 Institute of Agricultural Engineering and Informatics, University of Agriculture in Kraków, 30-149 Kraków, Poland
3 Department of Geography and the Environment, Rio Conservation and Sustainability Science Centre, Pontifícia Universidade Católica, Rio de Janeiro 22453-900, Brazil
4 Department of Manufacturing Engineering and Production Automation, Faculty of Mechanical Engineering, Opole University of Technology, 45-271 Opole, Poland
5 Institute of Soil Sciences and Environmental Protection, Wroclaw University of Environmental and Life Sciences, 53, 50-357 Wrocław, Poland
* Correspondence: iwona.gruss@upwr.edu.pl

Abstract: Biochar was reported to improve the chemical and physical properties of soil. The use of biochar as a soil amendment have been found to improve the soil structure, increase the porosity, decrease bulk density, as well increase aggregation and water retention. Knowing that springtails (Collembola) are closely related to soil properties, the effect of biochar on morphological diversity of these organisms was evaluated. The main concept was the classification of springtails to the life-form groups and estimation of QBS-c index (biological quality index based on Collembola species). We conducted the field experiment where biochar was used as soil amendment in oilseed rape and maize crops. Wood-chip biochar from low-temperature (300 °C) flash pyrolysis was free from PAH (polycyclic aromatic hydrocarbon) and other toxic components. Results showed that all springtail life-form groups (epedaphic, hemiedaphic, and euedaphic) were positively affected after biochar application. The QBS-c index, which relates to springtails' adaptation to living in the soil, was higher in treatments where biochar was applied. We can recommend the use of Collembola's morphological diversity as a good tool for the bioindication of soil health.

Keywords: biochar; biological soil quality; Collembola life-form groups; QBS-c index

---

1. Introduction

One of the major threats to global agriculture is soil degradation, including decreased fertility and increased erosion [1]. The common problem is acidification and soil organic matter depletion, which decreases soil aggregate stability [2]. Therefore, the development of methods is needed to sustain soil resources by different remediation strategies. The application of organic materials like manure, compost, and biomass waste seems very promising, but a lot of attention has been paid to stable forms of organic carbon like biochars [3,4]. The main feature of biochar is the porous carbonaceous structure, which can contain amounts of extractable humic-like and fluvic-like substances [5]. Biochar was reported to improve the chemical and physical properties of soil [6]. The use of biochar as a soil amendment has been found to improve the soil structure, increase the porosity, decrease bulk density, as well increase aggregation and water retention [7–9]. On the other hand, the main concerns with respect to biochar use as a soil amendment is its potential contamination with heavy metals (HMs) and polycyclic aromatic hydrocarbons (PAHs) [10].

Springtails (Hexapoda: Collembola) are a key group of soil arthropods with densities often reaching thousands of individuals per square meter [11,12]. They contribute mainly in substrate decomposition and nutrient cycling [12,13]. Moreover, these organisms are sensitive to environmental changes in soil and are therefore often used as indicators of soil quality [14]. For bioindication, Collembola species diversity is used [15,16]. The disadvantage of this method is the difficulty in the determination to the species level. An alternative could be the QBS-c index (biological quality index based on Collembola), which responds to the morphological diversity of springtails [17]. Using this index, each individual is evaluated in terms of different morphological traits, e.g., for antennal length, size of furca, presence of ocelli pigmentation, and the presence of hairs and/or scales along the body. The principle of this index is that the presence of individuals with better adaptation to live in soil (with reduced appendages or less pigmented) indicates better soil quality [17,18]. Also, on the basis of morphological traits, springtails can be divided into three main life-forms [19]. First, epedaphic Collembola are adapted to live on the soil surface. The major features of this group are a pigmented body, well developed eyes and appendages, as well a fast dispersal ability. In contrast to them, soil dwelling species (euedaphic), with a relatively small, less pigmented body and reduced eyes. Their dispersal ability is limited. Species showing adaptations between epedaphic and euedaphic species are classified as belonging to the hemiedaphic group [20]. The vertical stratification of springtails reflects their function in the ecosystem. For instance, only epedaphic springtails contribute in the early stages of organic matter decomposition [19]. Ponge et al. [21] suggested that Collembola living in the soil characterized by limited active dispersal, may suffer more from land use intensification, than species living on the soil surface. In contrast, Ellers et al. [22] showed stronger effects of intensive land use on epedaphic than on euedaphic Collembola. The majority of studies on biochar effect on springtails were conducted in laboratory conditions on one model species [23–25]. Considering the impact of biochar under field conditions some experiments have been made also on nematodes [26] and earthworms [27].

The potential of biochar for pH and nutrient availability changes or improvement of some physical properties like porosity, water retention, or temperature and impact on soil microbial life, are well documented [28–30]. Therefore, springtails can be affected directly by the changes in soil chemical properties [31,32] or indirectly from biochar-induced changes in microorganisms' biomass [33]. It has been reported that many Collembola species feed on bacteria or fungi [34,35]. Biochar particles might be considered analogous to soil aggregates in that their large internal surface areas and pores could be important for biological processes [36].

Within the presented study we aimed to estimate the effect of biochar on the morphological diversity of Collembola species and evaluate its potential for field application.

It was hypothesized that:

1. Biochar will increase the soil biological quality mainly through improvement of the physical and chemical soil properties.
2. From the analyzed life-form groups of springtails, the response of euedaphic assemblages will be most distinct after biochar amendment. On the other hand, springtails living on the soil surface and in the litter layer (epigeic and hemiedaphic) will be more sensitive to cover plants.
3. The QBS-c index will show higher values in crops where biochar was applied.

2. Materials and Methods

2.1. Experimental Design

The field experiment was set up in mid-April 2014 in the south of Poland (50.5740 N, 17.8908 E) and continued until October 2016. The soil type was poor (sandy and weakly acidic) agricultural soil [37]. The climate of the area is moderately warm with an average annual temperature of 8.4 °C and an average annual rainfall of 611 mm. The biochar effect on soil dwelling springtails was explored in two crops (oilseed rape and maize) compared to control (two crops with no biochar application). Within each treatment (plot), three replicates (subplots) were established. The area of each subplot was 3 × 3 m. The research area was previously (before 2014) used as conventionally agricultural field. The forecrop was maize. Biochar was applied up to a depth of 30 cm at a rate of 50 t/ha and was mixed by ploughing. No chemical protection was applied before or during experiment period. Weeds were removed manually upon occurrence. Weeds were harvested manually a few times during the vegetation season. The maize variety in two years of the study was P8745 (FAO 250, Pioneer Company) and oilseed rape variety Monolit. The only fertilizer used in oilseed rape was ammonium sulfate 34% in a dose of 300 kg/ha, and in maize ammonium phosphate (Polydap) in a dose of 25 kg/ha. The same amount of fertilizers was applied in biochar and control treatments.

2.2. Biochar Characteristic and Soil Properties

The biochar used in the experiment was industrial produced by Fluid S.A. Company (Poland). It was produced in the low-temperature flash pyrolysis (300 °C) of pine and spruce wood chips. Its heating value was 25 MJ/kg. During the experiment selected properties of biochar (pH, organic carbon content, cation exchange capacity, heavy metal content and total PAH's) were analyzed according to International Biochar Initiative (IBI) Standard Product Definition and Product Testing Guidelines [38]. The particle size fraction of biochar applied on the field was more than 2 mm (sieve method).

The tested biochar was alkaline (pH 8.2) and had 52.3% of carbon (Table 1). The surface area of the tested biochar was low (only 16.5 m^2/g) and cation exchange capacity was also lower—39.5 cmol/kg, compared with biochars produced at higher temperatures and from other feedstock, like wheat straw, giant miscanthus, rice husk, or sewage sludge [39,40]. It was free from polycyclic aromatic hydrocarbon and the concentration of all tested toxic compounds was very low or even under the level of detection, passing fixed recommendations for acceptable levels [38].

Table 1. The chemical characteristic of biochar properties used in the experiment (sourced from Gruss et al. [41]).

Parameter	Value
pH H_2O	8.2
CEC (cmol/kg)	16.8
Carbon content (% of DM)	52.3
H/Corg ratio	0.026
Pb (g/t DM)	1.57
Mn (g/t DM)	29.7
Cu (g/t DM)	0.50
Hg (g/t DM)	0.32
Zn (g/t DM)	13.04

For physicochemical analysis, soil samples were collected twice a year from topsoil, before each crop in rotation in five replicates from each plot. The pH, total organic carbon, CEC, exchangeable acidity, and water properties were measured. Soil was classified as Cambisol [38], with a typical sandy loam texture with the addition of medium fine gravel. Application of biochar significantly increased CEC values in both trials, due to the increase of exchangeable Ca2+, Mg2+, and H+ + Al3+ (exchangeable acidity) in the soil sorption complex and total organic carbon in biochar trials (Table 2).

Table 2. Soil properties after biochar application in oilseed rape and maize (sourced from Gruss et al. [41]).

Parameter	Oilseed Rape		Maize	
	Biochar	Control	Biochar	Control
C_{org} (%)	0.94	0.92	0.94	0.78
pH H_2O	6.88	7.26	6.49	7.22
Na^+ (cmol/kg)	0.20 *	0.12	0.12	0.18
Mg^{2+} (cmol/kg)	3.14	1.13	2.76	0.86
K^+ (cmol/kg)	0.30	0.19	0.34	0.31
Ca^{2+} (cmol/kg)	5.12	2.29	5.72	2.28
CEC (cmol/kg)	8.76	3.74	8.98	4.73

* The values in bold font differ significantly between treatments.

2.3. Collembola Studies

Soil samples for Collembola analysis were taken three times during each of the vegetation season (from May to July) in 2015 and 2016. The growth stages according to the BBCH (growth stages of plants) scale [42] in the sampling dates were: maize: 10–15, 32–37, and 61–67; oilseed rape: 60–69, 72–79, and 83–89. On each date, 12 samples were taken from each subplot (36 samples from one plot), and transported to the laboratory. The samples were taken with the use of a soil sampler (diameter 5 cm and depth 10 cm). The volume of one sample was 196 cm^2. Collembola were extracted over 24 h from the soil samples with the use of Tullgren funnels modified by Murphy [43]. After the extraction the springtails were kept in 75% ethyl alcohol.

Springtails from each sample were counted and identified to the species or genus. Each individual was placed on permanent microscope slide and determined to the species level with the use of following keys [44–46]. Springtails were classified to three life-form groups (euedaphic, hemiedaphic and epigeic) according to Karaban [20]. Epedaphic forms have strong pigmentation, fully developed furca and other appendages, and pigmented eyes (8 + 8). Hemiedaphic have reduced body pigmentation, eye numbers, and a reduced furca. Euedaphic forms are characterized by an unpigmented body (or eyes' pigmentation) with eyes and furca not developed. The QBS-c (biological quality index based on Collembola species) is calculated as the sum of EMI values in each sample (Table 3). The springtails species were evaluated for seven morphometric traits according to the scale. The results were the sums of scores (EMI) obtained for each trait. Species which are well adopted to live in soil obtain more EMI scores in comparison to those with adaptation to live on the soil layer [17].

Table 3. Description of the Collembola species identified in the experiment including their morphological description.

Species	Abbr. on the CCA Biplot	Life-form Group	Size *	Pigmentation	Structures on Cuticle	Ocelli	Antennae	Legs	Furcula	EMI Scores (QBS-c)
Bourietiella hortensis (Fitch)	Bou_hor	Epedaphic	4	0	1	0	0	0	0	5
Brachystomella parvula (Schaffer)	Bra_par	Hemiedaphic	4	0	1	0	3	0	3	14
Caprainea marginata (Schoett)	Cap_mar	Epedaphic	4	0	0	0	0	0	0	4
Desoria multisetis (Carpenter & Phillips)	Des_mul	Epedaphic	4	3	0	0	0	0	0	7
Desoria tigrina (Nicolet)	Des_tig	Hemiedaphic	4	6	0	0	0	0	0	10
Folsomia sexuolata (Tullberg)	Fol_sex	Hemiedaphic	4	6	3	6	2	2	2	25
Folsomides angularis (Axelson)	Fol_ang	Hemiedaphic	4	6	3	6	3	2	3	27
Folsomides parvulus (Stach)	Fol_par	Hemiedaphic	4	6	3	3	3	3	3	25
Friesea mirabilis (Tullberg)	Fri_mir	Hemiedaphic	4	3	1	0	2	2	2	14
Hypogastrura spp.	Hopogast	Hemiedaphic	4	0	3	0	2	2	2	13
Isotoma anglicana (Lubbock)	Iso_ang	Epedaphic	4	3	1	0	0	0	0	8
Isotoma antennalis (Bagnall)	Iso_ant	Epedaphic	4	1	0	0	0	0	0	5
Isotoma viridis (Bourlet)	Iso_vir	Epedaphic	4	3	0	0	0	0	0	7
Isotomiella minor (Schaeffer)	Iso_min	Hemiedaphic	4	6	3	3	3	3	3	25
Isotomodes productus (Axelson)	Iso_pro	Hemiedaphic	4	6	3	6	2	3	2	26
Isotomurus palustris (Mueller)	Iso_pal	Epedaphic	4	1	0	0	0	0	0	5
Isotomutus gallicus (Carapelli et al.)	Iso_gal	Epedaphic	4	1	0	0	0	0	0	5
Lepidocyrtus violaceus (Fourcroy)	Lep_vio	Epedaphic	4	0	0	0	0	0	0	4
Mesaphorura spp.	Mesaphor	Euedaphic	4	6	3	6	3	3	6	31
Parisotoma notabilis (Schaeffer)	Par_not	Hemiedaphic	4	6	1	3	0	2	0	16
Proisotoma minima (Absolon)	Pro_mini	Hemiedaphic	4	3	1	3	2	2	2	17
Proisotoma minuta (Tullberg)	Pro_minu	Hemiedaphic	4	3	1	0	2	2	2	14
Proisotoma tenella (Reuter)	Pro_ten	Hemiedaphic	4	3	1	0	2	2	2	14
Protaphorura spp.	Protapho	Euedaphic	4	6	3	6	3	3	6	31
Pseudosinnela sexoculata (Schott)	Pse_sex	Hemiedaphic	4	6	1	3	0	0	0	14
Sminthurides parvulus (Krausbauer)	Smi_par	Epedaphic	4	0	0	0	0	0	0	4
Sminthurinus alpinus (Gisin)	Smi_alp	Epedaphic	4	0	1	0	2	0	0	7
Sphaeridia pumilis (Krausbauer)	Sph_pum	Epedaphic	4	0	0	0	0	0	0	4
Stenacidia violacea (Reuter)	Ste_vio	Epedaphic	4	0	0	0	0	0	0	4
Stenaphorura spp.	Stenapho	Euedaphic	4	6	3	6	3	3	6	31

* Size: >3 mm = 0; 2–3 mm = 2; <2 mm = 4.

Pigmentation: Fully pigmented=0; only strips on the body = 1; r = Reduced to appendages = 3; none = 6.

Structures on cuticle: Well developed chaeta or scales, present trichobothria = 0; relatively low number of structures on cuticle=1; Reduced number of chaetae, presence of PSO (pseudocelli) on cuticle = 3; Low number of chaeteae, other structures present only in selected parts of the body = 6.

Number of ocelli in the eye spot: 8 + 8 = 0; 6 + 6 = 2; form 5 + 5 to 1 + 1 = 3; absence of ocelli = 6.

Antennae: antennae longer than the head = 0; antennae more or less the same length as the head = 2; antennae shorter than the head = 3; antennae much shorter than the head = 6.

Legs: Well developed = 0, Medium developed = 2; Short = 3; Reduced or with reduced claw and mucro =6.

Furcula: Well developed = 0; Medium developed = 2; Short with reduced number of chaetae = 3; the absence of mucro and modification of manubrium = 5; Furcula reduced in residual form = 6.

2.4. Data Analysis

The effect on springtails life-form groups and QBS-c index was analyzed with the mixed model in SAS University Edition (proc Mixed). In the analysis, the effect of crop and treatment, as well their interaction, were included. The year and term of the study were the random factors. The abundance of springtails per sample was relatively low (with the number of individuals of 10.5 per sample). Therefore, the abundance of springtails was calculated for 1 m^2, knowing that the area of one sample was 0.000785 m^2 (5 cm diameter).

The springtails abundance as well morphometric trails in relation to experimental treatments was analyzed using canonical correspondence analysis (CCA) in Canoco, Version 4.5 (Ithaca, New York, USA). Significance of the first canonical axis and all axes together was calculated with Monte Carlo Test.

3. Results

The abundance of Collembola per m^2 and the QBS-c index differed significantly between all tested factors (Table 4). Generally, the most abundant group was epedaphic, then hemiedaphic, and the least, euedaphic Collembola (Figure 1). Within the epedaphic and hemiedaphic groups, both the plant ($p < 0.0001$) and biochar ($p = 0.0009, 0.0058$) significantly differed with respect to springtails abundance. Springtails were significantly more abundant in oilseed rape in comparison to maize crop. At the same time more Collembola were found in biochar, but only in oilseed rape. The abundance of euedaphic Collembola differed between biochar and control plots ($p = 0.003$). In both crops, significantly more individuals were found in biochar treated soil.

Table 4. Results of repeated ANOVA (GML, $p \leq 0.05$) considering effects on treatment, plant and year, and its interactive effects on Collembola life-form groups and QBS-c index.

Dependent Variable	Treatment		Plant		Treatment × Plant	
	F *	p	F *	p	F	p
Epedaphic	11.10	0.0009	20.53	<0.0001	12.43	0.0004
Hemiedaphic	7.65	0.0058	22.04	<0.0001	11.53	0.0007
Euedaphic	13.09	0.0003	0.04	0.8501	2.15	0.1425
QBS-c	6.16	0.01132	4.77	0.0292	0.02	0.8943

* F = ratio of two different measure of variance for the data.

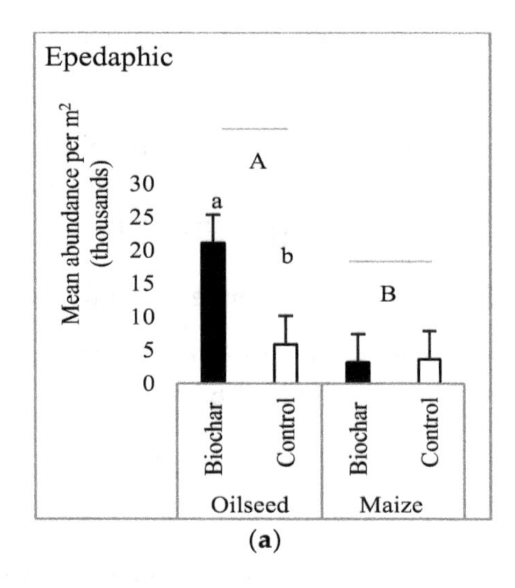

(a)

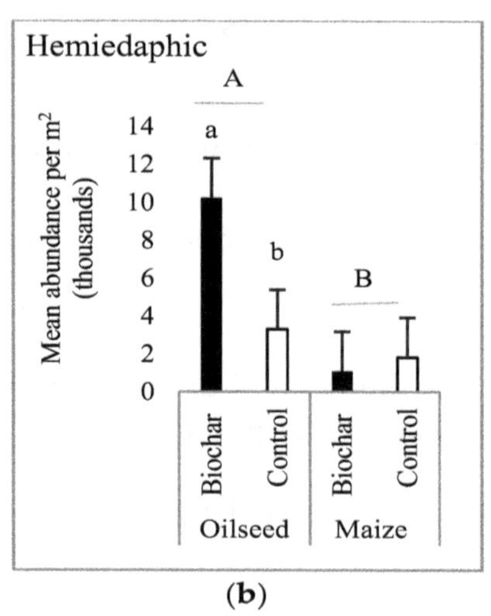

(b)

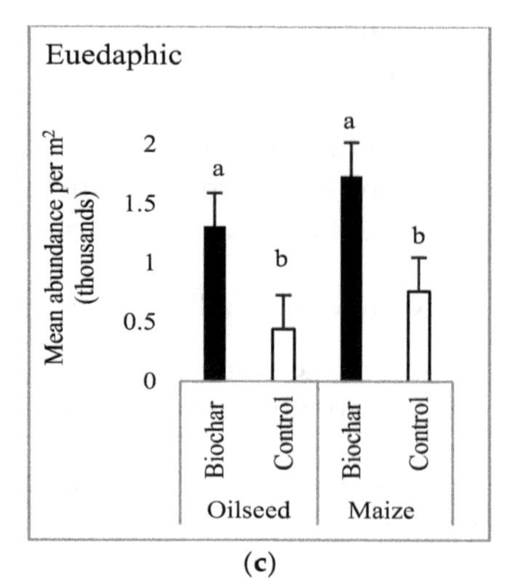

(c)

Figure 1. Effect of crop and biochar application on different Collembola life-form groups. Note: (**a**), (**b**), (**c**); indicate significant differences ($p \leq 0.05$) between biochar and control; A, B indicate significant differences ($p \leq 0.05$) between plants; the results of repeated ANOVA used for data analysis are given in the Table 4.

Considering the QBS-c index, it was significantly higher in oilseed rape crop in comparison to maize (Figure 2) ($p = 0.0292$). In both plants, the index was significantly higher in treatments, where biochar was applied ($p = 0.01132$). As shown by the life-form groups, significantly higher QBS-c index was found in oilseed rape only.

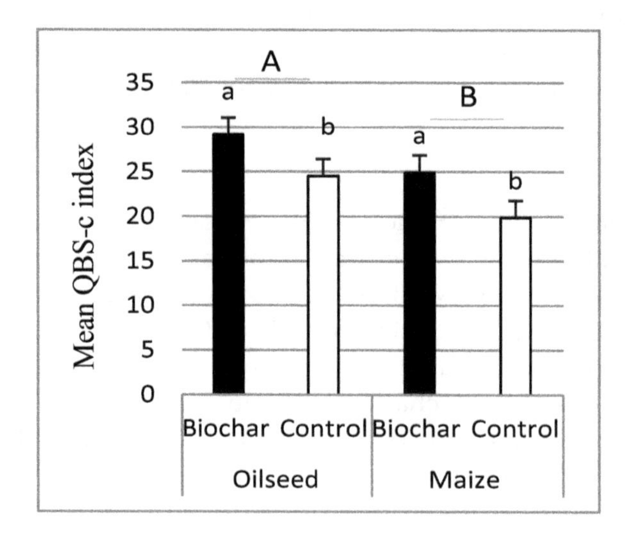

Figure 2. Effect of crop and biochar application on the QBS-c index. Note: (a), (b), (c); indicate significant differences ($p \leq 0.05$) between biochar and control; A, B indicate significant differences ($p \leq 0.05$) between plants; the results of repeated ANOVA used for data analysis are given in the Table 4.

The morphometric traits used for the calculation of QBS-C index were correlated with experimental treatments (Figure 3). The significance of the first canonical axis (CCA1), as well all axes together was $p = 0.002$ (Table 5). Biochar (in oilseed rape) was positively correlated with reduced legs, antennae and furcula, as well absence of specific structures on cuticle. In maize, where biochar was applied, springtails were characterized by a reduced number of ocelli. Size and pigmentation were positively correlated with oilseed rape and maize, both without biochar.

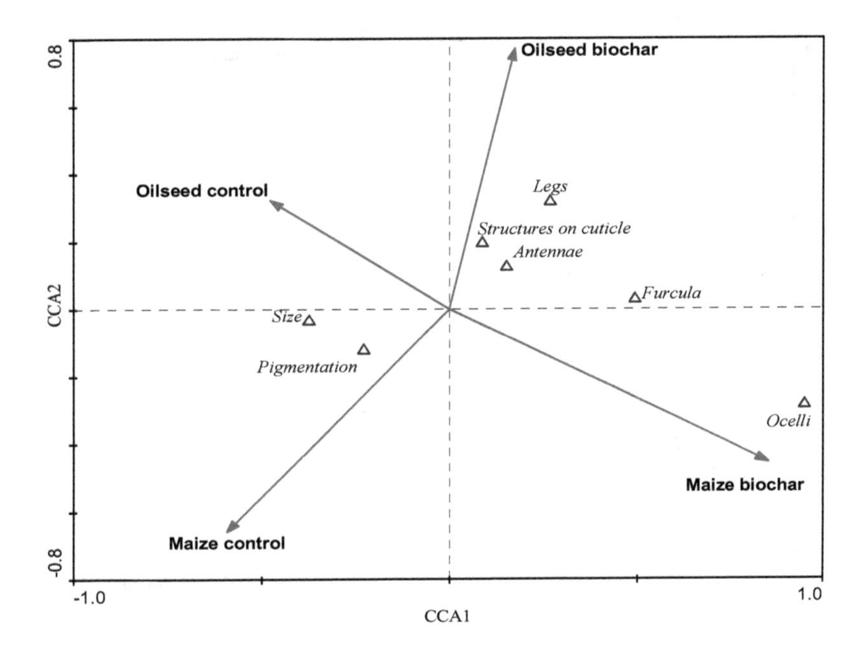

Figure 3. Canonical correspondence analysis (CCA) biplot on Collembola morphomentric traits in relation to experimental treatments. Notes: The morphometric traits are described in more detail in Table 1.

Table 5. Results of Canonical correspondence analysis (CCA) of Collembola morphometric traits correlated with experimental treatments.

CCA Axes	1	2	3	4
Eigenvalues	0.013	0.002	0.000	0.103
Morphometric traits-environment correlations	0.013	0.002	0.000	0.103
Significance of the first canonical axis		$F = 26.362, p = 0.002$		
Significance of all canonical axes		$F = 10.296, p = 0.002$		

The eigenvalues of the first two CCA axes were 0.0102 and 0.041, respectively (Table 6). Both the first canonical axis (CCA1), as well all axes were significant ($p = 0.002$, Monte Carlo test). As shown on the CCA biplot (Figure 4), the Collembola community was affected more by crop than by biochar. There was only minor effect of biochar in oilseed. In maize the group of species related to the control site (e.g., *Desoria tigrina* and *Pseusinella sexoculata*) differed from the species which preferred biochar (e.g., *Stenaphorura* spp., *Isotoma antenalis*). Considering life-form groups, most of the hemiedaphic species were found in the oilseed crops to oil seed (with similar effect for biochar and control). The epedaphic species were frequently found in all of the treatments, while euedaphic mostly in maize with biochar.

Table 6. Results of Canonical correspondence analysis (CCA) of Collembola species in relation to experimental treatment.

CCA Axes	1	2	3	4
Eigenvalues	0.102	0.041	0.013	0.476
Species-environment correlations	0.563	0.388	0.270	0.000
Significance of the first canonical axis		$F = 7.754, p = 0.002$		
Significance of all canonical axes		$F = 4.016, p = 0.002$		

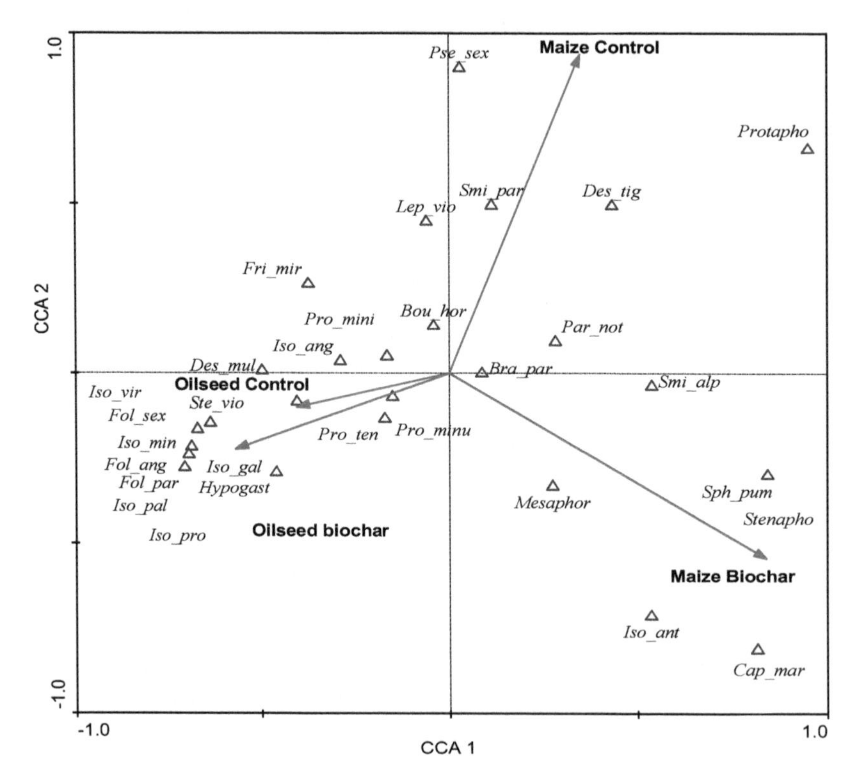

Figure 4. Canonical correspondence analysis (CCA) biplot on Collembola species in relations to the crop and biochar application. * Designation of the life-form groups: epedaphic, hemiedaphic, euedaphic. Note: the detailed description of the species is included in Table 1.

4. Discussion

The low temperature pine/wood chip biochar used in the experiment is characterized by the high carbon content but low surface area, and ability for nutrient storage, therefore the effect of its application to soil had little effect on soil properties. Tested pined wood biochar, was free from PAH and the concentration of all tested toxic compounds was very low or even under the level of detection. The release of toxic compounds like PAH's and heavy metals after biochar application seems to be crucial for survival and reproduction of soil fauna [47,48]. Some of the soil properties were improved, like total organic carbon content (only in maize trials) or CEC, as reported by other authors [49–53]. The liming effect, which is mostly expected [54,55], when biochar is applied to soil was not determined in our experimental trials.

Our estimations showed a higher QBS-c index value and higher individual number of particular life-form groups in biochar amended soils, confirming the hypothesis of soil biological properties improvement. We state that the positive response of Collembola to biochar addition was the result of improved soil properties. Some authors have found that soil mesofauna abundances are closely related to soil conditions, especially soil pH and organic matter content [56,57]. To compare, the significant increase of fungivorous nematodes was found after biochar application in the rates from 12 to 48 t/ha [26]. In contrast, no response of soil faunal feeding activity to biochar addition was found in the rates range from 0 to 30 t/ha [58]. Also, Castracani et al. [59] did not find any interaction between biochar and the abundance of epigeic macroarthropods in the rate of 14 t/ha.

The response of euedaphic springtails was predicted to be most distinct after biochar amendment. This would result from low dispersal ability and living in deeper soil layers [60]. Therefore, euedaphic springtails would be more sensitive to changes in the soil environment [61]. In our experiment the abundance of euedaphic springtails was relatively low compared to hemiedaphic and epedaphic groups. However, in both analyzed crops, its number increased after biochar application. Considering the two other groups living on upper soil layers, the effect was significant only for oilseed.

The main concept of the QBS-c index is that soil quality is positively correlated with the number of Collembola species that are well adapted to soil habitats [17]. Otherwise, numerous occurrences of euedaphic forms of springtails in a given habitat can indicate the better biological quality of the soil [43]. Based on the results, we can agree with the hypothesis that the QBS-c index will have higher values in crops where biochar was applied. For instance, in the study of Twardowski et al. [61], the QBS-c showed higher soil quality in the potato crop rotation in comparison to monoculture. Jacomini et al. [62] found decreased QBS-c index values in degraded soils. To confirm our last hypothesis, two springtails life-form groups (epigeic and hemiedaphic) differed significantly between crops. More springtails were found in oilseed rape in comparison to maize crop. Considering that maize and oilseed rape are plants which differ in their development, this should also affect organisms living on the soil surface. Similarly, Op Akkerhuis [63] found differences in mesofauna abundance between crops, i.e., specifically the mesofauna groups were much more abundant in cereals than in root and tuber crops. We state also that cover plant might modify the effect of biochar on soil fauna.

5. Conclusions

The main findings of the present study are that:

(1) Crop affected more Collembola community than the biochar application. More springtails occurred in oilseed rape.

(2) In each of the life-form groups, biochar caused a significant increase in individual number of Collembola in comparison to the no-biochar treatment. The effect was significant mainly for the oilseed rape crop.

(3) The QBS-C index (biological quality index based on Collembola species) was higher in treatments where biochar was applied.

(4) Collembola related to biochar were characterized by reduced appendages and the absence of specific structures on the cuticle, what indicates better adaptation to live in soil.

To conclude, biochar was found to increase springtails abundance and diversity in field conditions. A greater occurrence of species better adopted to life in the soil with biochar use indicated better soil quality. Thus, we can recommend the use of the morphological diversity of Collembola as a good tool for the bioindication of soil health.

Author Contributions: Conceptualization, I.G. and J.P.T.; Methodology, A.L., J.K. and A.M.-J.; Formal analysis, I.G. and J.P.T.; Writing, Original Draft Preparation, Review and Edition, I.G., and J.P.T.

Acknowledgments: We kindly thank the Fluid S.A. Company for free-of-charge delivery of biochar for the experiment. We are greatly indebted to Kamila Twardowska and Joanna Magiera-Dulewicz for support in laboratory experiments.

References

1. Gregory, A.S.; Ritz, K.; McGrath, S.P.; Quinton, J.N.; Goulding, K.W.; Jones, R.J.; Harris, J.A.; Bol, R.; Wallace, P.; Pilgrim, E.S.; et al. A review of the impacts of degradation threats on soil properties in the UK. *Soil Use Manag.* **2015**, *31*, 1–15. [CrossRef] [PubMed]
2. De Meyer, A.; Poesen, J.; Isabirye, M.; Deckers, J.; Rates, D. Soil erosion rate in tropical villages: A case study from Lake Victoria Basin, Uganda. *Catena* **2011**, *84*, 89–98. [CrossRef]
3. Lehmann, J.; Joseph, S. *Biochar for Environmental Management Science and Technology*, 1st ed.; Earthscan: London, UK, 2009; pp. 1–944.
4. Latawiec, A.E.; Królczyk, J.B.; Kubon, M.; Szwedziak, K.; Drosik, A.; Polańczyk, E.; Grotkiewicz, K.; Strassburg, B. Willingness to Adopt Biochar in Agriculture: The Producer's Perspective. *Sustainability* **2017**, *9*, 655. [CrossRef]

5. Lin, Y.; Munroe, P.; Joseph, S.; Henderson, R.; Ziolkowski, A. Water extractable organic carbon in untreated and chemical treated biochars. *Chemosphere* **2012**, *87*, 151–157. [CrossRef]
6. Herath, H.M.; Camps-Arbestain, M.; Hedley, M. Effect of biochar on soil physical properties in two contrasting soils: An Alfisol and an Andisol. *Geoderma* **2013**, *209–210*, 188–197. [CrossRef]
7. Baiamonte, G.; De Pasquale, C.; Marsala, V.; Cimò, G.; Alonzo, G.; Crescimanno, G.; Conte, P. Structure alteration of a sandy-clay soil by biochar amendments. *J. Soils Sediments* **2015**, *15*, 816–824. [CrossRef]
8. Ding, Y.; Liu, Y.; Liu, S.; Li, Z.; Tan, X.; Huang, X.; Zeng, G.; Zhou, L.; Zheng, B. Biochar to improve soil fertility. A review. *Agron. Sustain. Dev.* **2016**, *36*, 36. [CrossRef]
9. Latawiec, A.E.; Peake, L.; Baxter, H.; Cornelissen, G.; Grotkiewicz, K.; Hale, S.; Królczyk, J.B.; Kubon, M.; Łopatka, A.; Medynska-Juraszek, A.; et al. A reconnaissance-scale GIS-based multicriteria decision analysis to support sustainable biochar use: Poland as a case study. *J. Environ. Eng. Landsc.* **2017**, *25*, 208–222. [CrossRef]
10. Freddo, A.; Cai, C.; Reid, B.J. Environmental contextualization of potential toxic elements and polycyclic aromatic hydrocarbons in biochar. *Environ. Pollut.* **2012**, *171*, 18–24. [CrossRef]
11. Hopkin, S.P. *Biology of the Springtails (Insecta: Collembola)*; Oxford University Press: Oxford, UK, 1997; pp. 1–330.
12. Rusek, J. Biodiversity of Collembola and their functional role in the ecosystem. *Biodivers. Conserv.* **1998**, *7*, 1207–1219. [CrossRef]
13. Filser, J. The role of Collembola in carbon and nitrogen cycling in soil. *Pedobiologia* **2012**, *46*, 234–245. [CrossRef]
14. Frampton, G.K. The potential of Collembola as indicators of pesticide usage: Evidence and methods from the UK arable ecosystem. *Pedobiologia* **1997**, *41*, 179–184.
15. Fiera, C. Application of stable isotopes and lipid analysis to understand trophic interactions in springtails. *North West. J. Zool.* **2014**, *10*, 227–235.
16. Sousa, J.P.; Bolger, T.; da Gama, M.M.; Lukkari, T.; Ponge, J.-F.; Simón, C.; Traser, G.; Vanbergen, A.J.; Brennan, A.; Dubs, F.; et al. Changes in Collembola richness and diversity along a gradient of land-use intensity: A pan European study. *Pedobiologia* **2006**, *50*, 147–156. [CrossRef]
17. Parisi, V. The biological soil quality, a method based on microarthropods. *Acta Naturalia de L'Ateneo Parmense* **2001**, *37*, 97–106.
18. Machado, J.S.; Oliveira, F.L.; Sants, J.C.; Paulino, A.T.; Baretta, D. Morphological diversity of springtails (Hexapoda: Collembola) as soil quality bioindicators in land use systems. *Biota Neotrop.* **2019**, *19*, e20180618. [CrossRef]
19. Potapov, A.; Semeninaa, E.; Korotkevich, A.; Kuznetsova, N.; Tiunova, A. Connecting taxonomy and ecology: Trophic niches of collembolans as related to taxonomic identity and life forms. *Soil Biol. Biochem.* **2016**, *110*, 20–31. [CrossRef]
20. Karaban, K.; Karaban, E.; Uvarov, A. Determination of life form spectra in soil Collembola communities: A comparison of two methods. *Pol. J. Ecol.* **2012**, *59*, 381–389.
21. Ponge, J.F.; Dubs, F.; Gillet, S.; Sousa, J.P.; Lavelle, P. Decreased biodiversity in soil springtail communities: The importance of dispersal and land use history in heterogeneous landscapes. *Soil Biol. Biochem.* **2006**, *38*, 1158–1161. [CrossRef]
22. Ellers, J.; Berg, M.P.; Dias, A.T.; Fontana, S.; Ooms, A.; Moretti, M. Diversity in form and function: Vertical distribution of soil fauna mediates multidimensional trait variation. *J. Anim. Ecol.* **2018**, *87*, 933–944. [CrossRef]
23. Greenslade, P.; Vaughan, G.T. A comparison of Collembola species for toxicity testing of Australian soils. *Pedobiologia* **2003**, *47*, 171–179. [CrossRef]
24. Marks, E.A.N.; Mattana, S.; Alcañiz, J.M.; Domene, X. Biochars provoke diverse soil mesofauna reproductive responses in laboratory bioassays. *Eur. J. Soil Biol.* **2014**, *60*, 104–111. [CrossRef]
25. Domene, X.; Hanley, K.; Enders, A.; Lehmann, J. Short-term mesofauna responses to soil additions of corn stover biochar and the role of microbial biomass. *Appl. Soil Ecol.* **2015**, *89*, 10–17. [CrossRef]
26. Zhang, X.; Li, Q.; Liang, W.; Zhang, M.; Bao, X.; Xie, Z. Soil nematode response to biochar addition in a Chinese wheat field. *Pedosphere* **2013**, *23*, 98–103. [CrossRef]
27. Tammeorg, P.; Parviainen, T.; Nuutinen, V.; Simojoki, A.; Vaara, E.; Helenius, J. Effects of biochar on earthworms in arable soil: Avoidance test and field trial in boreal loamy sand. *Agric. Ecosyst. Environ.* **2014**, *191*, 150–157. [CrossRef]

28. Kolton, M.; Harel, Y.M.; Pasternak, Z.; Graber, E.R.; Elad, Y.; Cytryn, E. Impact of biochar application to soil on the root-associated bacterial community structure of fully developed greenhouse pepper plants. *Appl. Environ. Microbiol.* **2011**, *77*, 4924–4930. [CrossRef] [PubMed]
29. Lehmann, J.; Rillig, M.C.; Thies, J.; Masiello, C.A.; Hockaday, W.C.; Crowley, D. Biochar effects on soil biota—A review. *Soil Biol. Biochem.* **2011**, *43*, 1812–1836. [CrossRef]
30. Cole, E.; Zandvakili, O.R.; Blanchard, J.; Xing, B.; Hashemi, M.; Etemadi, F. Investigating responses of soil bacterial community composition to hardwood biochar amendment using high-throughput PCR sequencing. *Appl. Soil Ecol.* **2019**, *136*, 80–85. [CrossRef]
31. Van Straalen, N.M.; Verhoef, H.A. The development of a bioindicator system for soil acidity based on arthropod pH preferences. *J. Appl. Ecol.* **1997**, *34*, 217–232. [CrossRef]
32. Bardgett, R. *The Biology of Soil: A Community and Ecosystem Approach*; Oxford University Press: Oxford, UK, 2005; pp. 1–254.
33. Lehmann, J.; Gaunt, J.; Rondon, M. Bio-char Sequestration in Terrestrial Ecosystems—A Review. *MITIG ADAPT STRAT GL* **2006**, *11*, 403–427. [CrossRef]
34. Berg, M.P.; Stoffer, M.; van den Heuvel, H.H. Feeding guilds in Collembola based on digestive enzymes. *Pedobiologia* **2004**, *48*, 589–601. [CrossRef]
35. Chahartaghi, M.; Scheu, S.; Ruess, L. Sex Ratio and Mode of Reproduction in Collembola in an Oak-Beech Forest. *Pedobiologia* **2006**, *50*, 331–340. [CrossRef]
36. Tisdall, J.M.; Oades, J.M. Organic Matter and Water-Stable Aggregates in Soils. *Eur. J. Soil Sci.* **1982**, *33*, 141–163. [CrossRef]
37. FAO-WRB. *World Reference Base for Soil Resources. International Soil Classification System for Naming Soils and Creating Legends for Soil Maps*; World Soil Resources Reports 106; Food and Agriculture Organization of the United Nations: Rome, Italy, 2014.
38. *Standardized Product Definition and Product Testing Guidelines for Biochar that is Used in Soil*; IBI-STD-2.1; IBI (International Biochar Initiative): Canandaigua, NY, USA, 2014.
39. Liang, B.; Lehmann, J.; Solomon, D.; Kinyangi, J.; Grossman, J.; O'Neill, B.; Skjemstad, J.O.; Thies, J.; Luizao, F.J.; Petersen, J.; et al. Black carbon increases cation exchange capacity in soils. *Soil Sci. Soc. Am. J.* **2006**, *70*, 1719–1730. [CrossRef]
40. Liang, B.; Lehmann, J.; Sohi, S.P.; Thies, J.; O'Neill, B.; Truillo, L.; Gaunt, J.; Solomon, D.; Grossman, J.; Neves, E.G.; et al. Black carbon affects the cycling of non-black carbon in soil. *Org. Geochem.* **2010**, *41*, 206–213. [CrossRef]
41. Gruss, I.; Twardowski, J.P.; Latawiec, A.; Medyńska-Juraszek, A.; Królczyk, J. Risk assessment of low temperature biochar used as soil amendment on soil mesofauna. *Environ. Sci. Pollut. Res.* **2019**, *26*, 18230–18239. [CrossRef] [PubMed]
42. Meier, U. *Growth Stages of Mono-and Dicotyledonous Plants*; Federal Biological Research Centre for Agriculture and Forestry: Berlin, Germany, 2001; pp. 1–204.
43. Murphy, P.W. Extraction methods for soil animals. I. Dynamic methods with particular reference to funnel processes. In *Progress in Soil Zoology*; Butterworths: London, UK, 1962; pp. 75–114.
44. Zimdars, B.; Dunger, W. *Synopses on Palaearctic Collembola Part I. Tullbergiinae Bagnall, 1935*; Abhandlungen und Berichte des Naturkundemuseums: Görlitz, Germany, 1994; pp. 1–71.
45. Fjellberg, A. *The Collembola of Fennoscandia and Denmark. Part II: Entomobryomorpha and Symphypleona*; Fauna Entomologica Scandinavica; Brill Publishers: Leiden, The Netherlands, 2007; pp. 1–264.
46. Hopkin, S.P. *A Key to the Springtails (Collembola) of Britain and Ireland*; Field Studies Council (AIDGAP Project): London, UK, 2007; pp. 1–245.
47. Conti, F.; Visioli, G.; Malcevschi, A.; Menta, C. Safety assessment of gasification biochars using *Folsomia candida* (Collembola) ecotoxicological bioassays. *ESPR* **2017**, *25*, 6668–6679. [CrossRef] [PubMed]
48. Liesch, A.M.; Weyers, S.L.; Gaskin, J.W.; Das, K.C. Impact of two different biochars on earthworm growth and survival. *Ann. Environ. Sci.* **2010**, *4*, 1–9.
49. Sohi, S.P.; Krull, E.; Lopez-Capel, E.; Bol, R. *A Review of Biochar and Its Use and Function in Soil. Advances in Agronomy*; Academic Press: Cambridge, MA, USA, 2010; pp. 47–82.
50. Xie, T.; Reddy, K.R.; Wang, C.; Yargicoglu, E.; Spokas, K. Characteristics and applications of biochar for environmental remediation: A review. *Crit. Rev. Environ. Sci. Technol.* **2015**, *45*, 939–969. [CrossRef]
51. Vaccari, F.P.; Maienza, A.; Miglietta, F.; Baronti, S.; Di Lonardo, S.; Giagnoni, L.; Lagomarsino, L.; Pozzi, A.;

Pusceddu, E.; Ranieri, R.; et al. Biochar stimulates plant growth but not fruit yield of processing tomato in a fertile soil. *Agric. Ecosyst. Environ.* **2015**, *207*, 163–170. [CrossRef]
52. Głąb, T.; Palmowska, J.; Zaleski, T.; Gondek, K. Effect of biochar application on soil hydrological properties and physical quality of sandy soil. *Geoderma* **2016**, *28*, 11–20. [CrossRef]
53. Yang, D.; Yunguo, L.; Liu, S.; Huang, X.; Li, Z.; Tan, X.; Zeng, G.; Zhou, L. Potential benefits of biochar in agricultural soils: A review. *Pedosphere* **2017**, *27*, 645–661.
54. Baronti, S.; Vaccari, F.P.; Miglietta, F.; Calzolari, C.; Lugato, E.; Orlandini, S.; Pini, R.; Zulian Genesio, C.L. Impact of biochar application on plant water relations in *Vitis vinifera* (L.). *Eur. J. Agron.* **2014**, *53*, 38–44. [CrossRef]
55. Wang, L.; Butterly, C.R.; Wang, Y.; Herath, H.M.S.K.; Xi, Y.G.; Xiao, X.J. Effect of crop residue biochar on soil acidity amelioration in strongly acidic tea garden soils. *Soil Use Manag.* **2013**, *30*, 119–138. [CrossRef]
56. Obia, A.; Cornelissen, G.; Mulder, J.; Dörsch, P. Effect of Soil pH Increase by Biochar on NO, N_2O and N_2 Production during Denitrification in Acid Soils. *PLoS ONE* **2015**, *10*, e0138781. [CrossRef] [PubMed]
57. Hågvar, S. Reactions to soil acidification in microarthropods: Is competition a key factor? *Biol. Fertil. Soils* **1990**, *9*, 178–181. [CrossRef]
58. Domene, X.; Mattana, S.; Hanley, K.; Enders, A. Medium-term effects of corn biochar addition on soil biota activities and functions in a temperate soil cropped to corn. *Soil Biol. Biochem.* **2014**, *72*, 152–162. [CrossRef]
59. Castracani, C.; Maienza, A.; Grasso, D.A.; Genesio, L.; Malcevschi, A.; Miglietta, F.; Vaccari, F.P.; Mori, A. Biochar–macrofauna interplay: Searching for new bioindicators. *Sci. Total Environ.* **2015**, *536*, 449–456. [CrossRef]
60. Larsen, T.; Schjønning, P.; Axelsen, J. The impact of soil compaction on euedaphic Collembola. *Appl. Soil Ecol.* **2004**, *26*, 273–281. [CrossRef]
61. Twardowski, J.P.; Hurej, M.; Gruss, I. Diversity and abundance of springtails (Hexapoda: Collembola) in soil under 90-year potato monoculture in relation to crop rotation. *Arch. Agron. Soil Sci.* **2016**, *62*, 1158–1168. [CrossRef]
62. Jacomini, C.; Nappi, P.; Sbrilli, G.; Mancini, L. *Indicatori ed Indici Ecotossicologicie Biologici Applicati al Suolo: Stato Dell'arte*; Agenzia Nazionale per la Protezione dell'Ambiente (ANPA): Roma, Italy, 2000.
63. Op Akkerhuis, G.J.; de Ley, F.; Zwetsloot, H.; Ponge, J.-F.; Brussaard, L. Soil microarthropods (Acari and Collembola) in two crop rotations on a heavy marine clay soil. *Rev. Ecol. Biol. Sol.* **1988**, *25*, 175–202.

Impact of Integrated and Conventional Plant Production on Selected Soil Parameters in Carrot Production

Anna Szeląg-Sikora [1,*], Jakub Sikora [1], Marcin Niemiec [2], Zofia Gródek-Szostak [3], Joanna Kapusta-Duch [4], Maciej Kuboń [1], Monika Komorowska [5] and Joanna Karcz [1]

1 Faculty of Production and Power Engineering, University of Agriculture in Krakow, ul. Balicka 116B, 30-149 Kraków, Poland; Jakub.Sikora@ur.krakow.pl (J.S.); Maciej.Kubon@ur.krakow.pl (M.K.); jwieczorek2@poczta.onet.pl (J.K.)
2 Faculty of Faculty of Agriculture and Economics, University of Agriculture in Krakow, al. Mickiewicza 21, 31-121 Kraków, Poland; Marcin.Niemiec@ur.krakow.pl
3 Department of Economics and Enterprise Organization, Cracow University of Economics, ul. Rakowicka 27, 31-510 Krakow, Poland; grodekz@uek.krakow.pl
4 Faculty of Food Technology, University of Agriculture in Krakow, ul. Balicka 122, 30-149 Kraków, Poland; Joanna.Kapusta-Duch@ur.krakow.pl
5 Faculty of Biotechnology and Horticulture, University of Agriculture in Krakow, al. Mickiewicza 21, 31-121 Kraków, Poland; m.komorowska@ogr.ur.krakow.pl
* Correspondence: Anna.Szelag-Sikora@urk.edu.pl

Abstract: Currently, the level of efficiency of an effective agricultural production process is determined by how it reduces natural environmental hazards caused by various types of technologies and means of agricultural production. Compared to conventional production, the aim of integrated agricultural cultivation on commercial farms is to maximize yields while minimizing costs resulting from the limited use of chemical and mineral means of production. As a result, the factor determining the level of obtained yield is the soil's richness in nutrients. The purpose of this study was to conduct a comparative analysis of soil richness, depending on the production system appropriate for a given farm. The analysis was conducted for two comparative groups of farms with an integrated and conventional production system. The farms included in the research belonged to two groups of agricultural producers and specialized in carrot production.

Keywords: soil fertility; integrated agricultural production; conventional agricultural production; management

---

1. Introduction

In an agronomic sense, the agricultural system is defined as a way to manage the space for the production of plant and animal products. The agricultural system also includes the processing of primary products [1,2]. In modern agriculture, there are three basic management systems [3,4]: conventional, ecological, and integrated. The basis for this distinction is the extent of dependence of agriculture on the industrial means of production, mainly mineral fertilizers and pesticides, and the degree of their impact on the natural environment [5]. Conventional production is a management method aimed at maximizing profits. It is based on increasing the use of means of production and minimizing the number of agrotechnical operations in order to maximize profits [6,7]. Integrated agriculture is a form of alternative farming, which is based on harmonizing the premises of conventional agriculture with elements of biological plant protection in order to increase the safety of food products. This form of agriculture treats the farm as an agricultural ecosystem (agro-ecosystem). Its main goal is to maintain a high level of agricultural production, while protecting the population of beneficial organisms that inhabit the ecosystem and preventing soil degradation. The integrated production (IP) of plants is a more restrictive system, in terms of both environmental protection and product safety. To ensure compliance with the principles of IP, quantitative and qualitative restrictions on the use of pesticides, as well as quantitative and technological restrictions related to the use of fertilizers, are introduced. Restrictions related to the use of agrochemicals require more precise application that takes into account a wide range of agrotechnical, climatic, and habitat factors. This requires greater knowledge and experience of producers [8–10]. Due to the smaller number of restrictions related to fertilization and protection, conventional agriculture is a greater burden on the natural environment and is therefore much less effective in achieving ecological goals. However, it should be noted that in both integrated and conventional agriculture, production capacities are not yet fully utilized. Similarly, in both systems, there are opportunities to better achieve environmental goals [11,12].

Created in 2001, the European Initiative for the Sustainable Development of Agriculture (EISA) was developed to promote and defend consistent principles of integrated production in the European Union. One of the first tasks of this organization was to create the Common Codex of Integrated Farming. The document, which presented EISA's policy in terms of integrated agriculture, was published in 2006 and reviewed in 2012. FAO (Food and Agriculture Organization) used the latter version to determine sustainable practices in agriculture [13]. The state of research on implementations of integrated agriculture systems in many Western European countries is advanced [14]. Research by Dutch scientists shows that an integrated farm can achieve income at 94% of the income of a conventional farm. In Germany, where the average area of an integrated farm in Germany is approx. 17 ha [15], the implementation of an integrated system is carried out by, e.g., the Institute for Plant Protection in Stuttgart. In Poland, the integrated production system is currently regulated by the provisions of art. 5 of the Act on Plant Protection of 18 December 2003 (Journal of Laws [Dz. U.] of 2008 No. 133, item 849, as amended) and the Regulation of the Minister of Agriculture and Rural Development of 16 December 2010 on integrated production (Journal of Laws [Dz. U.] of 2010 No. 256, item 1722, as amended). Since 14 June 2007, due to the decision of the Minister of Agriculture and Rural Development, integrated production, as understood by art. 5, par. 1 of the Act on Plant Protection, has been recognized as a national food quality system. Detailed guidelines for the production technology of each plant species are included in methodologies developed by the Main Inspectorate of Plant Health and Seed Inspection. A producer wishing to join the integrated production system is obliged to continue agricultural production based on the methodologies approved by the Chief Inspector of Plant Health and Seed Inspection. Each methodology contains practical information about the planting, care, and harvesting of the crop. At the request of the plant producer, a certificate of integrated production is issued by the regional inspector consistent with the place of cultivation, along with an integrated production trademark signed with the producer's number.

The aim of this research was to assess the soil properties on farms using integrated crop production and on conventional farms.

2. Materials and Methods

2.1. Material

The research objects were two producer groups, varying in terms of the available land resources, direction of production, and the number of members. The main grouping factor was the type of production, i.e., integrated production (22 farms) and conventional production (8 farms). On conventional farms, fertilization was carried out without reference to actual nutritional needs under given agrotechnical and environmental conditions. Therefore, the amounts of biogen introduced into the soil were much higher than the nutritional needs of plants. Plant nutrient balance was not maintained on conventional farms.

The soil sampling scheme included taking 20 primary samples per pooled sample, with one pooled sample per max. 4 ha. The weight of a single sample was approx. 0.5 kg.

2.2. *Analytical Methods*

Within the assumed objective, in 2016, tests were carried out on 22 farms producing in accordance with the IP standard (Integrated Plant Production) and on eight conventional farms carrying out intensive production with no certified quality management system. All integrated farms were subject to the control of a certification body and were certified based on inspections. The examined farms were located in the Małopolskie (22) and Śląskie (8) provinces and their area ranged from 30 to 90 ha. A soil sample from the 0–20 cm layer was collected from each farm, in accordance with the PN-R-04031:1997 standard at the end of the growing season. The collected soil samples were dried and sieved with a 1 mm mesh sieve. Next, their parameters determining the greatest agricultural usefulness, including the pH, organic carbon content, assimilable forms of phosphorus, potassium, calcium, and magnesium, were designated. The content of assimilable forms of phosphorus and potassium was determined by the Egner Riehm method. The content of the remaining macro- and micronutrients was determined by atomic emission spectrometry (ICP-OES), following prior extraction with acetic acid at a concentration of 0.03 $mol \cdot dm^{-3}$. Soil pH was determined using the potentiometric method, in a KCl suspension at 1 $mol \cdot dm^{-3}$.

3. Results and Discussion

Carrot (*Daucus carota* L.) is a two-year plant belonging to the celery family (*Apiaceae*), formerly umbellifers (*Umbelliferae*). Carrots are characterized by a high capacity for an excessive accumulation of heavy metals, especially cadmium and lead [16,17]. For this reason, the soil with the lowest content of these elements should be selected for the cultivation of this plant. Growing carrots in the first year after fertilizing with manure is not recommended as it fosters an increased accumulation of nitrates, resulting in distorted and forking roots, which significantly worsen the quality of the produce. In rational pest management, the plant should not be cultivated in monoculture, as well as following other umbelliferous plants [18].

Phosphorus is the basic fertilizer macronutrient that must be delivered to agroecosystems. Intensification of plant production has led to a high demand for this element. As a consequence, the plants' ability to nourish with this element through the soil's ecosystem has become impaired. Phosphorus is taken in the form of phosphoric acid (V) ions. In the plant cell, the element is a component of nucleotides, phospholipids, and adenosine triphosphate (ATP), the latter of which plays a fundamental role as an energy carrier in the plant cell. It participates in the activation of enzymes by their phosphorylation or dephosphorylation. The availability of phosphorus in the initial stage of plant development affects the proper development of roots and thus results in drought and nutrient deficiency resistance. The deficiency of this macronutrient negatively affects the growth and development of plants, which leads to reduced crops and the deterioration of their quality, both sensory and technological. Very often, growth inhibition of lateral shoots is observed. Purple spots appear on the leaves, which, over time, become deformed and dry. The plant blooms; however, it does not bear fruit. Fertilization with phosphorus is carried out based on the soil's content of this element, or when its deficiency in the plant is observed. Phosphorus is a macronutrient, which is very often deficient in agriculturally used soils. The reason for the low content of phosphorus is due, on the one hand, to the insufficient level of fertilization with this element, and on the other hand, to processes related to the chemisorption of this element [11]. Therefore, in addition to application of the phosphate fertilizer itself, the management of the fertilization process includes the control and maintenance of soil properties at an optimum level. The most important parameters affecting the availability of phosphorus for plants are the soil's pH and its content of organic matter [19–21]. Phosphorus is best absorbed by plants at a soil pH of 6–7. In a strongly acidic environment, phosphorus is rebound by combining with aluminum, iron, and manganese cations. On the other hand, at a very high pH, calcium phosphate precipitates. In line with the principles of the development of sustainable agriculture, primary production management should aim not only at the intensification of production, but also at the quality of yield, as well as at reducing the negative impact of agriculture on individual elements of the environment [22,23]. The proper management of phosphorus in agrocenoses is associated with maintaining an adequate amount of assimilable forms of this element in the soil. Both a too high and too low content of the element in the soil is unfavorable for plants and the environment. Carrot is not a plant with a high demand for phosphorus. The optimum content of phosphorus in soils intended for carrot cultivation varies between 40 and 60 mg $P{\cdot}dcm^{-3}$ [18]. According to the methodology of integrated carrot production, the level of phosphate fertilization at average soil fertility should amount to about 60–80 kg of $P_2O_5{\cdot}ha^{-1}$.

The average phosphorus content in the studied soils from integrated carrot farms was 118.0 $mg{\cdot}kg^{-1}$, ranging from 47.87 to 275.2 $mg{\cdot}kg^{-1}$. In the soils of farms producing carrots using the conventional method, an average of 29.5 $mg{\cdot}kg^{-1}$ more of this element was found, i.e., 147.5 $mg{\cdot}kg^{-1}$. No low or very low phosphorus content was found in the soil samples collected from carrot producing farms. In the group of farms with integrated carrot production implemented, an average phosphorus content was observed in 18% of soil samples, whereas in 23% of cases, the content of this element was found at a high level. Approx. 60% of the studied soils contained bioavailable phosphorus compounds at a very high level. In the group of conventional vegetable farms, a high content of phosphorus was found in only one case. On the other hand, the remaining soils contained a very high amount of this element. From the point of view of the rationalization of phosphorus management, it is not beneficial to maintain very high concentrations of this element in the soil. The phosphorus not used by plants undergoes immobilization processes, and as a result of erosion, it enriches aquatic ecosystems, leading to the intensification of eutrophication processes. In addition, very high levels of phosphorus in the soil can lead to a reduced absorption of certain micronutrients, for example, zinc. According to the methodology of integrated carrot production, for carrot cultivation, the optimal content of assimilable phosphorus in the soil should be between 40 and 60 $mg{\cdot}dcm^{-3}$ [18], which gives approximately 30–75 $mg{\cdot}kg^{-1}$. At higher contents of this element, fertilization with phosphorus should be limited. The results of our own research indicate that on approx. 60% of farms carrying out integrated carrot production, and on all conventional farms studied, the content of available phosphorus forms was higher than recommended in the integrated carrot production methodology. On the other hand, in conventional farms, these quantities were much higher (Table 1).

Table 1. Content of available forms of phosphorus in the soils of the studied farms ($mg{\cdot}kg^{-1}$).

	Min.	Average	Max.	Median	Standard Deviation
Integrated vegetable farms	47.87	118.0	275.2	116.2	69
Conventional vegetable farms	134.6	147.5	174.5	151.4	13.4

Potassium belongs to the group of macronutrients. One of its most important functions in plants is the regulation of water management and maintenance of cellular turgor. As an activator of many enzymes, it is responsible for regulatory functions and is involved in the synthesis of both simple and complex proteins and sugars. This element has an active part in the transport of nitrate (NO_3^-) and phosphate (PO_4^{3-}) ions, as well as assimilates. It increases plant resistance to frost, diseases, and pests. Potassium is responsible for the proper growth of apple fruits, as well as their color and firmness. A good supply of this element strengthens plants' resistance to drought. Over 50% of agricultural land is characterized by a potassium deficit [24–26], which is why rational fertilization with this element plays a strategic role in the development of modern agriculture. According to the principles of integrated carrot production, the optimal content of potassium in the soil should be 120–150 $mg{\cdot}dm^3$. The number of doses of fertilization with this element should be determined on the basis of a chemical analysis of the soil. However, when the analysis shows that the potassium content is equal to or higher than the optimal content, then it is reasonable to reduce the level of fertilization with this element, or in the case of a very high content, to discontinue fertilization altogether in a given year [24]. With a potassium content below optimal, the element should be supplemented with doses higher than that required by the plants. For a fruiting orchard, the level of fertilization should amount to 50–80 kg $K_2O \cdot ha^{-1}$ [18]. The greatest demand for this nutrient occurs in the period of setting of the fruit and its intensive growth. Later, potassium takes part in retarding the growth of tree shoots and entering winter dormancy.

According to the methodology of integrated carrot production, the optimal content of assimilable phosphorus in the soil for carrot cultivation should be between 120 and 150 $mg \cdot dcm^{-3}$ [18], which gives approx. 80–100 $mg \cdot kg^{-1}$. The results of our own research show that on 18% of farms carrying out integrated carrot production, and on all conventional farms, the content of assimilable forms of potassium in soil was much higher than recommended by the methodology (Table 2).

Table 2. Content of assimilable forms of potassium in the soils of the studied farms ($mg \cdot kg^{-1}$).

	Min.	Average	Max.	Median	Standard Deviation
Integrated vegetable farms	48.35	57.9	148.6	52.10	31.32
Conventional vegetable farms	86.2	100.9	248.3	109.6	74.80

Boron is a common element worldwide; however, in agroecosystems, its deficit is observed more and more often [27]. It is most common in the form of boric acid and belongs to the group of micronutrients essential for living organisms [28]. Plants collect boron from the soil through the root system, in the form of anion $H_2BO_3^-$ or in the form of undissociated boric acid molecules (H_2BO_3). Only a part of this element is fully available to plants; usually no more than 5%–6% of the total boron content in the soil, and sometimes even less than 1%. Carrot is very sensitive to the deficiency of some micronutrients, especially boron. The deficiency of this element is most often observed in alkaline soils. The effect of the boron deficit is the stunting of the plant growth cone and the appearance of black spots on carrot roots after washing.

The average content of available boron forms in soil on integrated carrot farms was 2.960 $mg \cdot kg^{-1}$, whereas on conventional farms, it was 1.70 $mg \cdot kg^{-1}$. Soil analysis carried out on integrated fruit farms demonstrated that the average content of available boron was 0.70 $mg \cdot kg^{-1}$, while for conventional production farms, this value fluctuated at 0.66 mg $B \cdot kg^{-1}$. The boron contents indicated the possibility of a deficit of this element in the agroecosystem [29].

Calcium plays a very important role in the production of vegetables and fruits. As a nutrient that is not very mobile in the plant, it is absorbed into fruits and vegetables in small quantities, causing a need for its urgent replenishment [30]. For this reason, even a high content of calcium in the soil may not provide a sufficient level of plant nutrition. Therefore, foliar feeding of the apple tree is a necessary element of integrated production, and is an integral part of the full fertilization program. Symptoms of calcium deficiency on vegetative parts of plants are rare, appearing in the form of brightening apical leaves with yellow spots. With a deficit of calcium, apples are small and tend to crack and cork. They are sensitive to sunburn. Regardless of the use of calcium in apple cultivation, its optimal content in soil is a prerequisite for cultivating fruit plants. It improves the soil structure and prevents it from crusting, regulates its pH, and supports soil microbes by accelerating the distribution of organic matter and facilitating the development of the root system. According to the methodology of integrated carrot production, the optimum level of this component in the soil is 1000–2000 $mg \cdot dm^{-3}$ [18], which amounts to approx. 666–1333 Ca $mg \cdot kg^{-1}$ of soil. The available amount of this element in the soil on farms carrying out integrated carrot production was 1095 $mg \cdot kg^{-1}$ on average, whereas for farms from the conventional group, the average amount of available calcium forms in the soil was 255.9 $mg \cdot kg^{-1}$ (Table 3). An appropriate content of calcium in carrot increases the strength of cell walls, making the roots less susceptible to cracking.

Table 3. Calcium content in the tested soil samples ($mg \cdot kg^{-1}$).

	Min.	Average	Max.	Median	Standard Deviation
Integrated vegetable farms	416.4	1095	2119	768.4	536
Conventional vegetable farms	174.6	255.9	474.1	204.4	300.8

Copper plays an important role in plant organisms, impacting the regulation of cellular metabolism [29]. This element controls the transport of electrons in the process of photosynthesis, takes an active part in nitrogen transformations and in the synthesis of proteins and vitamin C, and binds and neutralizes free radicals. Plants require small amounts of copper for proper development. In the majority of soils in Poland, there are shortages of copper. On average, the largest amount of assimilable forms of copper was recorded in the soil from vegetable farms with an integrated production profile (0.54 $mg \cdot kg^{-1}$). Soils collected from vegetable and conventional farms contained 0.39 mg $Cu \cdot kg^{-1}$ on average (Table 4).

Table 4. Copper content in the tested soil samples ($mg \cdot kg^{-1}$).

	Min.	Average	Max.	Median	Standard Deviation
Integrated vegetable farms	0.25	0.54	1.60	0.50	0.32
Conventional vegetable farms	0.23	0.39	0.73	0.35	0.20

The iron content in the soils of Poland is very diverse, ranging from 0.8% to 1.8% [31]. As a rule, heavy soils contain much more of this element than sandy soils. The physiological functions of this metal in plants are related to the activation of oxidation-reduction reactions associated with many metabolic processes, such as respiration, photosynthesis, or the transformation of nitrogen compounds. The symptom of iron deficiency in plants is iron chlorosis of leaves, which first appears on the youngest leaves. The most important causes of iron deficiency in agroecosystems include intensive cultivation, large temperature fluctuations in the growing season, and intensive exposure. The iron content varies considerably in the individual organs. Its concentration in plant tissues ranges from 50 to 300 ppm [32]. The iron is taken up by plants in the ionic form of Fe^{2+} and the form of chelates. In the case of this micronutrient, its deficiencies are most often associated with soil properties. At a pH level above 6, and with the presence of large quantities of other macro- and micronutrients, the ability of its assimilation by plants may be limited. The average content of available forms of iron in the soil (Table 5) on the farms producing carrots using the integrated method was 0.750 $mg \cdot kg^{-1}$, while on conventional farms, the average was 1.330 $mg \cdot kg^{-1}$.

Table 5. Iron content in the tested soil samples ($mg \cdot kg^{-1}$).

	Min.	Average	Max.	Median	Standard Deviation
Integrated vegetable farms	0.28	0.75	1.65	0.67	0.42
Conventional vegetable farms	0.73	1.33	2.43	1.01	0.49

Most Polish soils are characterized by a low magnesium content [33,34]. The reason for the deteriorating deficit of this element in the soils is their acidification and low content of organic matter [35,36]. Magnesium is an element that is easily washed into deeper soil layers. It is estimated that its annual leaching oscillates between 10 and 40 kg $MgO \cdot ha^{-1}$ [37]. Magnesium is taken up by plants in accordance with its osmotic concentration, i.e., passive movement from the soil water. A high content of dissolved magnesium in the soil water allows it to be better absorbed by the roots. In order to prevent soil degradation and to ensure the supply of this element to plants, its content in soil should be maintained at the level of 60 to 80 mg·dm^3, i.e., approx. 40–53 mg·kg^{-1} [18]. Magnesium is the basic ingredient of chlorophyll. It determines the course of photosynthesis and energy transformations taking place in the plant, as well as the synthesis of proteins, carbohydrates, and fats. Magnesium is an activator of many enzymes. It plays an important role in the construction of cell walls, and thus increases the resistance of plants to diseases. Magnesium deficiency is most often observed in young trees, as chlorosis and necrosis between the main nerves of the leaves, or yellow-purple spots on the lamina. Magnesium deficiency accelerates the generative development of plants and early maturing of fruits. In addition, plants are less resistant to low temperatures. The average magnesium content in the soils in which the carrots were grown in the integrated system was 33.63 mg·kg^{-1} (Table 6). On conventional farms, however, a slightly smaller amount of this element was found, i.e., 31.90 mg·kg^{-1}.

Table 6. Magnesium content in the tested soil samples (mg·kg^{-1}).

	Min.	Average	Max.	Median	Standard Deviation
Integrated vegetable farms	20.85	33.63	36.95	32.07	3.30
Conventional vegetable farms	29.56	31.90	36.15	3.02	2.05

Manganese is a micronutrient that is essential for the life of plants as it contributes to the processes of nitrogen absorption and protein synthesis, vitamin C synthesis, respiration, and photosynthesis. Manganese deficiency leads to excessive iron uptake by plants. The range of manganese content in the soil varies from 20 to 5000 mg·kg^{-1} and it occurs in several forms of mineral and organic compounds [37]. The absorption of manganese for plants is strongly correlated with the pH of the soil. Acidic soil promotes solubility of this element. In most cases, acidic soils demonstrate no need for manganese fertilization. Symptoms of manganese deficiency are similar to those of iron deficiency; however, yellowing of leaves starts from the margin of the lamina and develops in a V-shaped direction towards the midrib. In the case of an iron deficit, all veins remain green, while with manganese deficiency, the final vein segments discolor. Most often, manganese deficiency is observed in older leaves. In the case of apple trees, the symptoms of manganese chlorosis occur on long and short shoots. The fruits are small and not very juicy and they quickly lose the green color of the skin. In the case of carrot cultivation, manganese is not as important an element as in the cultivation of apple trees. However, the deficiency of manganese in carrot causes retarded growth, and thus a reduction in yield [38].

The average content of assimilable manganese in the soil recorded for samples from conventional vegetable farms was 41.02 mg·kg^{-1}, while in the integrated production group, it was 33.66 mg·kg^{-1} (Table 7).

Table 7. Manganese content in the tested soil samples (mg·kg^{-1}).

	Min.	Average	Max.	Median	Standard Deviation
Integrated vegetable farms	5.20	33.56	87.60	29.90	80.0
Conventional vegetable farms	10.95	41.02	97.83	37.54	66.0

Zinc and its compounds are characterized by a good solubility. Its best solubility occurs in acidic and slightly acidic soils [39]. Organic matter and soil minerals bind zinc ions present in the soil. On Polish farms, zinc is an element often overlooked in the process of fertilization, because farmers believe this element has little impact on the yield increase [40,41].

The results of the conducted research indicate that the average content of available zinc forms in the studied soil samples on conventional vegetable farms was 0.710 mg·kg^{-1} (Table 8), while in soils from integrated farms, this value fluctuated at 0.580 mg·kg^{-1}.

Table 8. Zinc content in the tested soil samples (mg·kg^{-1}).

	Min.	Average	Max.	Median	Standard Deviation
Integrated vegetable farms	0.10	0.58	1.85	0.52	0.41
Conventional vegetable farms	0.20	0.71	1.78	0.56	0.65

Zinc deficiency in agroecosystem plants is often observed at a very high level of phosphorus fertilization, which has been pointed out by many researchers who studying this problem [42]. On the studied farms, the results of our own research indicate a too high level of phosphorus fertilization, inconsistent with the plants' demand. This may lead to an insufficient supply of zinc, especially under intense production conditions.

pH is one of the most important soil parameters, determining its fertility. It is influenced by external and internal factors determined by the applied production techniques. External factors include the type of parent rock, acid rain, and the leaching of alkaline cations, whilst internal factors are fertilization and liming treatments [19]. The pH measured in the aqueous soil suspension indicates the active acidity created by the hydrogen ions found in the soil solution, while the pH measured in the potassium chloride suspension (KCl) also takes into account the acid ions associated with the sorption complex [43].

This parameter determines the conditions of plant growth and development, as well as the direction and speed of biological and physicochemical processes in the soil [43–45]. It is the basic factor affecting the uptake of nutrients by plants, as well as the transformation of nitrogen and phosphorus compounds in the soil. The optimal pH of mineral soil for carrot cultivation is within the range of pH 6–7 [18]. The results of our own research (Table 9) indicate that the soil pH on 27% of farms producing carrot with the integrated methodology was below the optimal values. As a result, on some farms, the necessity of soil liming was identified, while on others, liming was only recommended. Almost 90% of conventional farms also had soil pH below optimal values, thus the need for liming was identified. Only one farm in this group was characterized by a soil with a pH level optimal for carrot production.

The basic element that significantly impacts the formation of soil properties is organic matter consisting of carbohydrates, proteins, fats, and humus. Hummus is part of the organic matter that impacts soil fertility, and is characteristic for each soil [46,47]. The content of organic matter in the soil depends on, e.g., the climate, terrain, parent rock, and water conditions prevailing in the area. In addition, the amount and type of organic matter in the soil are impacted by anthropogenic factors such as indirect or direct human influence on the environment, as well as its flora and fauna. The classification of soils according to the content of humus in soil is presented in Table 10.

The results of tests for humus content in the soils of vegetable and fruit farms indicate large differences between them (Table 11). In the group of integrated vegetable farms, 13.6% of the samples were classified as humus-deficient, 63.6% were low-humus soils, and 22.7% were medium-humus soils. In the above group of farms, there was not a single farm with humus soil. On conventional farms, no humus-deficient soil samples were identified: 62.5% of samples were low-humus soils, 25% were medium-humus soils, and 12.5% were humus soils.

Table 9. Liming demand of individual farms.

	Type of Soil	Reaction	
		pH	Liming
Farms producing carrots in the integrated system			
1	III	4.24	necessary
2	III	6.66	unnecessary
3	III	6.3	unnecessary
4	IV	4.89	necessary
5	III	6.85	unnecessary
6	III	5.21	necessary
7	III	6.65	unnecessary
8	III	6.94	unnecessary
9	III	7.63	unnecessary
10	III	6.4	unnecessary
11	IV	7.05	unnecessary
12	III	5.14	necessary
13	III	6.54	unnecessary
14	IV	6.94	unnecessary
15	IV	6.33	unnecessary
16	III	6.75	unnecessary
17	IV	7.06	unnecessary
18	IV	6.92	unnecessary
19	IV	5.28	necessary
20	III	6.68	unnecessary
21	IV	4.69	necessary
22	III	6.94	unnecessary
Farms producing carrots in the conventional system			
1	III	5.65	necessary
2	III	4.89	necessary
3	IV	4.92	necessary
4	III	6.93	unnecessary
5	III	5.16	necessary
6	III	4.85	necessary
7	IV	4.99	necessary
8	III	4.8	necessary

Table 10. Soil classification according to the content of humus in the soil [48].

Humus-Deficient Soils	less than 1%
Soils with Low Humus Level	1.0%–2.0%
Soils with Medium Humus Level	2.1%–3.0%
Humus Soils	above 3.0%

Table 11. The content of humus in the soils of vegetable farms, based on the obtained test results.

	Farms Producing Carrots in the Integrated System	Farms Producing Carrots in the Conventional System
Humus-deficient soils <1%	3 farms (13.6%)	-
Soils with low humus level 1.0%–2.0%	14 farms (63.6%)	5 farms (62.5%)
Soils with medium humus level 2.1%–3.0%	5 farms (22.7%)	2 farms (25%)
Humus soils (3.0%)	-	1 farm (12.5%)

Research carried out in recent years has shown a decrease in humus content in Polish soils [46]. This is related to the disturbance of hydrographic conditions and intensive soil use. Low levels of humus in Polish soils, as well as risks associated with mineralization, can cause large emissions of carbon dioxide from soils [1,46]. The mechanism that allows humus depletion to be counteracted is the development of agri-environmental programs, under which farmers receive subsidies for the cultivation of after and intercrops improving the balance of organic matter in the soils of their farms.

4. Conclusions

The following conclusions can be drawn from this study:

1. The rational management of plant nutrients and maintenance of appropriate soil parameters are strategic elements of the quality management system in plant production;
2. The results of our own research indicate that on approx. 60% of farms carrying out integrated carrot production, and on all conventional farms studied, the content of available phosphorus forms was higher than recommended in the integrated carrot production methodology. On the other hand, on conventional farms, these quantities were much higher (Table 1);
3. One of the goals of integrated production is to improve soil properties. In the majority of both integrated and conventional farms, balanced fertilization was not implemented due to irrational fertilization with potassium. The result of such a management strategy may negatively impact both the soil and economic efficiency of production [11,14,31,49]. The calcium content in the tested soil samples varied significantly within the compared production systems. Unfavorable values, i.e., below 1000 (mg·kg^{-1}), were observed on farms with the conventional production system;
4. The results of our own research indicate a very small variability in the amount of available forms of iron in individual samples within the research groups. The value of the coefficient of variation in the group of vegetable farms carrying out production in accordance with the principles of integrated and conventional agriculture was 56% and 36%, respectively;
5. The results of the conducted research indicate that on each of the studied farms, both the integrated and conventional group, the soil had a magnesium deficit. A too low magnesium content in the soil can cause plant metabolism disorders;
6. Comparative analysis indicates an insufficient effectiveness of the integrated production system in terms of soil resource management. However, compared to conventional farms, soil resource management on integrated farms follows the concept of sustainable agriculture more closely. The implementation of obligatory consulting on practical aspects of fertilization should impact optimization of the production process.

Author Contributions: Conceptualization, A.S.-S., M.N., and J.S.; methodology, A.S.-S., M.N., and J.K.-D.; resources, J.K., A.S.-S., M.N., and J.S.; formal analysis, A.S.-S., M.N., J.S., and M.K.; investigation, A.S.-S., Z.G.-S., and M.K.; data curation, A.S.-S. and M.N.; writing, A.S.-S. and M.N.; visualization, J.K.-D., J.S., and M.K.

References

1. Szeląg-Sikora, A.; Niemiec, M.; Sikora, J.; Chowaniak, M. Possibilities of Designating swards of grasses and small-seed legumes from selected organic farms in Poland for feed. In Proceedings of the IX International Scientific Symposium "Farm Machinery and Processes Management in Sustainable Agriculture", Lublin, Poland, 22–24 November 2017; pp. 365–370.
2. Holzapfel, S.; Wollni, M. Global GAP Certification of small-scale farmers sustainable? Evidence from Thailand. *J. Dev. Stud.* **2014**, *50*, 731–746. [CrossRef]
3. Mzoughi, N. Farmers adoption of integrated crop protection and organic farming: Do moral and social concerns matter? *Agric. Econ.* **2011**, *70*, 1536–1545. [CrossRef]
4. Niemiec, M.; Komorowska, M.; Szeląg-Sikora, A.; Sikora, J.; Kuboń, M.; Gródek-Szostak, Z.; Kapusta-Duch, J. Risk Assessment for Social Practices in Small Vegetable farms in Poland as a Tool for the Optimization of Quality Management Systems. *Sustainability* **2019**, *11*, 3913. [CrossRef]
5. Szeląg-Sikora, A.; Niemiec, M.; Sikora, J. Assessment of the content of magnesium, potassium, phosphorus and calcium in water and algae from the black sea in selected bays near Sevastopol. *J. Elem.* **2016**, *21*, 915–926. [CrossRef]
6. Rivas, J.; Manuel Perea, J.; De-Pablos-Heredero, C.; Angon, E.; Barba, C.; García, A. Canonical correlation of technological innovation and performance in sheep's dairy farms: Selection of a set of indicators. *Agric. Syst.* **2019**, *176*, 102665. [CrossRef]
7. Erbaugha, J.; Bierbaum, R.; Castillejac, G.; da Fonsecad, G.A.B.; Cole, S.; Hansend, B. Toward sustainable agriculture in the tropics. *World Dev.* **2019**, *121*, 158–162. [CrossRef]
8. Niemiec, M.; Komorowaka, M.; Szeląg-Sikora, A.; Sikora, J.; Kuzminova, N. Content of Ba, B, Sr and as in water and fish larvae of the genus Atherinidae, L. sampled in three bays in the Sevastopol coastal area. *J. Elem.* **2018**, *23*, 1009–1020. [CrossRef]
9. Niemiec, M.; Sikora, J.; Szeląg-Sikora, A.; Kuboń, M.; Olech, E.; Marczuk, A. Applicability of food industry organic waste for methane fermentation. *Przem. Chem.* **2017**, *69*, 685–688. [CrossRef]
10. Kuboń, M.; Krasnodębski, A. Logistic cost in competitive strategies of enterprises. *Agric. Econ.* **2010**, *56*, 397–402.
11. Niemiec, M. Efficiency of slow-acting fertilizer in the integrated cultivation of Chinese cabbage. *Ecol. Chem. Eng.* **2014**, *21*, 333–346.
12. Cassman, K.; Dobermann, A.; Walters, D. Agroecosystems, nitrogen-use efficiency, and nitrogen. *J. Hum. Environ.* **2002**, *31*, 132–140. [CrossRef] [PubMed]
13. Meziere, D.; Lucas, P.; Granger, S.; Colbach, N. Does Integrated Weed Management affect the risk of crop diseases? A simulation case study with blackgrass weed and take all disease. *Eur. J. Agron.* **2013**, *47*, 33–43. [CrossRef]
14. Helander, A.; Delin, K. Evaluation of farming systems according to valuation indices developed within a European network on integrated and ecological arable farming systems. *Eur. J. Agron.* **2004**, *21*, 53–67. [CrossRef]
15. Musshoff, O.; Hirschauer, N. Adoption of organic farming in Germany and Austria: An integrative dynamiv investment perspective. *Agric. Econ.* **2008**, *39*, 135–145. [CrossRef]
16. Bizkarguenaga, E.; Zabaleta, I.; Mijangos, L.; Iparraguirre, A.; Fernández, L.A.; Prieto, A.; Zuloaga, O. Uptake of perfluorooctanoic acid, perfluorooctane sulfonate and perfluorooctane sulfonamide by carrot and lettuce from compost amended soil. *Sci. Total Environ.* **2016**, *571*, 444–451. [CrossRef]
17. Kapusta-Duch, J.; Szeląg-Sikora, A.; Sikora, J.; Niemiec, M.; Gródek-Szostak, Z.; Kuboń, M.; Leszczyńska, T.; Borczak, B. Health-Promoting Properties of Fresh and Processed Purple Cauliflower. *Sustainability* **2019**, *11*, 4008. [CrossRef]
18. *Methodology of Integrated Carrot Production*; Polish Institute of Plant Protection and Fertilization (PIORIN): Warszawa, Poland, 2016. Available online: www.piorin.gov.pl (accessed on 15 June 2019).

19. Qian, X.; Gu, J.; Sun, W.; Li, D.; Fu, X.; Wang, J.; Gao, H. Changes in the soil nutrient levels, enzyme activities, microbial community function, and structure during apple orchard maturation. *Appl. Soil Ecol.* **2014**, *77*, 18–25. [CrossRef]
20. Higgs, B.; Johnston, A.E.; Salter, J.L.; Dawson, C.J. Some Aspects of Achieving Sustainable Phosphorus Use in Agriculture. *J. Environ. Qual.* **2000**, *17*, 80–87. [CrossRef]
21. Ayaga, G.; Todd, A.; Brookes, P.C. Enhanced biological cycling of phosphorus increases its availability to crops in low-input sub-Saharan farming systems. *Soil Biol. Biochem.* **2006**, *38*, 81–90. [CrossRef]
22. Skafa, L.; Buonocorea, E.; Dumonteta, S.; Capone, R.; Franzesea, P.P. Food security and sustainable agriculture in Lebanon: An environmental accounting framework. *J. Clean. Prod.* **2019**, *209*, 1025–1032. [CrossRef]
23. Yu, J.; Wu, J. The Sustainability of Agricultural Development in China: The Agriculture–Environment Nexus. *Sustainability* **2018**, *10*, 1776. [CrossRef]
24. Malik, M.A.; Marschner, P.; Khan, K.S. Addition of organic and inorganic P sources to soil e Effects on P pools and microorganisms. *Soil Biol. Biochem.* **2012**, *49*, 106–113. [CrossRef]
25. Gródek-Szostak, Z.; Malik, G.; Kajrunajtys, D.; Szeląg-Sikora, A.; Sikora, J.; Kuboń, M.; Niemiec, M.; Kapusta-Duch, J. Modeling the Dependency between Extreme Prices of Selected Agricultural Products on the Derivatives Market Using the Linkage Function. *Sustainability* **2019**, *11*, 4144. [CrossRef]
26. Kocira, S.; Kuboń, M.; Sporysz, M. Impact of information on organic product packagings on the consumers decision concerning their purchase. *Int. Multidiscip. Sci. GeoConf. SGEM* **2017**, *17*, 499–506. [CrossRef]
27. Brown, P.; Bellaloui, N.; Wimmer, M.A.; Bassil, E.S.; Ruiz, J.; Hu, H.; Pfeffer, H.; Dannel, F. Boron in plant biology. *Plant Biol.* **2002**, *4*, 205–223. [CrossRef]
28. Itaktura, T.; Sasai, R.; Itoh, H. Precipitation recovery of boron from wastewater by hydrothermal mineralization. *Water Res.* **2005**, *39*, 2543–2548. [CrossRef]
29. Dirceu, M.; Hippler, F.; Boaretto, R.; Stuchi, E.; Quaggio, J. Soil boron fertilization: The role of nutrient sources and rootstocks in citrus production. *J. Integr. Agric.* **2017**, *16*, 1609–1616.
30. Danis, T.G.; Karagiozoglou, D.T.; Tsakiris, I.N.; Alegakis, A.K.; Tsatsakis, A.M. Evaluation of pesticides residues in Greek peaches during 2002–2007 after the implementation of integrated crop management. *Food Chem.* **2011**, *126*, 97–103. [CrossRef]
31. Courtney, R.G.; Mullen, G.J. Soil quality and barley growth as influenced by the land application of two compost types. *Bioresour. Technol.* **2008**, *5*, 2913–2918. [CrossRef]
32. Lemberkovics, E.; Czinner, E.; Szentmihalyi, K.; Bals, A.; Szoke, E. Comparitive evaluation of Helichrysi flos herbat extracts as dietary sources of plant polyphenols, and macro- and microelements. *Food Chem.* **2002**, *78*, 119–127. [CrossRef]
33. Wang, H.Q.; Zhao, Q.; Zeng, D.H.; Hu, Y.L.; Yu, Z.Y. Remediation of a Magnesium-Contaminated Soil by Chemical Amendments and Leaching. *Land Degrad. Dev.* **2015**, *15*, 613–619. [CrossRef]
34. Qadir, M.; Schubert, S.; Oster, J.D.; Sposito, G.; Minhas, P.S.; Cheraghi, S.M.A.; Murtaza, G.; Mirzabaev, A.; Saqib, M. High-magnesium waters and soils: Emerging environmental and food security constraints. *Sci. Total Environ.* **2018**, *642*, 1108–1117. [CrossRef] [PubMed]
35. Kuboń, M.; Sporysz, M.; Kocira, S. Use of artificial of clients of organic farms. In Proceedings of the 17th International Multidisciplinary Scientific GeoConference SGEM 2017, Bulgaria, Balkans, 27 June–6 July 2017; Volume 17, pp. 1099–1106.
36. Gródek-Szostak, Z.; Szeląg-Sikora, A.; Sikora, J.; Korenko, M. Prerequisites for the cooperation between enterprises and business supportinstitutions for technological development. In *Business and Non-Profit Organizations Facing Increased Competition and Growing Customers' Demands*; Wyższa Szkoła Biznesu—National-Louis University: Nowy Sącz, Poland, 2017; Volume 16, pp. 427–439.
37. Wanli, G.; Hussain, N.; Zongsuo, L.; Dongfeng, Y. Magnesium deficiency in plants: An urgent problem. *Crop J.* **2016**, *4*, 83–91.
38. Serpil, S. Investigation of Effect of Chemical Fertilizers on Environment. *Apcbee Procedia* **2012**, *1*, 287–292.
39. Chen, C.; Zhang, H.; Gray, E.; Boyd, S.; Yang, H.; Zhang, D. Roles of biochar in improving phosphorus availability in soils: A phosphate adsorbent and a source of available phosphorus. *Geoderma* **2016**, *276*, 1–6.
40. Domańska, J. Soluble forms of zinc in profiles of selected types of arable soils. *J. Elem.* **2009**, *14*, 55–62. [CrossRef]

41. Van Oort, F.; Jongmans, A.G.; Citeau, L.; Lamy, I.; Chevallier, P. Microscale Zn and Pb distribution patterns in subsurface soil horizons: An indication for metal transport dynamics. *Europ. Soil Sci.* **2006**, *57*, 154–166. [CrossRef]
42. Zhu, G.; Smith, E.; Smith, A. Zinc (Zn)—Phosphorus (P) interaction in two cultivars of spring wheat (*Triticum aestivum* L.) differing in P uptake efficiency. *Ann. Bot.* **2001**, *88*, 941–945. [CrossRef]
43. Han, T.; Cai, A.; Liu, K.; Huang, J.; Wang, B.; Li, D. The links between potassium availability and soil exchangeable calcium, magnesium, and aluminum are mediated by lime in acidic soil. *J. Soils Sediments* **2019**, *19*, 1382–1392. [CrossRef]
44. Ji, C.-J.; Yang, Y.-H.; Han, W.-X.; He, Y.-F.; Smith, J.; Smith, P. Climatic and Edaphic Controls on Soil pH in Alpine Grasslands on the Tibetan Plateau, China: A Quantitative Analysis. *Pedosphere* **2014**, *24*, 39–44. [CrossRef]
45. Sikora, J.; Niemiec, M.; Szeląg-Sikora, A.; Kuboń, M.; Olech, E.; Marczuk, A. Biogasification of wastes from industrial processing of carps. *Przem. Chem.* **2017**, *96*, 2275–2278.
46. Klimkowicz-Pawlas, A.; Smreczak, B.; Ukalska-Jaruga, A. The impact of selected soil organic matter fractions on the PAH accumulation in the agricultural soils from areas of different anthropopressure. *Environ. Sci. Pollut. Res.* **2017**, *24*, 10955–10965. [CrossRef] [PubMed]
47. Li, L.-J.; Zhu-Barker, X.; Ye, R.; Doane, T.A.; Horwath, W.R. Soil microbial biomass size and soil carbon influence the priming effect from carbon inputs depending on nitrogen availability. *Soil Biol. Biochem.* **2018**, *119*, 41–49. [CrossRef]
48. Mocek, A.; Drzymała, S.; Maszner, P. *Geneza, Analiza i Klasyfikacja Gleb*; Wydawnictwo Akademia Rolnicza w Poznaniu: Poznan, Finland, 1997; p. 416.
49. Domagała-Świątkiewicz, I.; Gąstoł, M. Soil chemical properties under organic and conventional crop management systems in south Poland. *Biol. Agric. Hortic.* **2013**, *29*, 12–28. [CrossRef]

6

The Review of Biomass Potential for Agricultural Biogas Production

Katarzyna Anna Koryś [1,2], Agnieszka Ewa Latawiec [1,2,3,4,*], Katarzyna Grotkiewicz [3] and Maciej Kuboń [3,5]

1 International Institute for Sustainability, Estrada Dona Castorina 124, Rio de Janeiro 22460-320, Brazil; k.korys@iis-rio.org
2 Rio Conservation and Sustainability Science Centre, Department of Geography and the Environment, Pontifícia Universidade Católica, Rio de Janeiro 22453900, Brazil
3 Department of Production Engineering, Logistic and Applied Computer Sciences, University of Agriculture in Kraków, Balicka 116B, 30-149 Kraków, Poland; katarzyna.grotkiewicz@urk.edu.pl (K.G.); maciej.kubon@urk.edu.pl (M.K.)
4 School of Environmental Science, University of East Anglia, Norwich NR4 7TJ, UK
5 Institute of Technical Sciences, State Vocational East European Higher School in Przemyśl, Książąt Lubomirskich 6, 37-700 Przemyśl, Poland
* Correspondence: a.latawiec@iis-rio.org

Abstract: Adequate management of biomass residues generated by agricultural and food industry can reduce their negative impacts on the environment. The alternative use for agricultural waste is production of biogas. Biomass feedstock intended as a substrate for the agricultural biogas plants may include energy crops, bio-waste, products of animal and plant origin and organic residues from food production. This study reviews the potential of selected biomass residues from the agri-food industry in terms of use for agricultural biogas production in Poland. The most common agri-food residues used as substrates for biogas plants in Poland are maize silage, slurry, and distillery waste. It is important that the input for the agricultural biogas installations can be based on local wastes and co-products that require appropriate disposal or storage conditions and might be burdensome for the environment. The study also discusses several limitations that might have an unfavourable impact regarding biogas plants development in Poland. Given the estimated biomass potential, the assumptions defining the scope of use of agricultural biogas and the undeniable benefits provided by biogas production, agricultural biogas plants should be considered as a promising branch of sustainable electricity and thermal energy production in Poland, especially in rural areas.

Keywords: biomass; agricultural biogas plants; agricultural waste; sustainable and renewable energy; organic residue management; Poland

---

1. Introduction

Global environmental change caused by excessive exploitation of natural resources and combustion of fossil fuels causes a range of negative impacts on human health and functioning of ecosystems humans ultimately depend on. Generation of solid waste is increasing rapidly as a result of industrialization, global urbanization and economic development [1,2]. Projections indicate that by the 2050 the word population will reach 9.7 billion [3]. There is therefore an increasing pressure on land and water resources to supply food and industrial products. Inability to effectively manage wastes as recently highlighted by scientists and decision-makers lead to serious environmental and socio-economic problems that need urgent and reliable solutions [4,5]. In response to this problem, a number of scientific initiatives were founded with the goal of creating new usages for organic residues [6]. One of the solutions of organic waste management can be their use for production of biogas, during anaerobic digestion (AD) [7].

Biomass [8] represents an important source of alternative energy and provides an opportunity to decrease environmental problems such as pollution and depletion of natural resources [6]. It is widely available and regenerates in a relatively short time [9]. There are many possibilities of biomass processing. For example, agricultural residues and municipal solid residues for biogas and biofuel production [10,11], the use of organic waste for vermicomposting [12], as a supplement during combustion with hard or brown coal or use of agricultural, urban, woody and industrial pyrolyzed residues to improve soil quality [13,14]. This paper discusses the potential of using residual biomass in Poland for the energy sector. It focuses on the assessment of the suitability of biomass from agricultural production and by-products or residues from the agri-food industry as a substrate for biogas production.

2. Methods

The first stage of the study was the literature review, which was performed in English and in Polish languages. The search was conducted on the online databases such as Google Scholar, Scopus, Web of Science, and Science Direct (without restriction to year). For this purpose, we provide an extensive combination of keywords in English and Polish (Table S1). The second stage was the selection of the scientific papers (articles, conference papers and PhD theses), based on title and abstract content. The selected documents were archived in the literature database Mendeley. Additionally, we reviewed the bibliography in the key documents with the aim of finding other relevant titles on the specific subject ("snowball" search method).

3. The Role of Biomass in Energy Production in Poland

The rational use of renewable energy sources is essential for sustainable development. Nowadays, Poland is the second largest coal consumer in the EU (after Germany), and the 10th largest coal (here referred to both: hard coal and lignite) consumer in the world, with consumption of 77 million tonnes of coal per year. In addition, the country is a leader in hard coal extraction and import. For example, in 2016 Poland produced 70.7 million tonnes of hard coal and imported 8.3 million tonnes [15]. Current data provided by the World Energy Council from 2017 showed that 92% of electricity and 89% of heat in Poland is generated from coal [16]. Biomass co-firing with fossil fuels might be a promising solution, allowing an increase in the share of renewable energy sources (RES) in the total energy in Poland [17]. Supporting energy production from renewable energy sources has become an essential objective of the European Union's policy as a strategy to pursue sustainable development [18,19]. In the case of Poland, the need for RES development results from the commitments of the "3 × 20" climate package imposed by the EU. By year 2020 Poland is obliged to obtain 15% of RES in gross final energy use and reduce emissions of air pollution. The use of biomass as the renewable energy source might be an essential aspect to achieve these obligations [20].

Biomass is one of the oldest as well as the most promising development of RES in Poland, with a great potential to be used for energy purposes [21]. It is mostly related to favourable geographical and climatic conditions for biomass production, wide range of its application, and its large resources [21–23]. In case of other RES, the limiting factors may result from unfavourable topography or insufficient resources (hydropower, wind energy), as well as the costs of production: for example, the high price of the solar cells for solar energy [21]. The energy obtained from biomass in Poland comprise approximately 80% of the entire energy pool, which is obtained from RES [20]. The technical potential of biomass in Poland is around 900 PJ/year [24]. The structure of primary energy production from renewable energy sources in Poland from 2016 is shown in Figure S1 [25]. Additionally, Poland is one of the largest exporters of biomass in Europe [21]. The annual biomass energy potential that can be managed is estimated as follows: over 20 million tons of waste straw, 4 million tons of wood waste (sawdust, tree cortex, sawdust, pellets), 6 million tons of sewage sludge from the paper industry and pulp, food and municipal waste [26].

From an environmental point of view, biomass can be considered better than coal. Previous studies demonstrated that combustion of biomass emits less SO_2 compared to coal and has zero-balance carbon dioxide emission [27,28]. Furthermore, the ash gained from the combustion of biomass is returned to the soil, where the plants used for thermal process were cultivated and collected, is consistent with the principles of sustainable development [21]. Despite the fact that electric energy in Poland is obtained mainly from coal [17], it has been demonstrated that using biomass from agriculture co-products for energy purposes is fully justified [23].

4. The Potential of Use of Agricultural Residues for Biogas Production in Poland

In Poland, the production of agricultural biogas has a great potential to grow given increasing demand for heat and electricity from renewables. Moreover, availability of feedstock (substrates) presents interesting opportunities for potential investors, farmers in particular [29]. Because the production of agricultural biogas requires a daily input of substrates [30], location of agricultural biogas installation and its capacity to process the biomass is determined by the constant supply of raw material [31]. This ensures high biogas yield, stability of the fermentation process and possibility of formed digestate utilization [32]. The composition and capacity of biogas obtained from biomass depends on many indicators, including moisture and physical state of the feedstock, technology used, temperature and pressure. As the input for agricultural power plant, both animal and plant origin substrates can be applied as well as waste from the agri-food industry [30,33].

4.1. Utilization of Biomass from Animal and Plant Production

Large breeding farms and agri-food processing generate a significant amount of organic waste [34]. In accordance with the requirements of environmental protection and waste management, if the breeding farm presents more than 40,000 places of poultry, or pig stock more than 2000, farmers are obliged to dispose at least 70% of animals manure on their own farmlands [35]. Semi-liquid and liquid slurry from animal farming are valuable fertilizer, however, their improper storage, application and spillage can lead to environmental pollution and cause odour problems [36]. Their utilization as the substrate for agricultural biogas plants could be one of the options. This solution might be practically attractive for farmers that produce large amounts of organic waste characterized by high energy value. In Poland, approximately 80,750 tons of manure and 35 million m^3 of slurry is provided as the organic waste of agricultural production. About one third of this amount could be processed into biogas for local use [34]. Banaszkiewicz and Wysmyk [37] estimated that the total technical potential to produce agricultural biogas from livestock excreta in Poland is 674 million m^3, i.e., 26.2 PJ. Furthermore, the anaerobic digestion conducted in agricultural biogas plants provides stabilization and deodorization of raw manure, also changing the category of fertilizer from "natural" to "organic" as the final product, so it can be disposed-of easier [36]. Nevertheless, digesting of raw manure as the only substrate under thermophilic conditions might be unprofitable due to some exploitation issues and inhibition of biogas production due to higher N amounts in animal excreta than in other organic waste [38,39]. A mixture of slurry with plant-biomass enhances C/N (carbon to nitrogen) ratio and nutrient balance, contributing to the improvement of biogas quality and lowering production costs [40,41].

In the case of farming, not only animal excrements can be used as a substrate for an agricultural biogas plant. Straw is commonly used in agriculture as feed for farm animals, however, its surplus is not suitable for agricultural purposes and may be burdensome for the environment. Currently, the surplus of straw is estimated to be 10–11 million tonnes per year. [42]. Nevertheless, straw has been recognized as a valuable biomass for the energy sector. This raw material can provide 934 TJ of energy. Assuming that the average calorific value of coal is 24 MJ kg^{-1}, the evaluated biomass could replace over 9.16 million tonnes of coal. However, it should be considered that straw combusting arouses controversy due to CO_2 emissions [43].

In Poland, approximately 25% of arable lands can be used for the growing of annual energy crops with the aim of agricultural biogas production [44]. Among them, maize, more precisely maize silage combined with slurry is used often for the co-fermentation process [33]. In comparison with other grain plants, maize shows higher biogas efficiency (Table S2), high yielding potential, harvesting and silage [45–47]. It is worth mentioning that silages of various plants, especially corn silage, are used for large-scale energy purposes. In Poland, maize growing is concentrated mainly in the western and northern part of the country [48]. The spatial analysis conducted by Jędrejek and Jarosz [31] demonstrated that the most suitable voivodeships for the construction of 50–100 kW micro-biogas plant using 100% maize silage are: Lower Silesian, Pomerania, Lubusz, Greater Poland, Warmian-Masurian and West Pomeranian. In the communities located in the voivodeships mentioned above, there is also a prospect of building micro-biogas plants obtaining biogas from a blend of maize silage (70%) and slurry (30%). Nonetheless, cultivation of monoculture—as in the case of maize intended for silage—might have an unfavourable impact for the environment, especially for soil [31]. Thus, due to the growing interest of advantages resulting from agricultural biogas production, the number of installations built in recent years and high costs of the feedstock, more attention is paid for alternative substrates for biogas plants.

4.2. *Biomass from the Agri-Food Industry*

The Polish agri-food industry generates large amounts of organic waste [49]. Replacing or co-firing of biomass obtained from agricultural crops with the raw material from agri-food production can be a promising solution for biogas production in Poland [50]. The usage applies to processed and unprocessed waste from the agri-food production, e.g., fruit processing [51], residues from diary industry, [52,53], distillery waste [54], meat processing [55] or fresh vegetables and fruits [56,57]. Theoretically, any biodegradable biomass that contains carbohydrates, proteins and fats can be used as the substrate for the agricultural biogas plant, however the prerequisite for the profitability of using raw material is the content of the organic dry matter amount above 30% [58,59]. It should be emphasized that regarding the methane fermentation process, it is important to use technologies based on the use of by-products from agriculture that do not compete with the production of food or forage [44].

4.2.1. Fruits Residues

During extraction of juice from fruits, a by-product called fruit pomace is separated. A certain amount of leftover goes to landfill, contributing to environmental pollution, while a significant part can be applicable as an energy source [50]. The use of fruit pomace as an input for a biogas facility has been well-described in literature. For example, Pilarska et al. [60] showed that the quantity of biogas and methane obtained from apple pomace was as follow: 203.64 m^3/ton of fresh substance and 101.36 m^3/ton of fresh substance. As a comparison, the results for activated sludge in the current study is: 4.38 m^3/ton of fresh substance for biogas and 2.21 m^3/ton of fresh substance for methane. Moreover, Prask et al. [61] demonstrated that grape pomace obtained during wine production might be successfully used in agricultural biogas plants, both as a substrate or co-substrate. It is important to highlight that fruit residues contain low concentrations of heavy metals [62]. Thus, as the result of processing the fruit residues during the methane fermentation, valuable organic fertilizer which is nutrient-rich and free from heavy-metals can be obtained [63].

4.2.2. Dairy Industry

The main by-product of the dairy industry is whey, whose annual production in Poland is 2–3 million m^3 [64]. Utilization of whey may be problematic, because it contains chemical substances, resistant for biodegradation in the conventional wastewater treatment [64]. Due to the content of lactose, which is the source of energy for many microbial groups like lactic acid bacteria, whey can be used in numbers of biotechnological processes, including methane fermentation, instead. Wesołowska-Trojanowska and Targoński [52] demonstrated that from one ton of substrate, up to 55 m^3 of biogas, containing about 78% of methane, can be obtained. In addition, the product does not contain sulphur compounds and can be directly used for combustion in steam boilers without prior desulphurisation. Nevertheless, it should be noticed that depending on the milk processing technology used in the dairy plant, acid, sweet and casein whey may be formed, differing mainly in pH. Whey is characterized by extremely high chemical oxygen demand (COD)—approximately 50,000 mg O_2/dm^3, and nitrogen (Kjeldahl nitrogen (Nog) 600 mg N/dm^3; nitrate nitrogen ($N\text{-}NO_3$) 2.5 mg/dm^3; $N\text{-}NH_4$ + 60 mg/dm^3). Therefore, despite the high biogas potential estimated, it is not recommended to use whey as the only substrate in the methane fermentation process, mainly due to the low C/N ratio required for the correct course of the process [53].

4.2.3. Meat Industry and Post-Slaughter Waste: Inconveniences Continuation

Meat industry, pork particularly, is one of the most important products of Polish agriculture [55]. However, the meat processing generates annually about 18 million tons of waste [65]. It poses a serious environmental and epidemiological threat and should be disposed of properly [44,65,66]. Due to the restrictive regulation regarding disposal of slaughterhouse waste, meat producers are struggling with the high costs of animal waste management [67]. Methane fermentation in purpose of their utilization seems to be an optimal solution [55]. The research conducted in the Institute of Biosystems Engineering (University of Life Sciences in Poznań) showed that the waste from the meat industry might be promising for biogas efficiency, due to the high content of protein and fats [68]. For example, the amount of biogas obtained from 1 ton of the content of the digestive tract was 275.77 m^3, including methane: 194.38 m^3/t fresh matter, which is 70.48% of gas. Also, it has been demonstrated that the mixture of slurry and digestive tract content of pigs is an energetically effective input in the methane fermentation process, producing high-energy biogas with a methane content exceeding 60% [55].

Nevertheless, the use of waste from slaughterhouse and meat processing in Poland as a substrate for biogas plants is associated with certain problems. First, this type of waste requires the prior sanitary treatment (thermal), in accordance with the Regulation of the European Parliament and Council Regulation (EC) No 1069/2009 of 21 October 2009. The exception applies to the content of the digestive tract [65]. Second, the technological issues related with processing of selected organs. For instance, brains and spinal cords that cannot be used as a substrate for a biogas plant, as they might be the potential source of pathogenic prions [69]. Third: regarding the UE legislation, post-slaughterhouse wastes are divided into three categories (Animal by-product, ABPs), based on the risk they pose: category I ABPs classed as "particular risk" (SRM—from polish language: "material szczególnego ryzyka"), category II ABPs—"high-risk", category III—"low-risk". Only waste classified to category II and III can be used to produce agricultural biogas, after earlier treatment [70].

An appropriate and efficient fermentation process depends on the quality and proper balance of the supplied substrate. In the case of meat waste, the other, considerable issue is the high amount of nitrogen which makes a lot of difficulties in running the anaerobic fermentation process properly. Balance of the C/N components ratio is an important element because in the fermentation process the organic nitrogen from the substrate is converted into ammonium nitrogen, which is partly used for the synthesis of protein of newly emerging bacterial cells. In addition, ammonia is formed with excess nitrogen, which inhibits bacterial growth at low concentrations, while higher carbon to nitrogen ratio causes a decrease in the amount of methane due to disruption of the carbon metabolism [63].

Despite of mentioned inconveniences, a number of studies showed that waste of meat processing has a higher biomethane yield, i.e., compared to maize silage, the basic biogas input used in Europe. Furthermore, the biomass fermentation can be an effective tool to reduce the unfavourable effects of improper waste management from the meat industry, and the electricity and heat obtained can be an additional source of income for meat processing plants or used for their technological purposes [55,69].

4.2.4. Distillery Waste

Distillery stillage is a main by-product obtained during ethanol production [71,72]. Poland is one of the largest spirits producers worldwide and the amount of distillery stillage exceeds up to 12 times of alcohol, resulting in an estimated several million tons per year [73]. It creates a great problem with disposal of its surplus. Distillers contain, in addition to organic carbon (Ct) compounds, mineral nutrients necessary for plants and can be used to fertilize or improve the soil quality. A common feature of decoctions is too low phosphorus content in relation to nitrogen and potassium and relatively high Ct content [74]. Grain stillage is characterized by high content of B vitamins, minerals, exogenous amino acids, dietary and lactogenic value and favourable protein–oat ratio; therefore, it is used in the forage industry [75]. In turn, molasses stillage is not suitable for the forage purpose, however, for the economic reasons, is often used by biofuel producers as a substrate to produce ethanol [76]. Though molasses stillage may be used as fertilizer, the application on the field might be problematic due to its polluting potential and would generate further costs related with the constructions of the appropriate tanks for its storage [71,76]. At this point, attention should be paid to the hazard associated with excess potassium accumulation in the case of fertilization with this waste of crops grown for feed purposes [75]. Moreover, the period of usefulness of stillage is relatively short because of the risk of microbiological development [73].

Nevertheless, a number of studies were conducted with the aim of finding an optimal solution, regarding the difficulties associated with the large loads of distillery waste. The anaerobic digestion with biogas production is one of the alternative methods proposed for the utilization of stillage [49,54,77,78]. Biogas efficiency from stillage has been estimated in the range between 430–725 m^3 Mg^1 of dry organic matter, with methane content of 55% and is comparable to other substrates from the agri-food industry [79]. From the other hand, the use of stillage as the input for biogas plants can be justified in the case of a stable situation of the domestic ethanol industry and when the installation is located in vicinity of a distillery that generates great quantities of this waste. This would minimize the costs related with transport which might contribute to unfavourable economic stability [80,81]. Additionally, the high amount of potassium in molasses and accordingly in distillery waste, not only contaminates fields when it is used as fertilizer but may also inhibit bacteria involved in the anaerobic digestion. Therefore, when supplying the fermentation chamber, special attention should be paid to quality as well as chemical and elemental composition of the substrate [74,82].

4.2.5. Fresh Fruits and Vegetables: The Controversial Case

Utilization of fresh agricultural products for energy purposes seems to be unjustified and controversial. On other hand, in certain cases might be necessarily. Fresh fruits and vegetables require appropriate storage conditions, which most of the farms are deprived of. The amount of unused fresh biomass might be problematic for many food producers, as the product that stays with the farmers, become a waste and need to be disposed. For instance, the embargo imposed by the Russian Federation in 2014, had significant influence on Poland´s economic situation, causing the saturation of local markets with fresh agricultural products [56,57]. Thus, the farmers were recommended to use the waste as the substrate for biogas plant, because it would be profitable from the economic point of view [83]. The study conducted by Smurzyńska et al. [57], with the aim to determine the biogas yield and dynamic of the fermentation process of surplus of fresh vegetables and fruits (biogas and methane productivity tests were carried out for the following vegetables and fruits: eggplants, pumpkins, cauliflower, cabbage, peppers, tomatoes, cucumbers) demonstrated that the process of methane production has not been impaired by any inhibitory factors and biogas yield obtained from individual vegetables and fruits tested was comparable. Furthermore, a large number of polysaccharides in substrates tested contributed to a short but intense process of biogas production. A comparable study (also with the fruits covered by embargo, such as: apples, pears, nectarines, peaches and plums) showed that the biogas yield obtained from the tested fruits were also similar (653 m^3/Mg peaches, 829.66 m^3/Mg apples on organic matter) [56]. Based on these studies it can be concluded that biomass obtained from fresh non-traded agricultural products are a suitable substrate for biogas production, however, should be used only in case of exceptional situations.

5. Limitation for Development of Agricultural Biogas Plants in Poland

Agricultural biogas plants have huge potential that can positively influence both the environmental aspect and socio-economic development of a given area [34,84]. Currently there are 96 agricultural biogas plants in Poland under operation [85]. In Germany, for example, the number of installations is over 9400 and the country has similar potential for biogas production, when comparing arable land [85]. This difference can be explained by limiting factors such as the location. Construction of the installation may cause social resistance of the local communities. For instance, the surveys conducted with the residents of rural commune Kamionka in Lublin province, Poland, demonstrated that most of the residents were concerned about the unpleasant odour (60% of respondents) [86]. Other fears of the respondents were related to pollution, noise and risk of explosion. On the other hand, over 80% of the respondents indicated the benefits from an agricultural biogas facility, such additional income for the farmers, provision of the cheap energy for the community and positive impact on local environment. Moreover, some local producers were willing to cooperate with the owners of biogas facilities, by buying the energy, providing the biomass and using the post-fermentation pulp (digestate) for fertilizing purposes. The inconvenience related with cost of transport or seasonality of agri-food waste availability, characteristic for the areas with small farms that may cause input instability can be clarified by cooperation between farmers, producing different biomass [87]. Therefore, the decision regarding the selection of the place for the biogas installation should not only consider technological, environmental and legal issues, but also the social aspect should be taken into account [86].

The utilization of digestate might be another limiting factor for the development of agricultural biogas plants in Poland. The operating of agricultural biogas installations is associated with the generation of a large amount of post-fermentation pulp. This product, resulted from the digestion process contains more inorganic nitrogen than non-digested organic fertilizers, and, in consequence, more nitrogen in a form available for plants [88]. Previous studies conducted in EU countries demonstrated the possibility of using the digestate as a replacement for the traditional fertilizer or soil amendment, with the benefits both for the farmers (impact on the crop yields) and soil properties [89–93]. Nevertheless, in some cases the digestate management can be problematic for the biogas producers. When using digestate as the organic fertilizer, it is necessary to comply with several legal requirements regarding both storage methods of biomass intended for methane fermentation as the input and for fermentation pulp on the premises [93,94]. The other important issue of utilizing digestate for fertilizing purposes concerns is its classification. According to the law, digestate may be considered waste or by-product. However, numerous legal regulations for the digestate to be classified as a by-product and therefore qualify it for use as an organic fertilizer (or soil amendment), may be inconvenient for the producers (e.g., farmers, owners of the installation, etc.). Furthermore, the bio-fertilizers based on digestate require accurate physicochemical and microbiological tests in specialized research institutions and need to meet the procedures set by decision-makers [95]. These procedures apply to the use of digestate in agriculture as well as in gardening and forestry. The most important legal acts regarding the utilization and the requirements for the classification of digestate as a by-product, as well as legal requirements regarding methods of storage of substrates and digestate on biogas plant areas were discussed by Czekała et al. [94] and Łagocka et al. [96]. Nonetheless, due to the growing interest in energy obtained from the agricultural biogas installations in Poland, and consequently, an increased amount of post-digestate pulp attention should be given to the alternative methods of its management. In Italy for instance, the usage of digestate became a key factor to maintain profitability of biogas plants and to promote bioeconomy [88,97].

For the further development of agricultural biogas plants in Poland it is crucial to show farmers and residents the benefits of this type of investment. Promotion of bioeconomy as an important element of environmental sustainability and usage of renewable biological resources [98] as well as creating favourable conditions for research on cost-effective and implementing practical solutions [99].

6. Final Considerations

Agriculture plays an important role in the Polish economy and Poland is considered a producer and exporter of good quality products. The country has also considerable potential for biomass processing using agricultural, forest and municipal waste. Biomass from the residues of agri-food production and agricultural, especially bovine slurry, maize silage and distilleries has a great energy potential and is a valuable substrate for agricultural biogas production. Simultaneously it would be essential to implement and develop available and cost-effective technologies that convert biomass of agricultural origin into energy, while not competing with the food and forage market. The use of controversial products, for example fresh fruits and vegetables as a substrate should be considered with caution. Regarding the use of biomass for energy purposes, factors such an economic aspect, substrate availability and substrate storage should be taken into account. Utilization of digestate as a bio-fertilizer or soil amendment and its effect on crop yields is a priority for farmers. Nevertheless, here we propose further research on the impact of digestate on soil carbon sequestration, greenhouse gas emissions and on the use of digestate in degraded areas in order to restore soil ecosystem services. Agricultural biogas installations have the potential to contribute to the greening of the Polish energy sector but unless certain restrictions are overcome, the share of biomass for energy production might be limited.

Author Contributions: Conceptualization, K.A.K. and M.K.; methodology, K.A.K. and A.E.L.; literature review, K.A.K. and K.G., database creation, K.A.K.; writing-original draft preparation, K.A.K.; writing-review end editing, A.E.L., K.G., and M.K.; funding acquisition, M.K.

Acknowledgments: Anonymous reviewers are gratefully acknowledged for their constructive review that significantly improved this manuscript.

References

1. Suocheng, D.; Tong, K.W.; Yuping, W. Municipal solid waste management in China: Using commercial management to solve a growing problem. *Util. Policy* **2011**, *10*, 7–11. [CrossRef]
2. Rosik-Dulewska, C.; Karwaczyńska, U.; Ciesielczuk, T. Możliwości wykorzystania odpadów organicznych i mineralnych z uwzględnieniem zasad obowiązujących w ochronie środowiska (Possibilities of using organic and mineral waste, regarding the principles applicable in environmental protection). *Rocz. Ochr. Srodowiska* **2011**, *13*, 361–376. (In Polish)
3. Department of Economic and Social Affairs, Population Division. *World Population Prospects: The 2015 Revision, Key Findings and Advance Tables*; United Nations: New York, NY, USA, 2015; p. 241.
4. Kostecka, J.; Koc-Jurczyk, J.; Brudzisz, K. Waste management in Poland and European Union. *Arch. Waste Manag. Environ. Prot.* **2014**, *16*, 1–10.
5. Jambeck, J.R.; Geyer, R.; Wilcox, C.; Siegler, T.R.; Perryman, M.; Andrady, A.; Narayan, R.; Law, L.K. Plastic waste inputs from land into the ocean. *Science* **2015**, *347*, 768–771. [CrossRef] [PubMed]
6. Rodriguez, A.; Latawiec, A.E. Rethinking Organic Residues: The Potential of Biomass in Brazil. *Mod. Concepts Dev. Agron.* **2018**, *1*, 1–5.
7. Grando, R.L.; De Souza Antune, A.M.; Da Fonseca, F.V.; Sánchez, A.; Barrena, R.; Font, X. Technology overview of biogas production in anaerobic digestion plants: A European evaluation of research and development. *Renew. Sustain. Energy Rev.* **2017**, *80*, 44–53. [CrossRef]
8. Renewable Energy Law of Poland [Dz.U. 2012 poz. 1229]. Available online: http://prawo.sejm.gov.pl/isap.nsf/download.xsp/WDU20120001229/O/D20121229.pdf (accessed on 8 October 2019).
9. Nowacka-Blachowska, A.; Resak, M.; Rogosz, B.; Tomaszewska, H. Zrównoważone wykorzystanie biomasy na terenie Dolnego Śląska (Sustainable use of biomass in Lower Silesia). *Gor. Odkryw.* **2016**, *6*, 48–53. (In Polish)

10. Roszkowski, A. Biomasa i bioenergia—Bariery technologiczne i energetyczne (Biomass and bioenergy—Technological and energy limitations). *Probl. Inz. Rol.* **2012**, *3*, 79–100. (In Polish)
11. Kacprzak, A.; Krzystek, L.; Ledakowicz, L. Badania biochemicznego potencjału metanogennego wybranych roślin energetycznych (Analysis on the biochemical methanogenic potential of selected energy plants). *Inżynieria I Aparatura. Chemiczna* **2010**, *49*, 32–33. (In Polish)
12. Ansari, A.A. Effect of Vermicompost on the Productivity of Potato (*Solanum tuberosum*), Spinach (*Spinaciaoleracea*) and Turnip (*Brassica campestris*). *World J. Agric. Sci.* **2008**, *4*, 333–336.
13. Castro, A.; Da Silva Batista, N.; Latawiec, A.E.; Rodrigues, A.; Strassburg, B.B.N.; Silva, D.; Araujo, E.; De Moraes, L.F.D.; Guerra, J.G.; Galvão, G.; et al. The effects of *Gliricidia*-Derived Biochar on Sequential Maize and Bean Farming. *Sustainability* **2018**, *10*, 578. [CrossRef]
14. Yuan, H.; Lu, T.; Wang, Y.; Chen, Y.; Lei, T. Sewage sludge biochar: Nutrient composition and its effect on the leaching of soil nutrients. *Geoderma* **2016**, *267*, 17–23. [CrossRef]
15. Climate and Energy Policies in Poland. 2017. Available online: http://www.europarl.europa.eu/RegData/etudes/BRIE/2017/607335/IPOL_BRI(2017)607335_EN.pdf (accessed on 8 October 2019).
16. World Energy Council. Available online: https://www.worldenergy.org/data/resources/country/poland/coal/ (accessed on 8 October 2019).
17. Dzikuć, M.; Piwowar, A. Ecological and economic aspects of electric energy production using the biomass co-firing method: The case of Poland. *Renew. Sustain. Energy Rev.* **2016**, *55*, 856–862. [CrossRef]
18. Dec, B.; Krupa, J. Wykorzystanie odnawialnych źródeł energii w aspekcie ochrony środowiska (The use of renewable energy sources in the aspect of environmental protection). *Przegląd Nauk. Metod. Eduk. Bezpieczenstwa* **2007**, *3*, 722–757. (In Polish)
19. Zuwała, J. Life cycle approach for energy and environmental analysis of biomass and coal co-firing in CHP plant with backpressure turbine. *J. Clean. Prod.* **2012**, *35*, 164–175. [CrossRef]
20. Ciepielewska, M. Development of Renewable Energy in Poland in the Light of the European Union Climate-Energy Package and the Renewable Energy Sources Act. 2016. Available online: http://dx.doi.org/10.18778/1429-3730.43.01 (accessed on 20 September 2018).
21. Gołuchowska, B.; Sławiński, J.; Markowski, G. Biomass utilization as a renewable energy source in polish power industry-current status and perspectives. *J. Ecol. Eng.* **2015**, *16*, 143–154. [CrossRef]
22. Ross, A.B.; Jones, J.M.; Chaiklangmuang, S.; Pourkashanian, M.; Williams, A.; Kubica, K.; Andersson, J.T.; Kerst, M.; Danihelka, P.; Bartle, K.D. Measurement and prediction of the emission of pollutants from the combustion of coal and biomass in a fixed bed furnace. *Fuel* **2002**, *81*, 571–582. [CrossRef]
23. Jasiulewicz, M. Potencjał energetyczny biomasy rolniczej w aspekcie realizacji przez Polskę Narodowego Celu Wskaźnikowego OZE i dyrektyw w UE w 2020 (Energy potential of agricultural biomass in the aspect of Poland's implementation of the National RES Target and EU directives in 2020). *Rocz. Nauk. Stowarzyszenia Ekon. Rol. Agrobiz.* **2014**, *16*, 70–76. (In Polish)
24. Bartosiewicz-Burczy, H. Potencjał i energetyczne wykorzystanie biomasy w rajach Europy Srodkowej (Biomass potential and its energy utilization in the Central European countries). *Energetyka* **2012**, *7*, 860–866. (In Polish)
25. Energy 2018. Available online: https://stat.gov.pl/files/gfx/portalinformacyjny/pl/defaultaktualnosci/5485/1/6/1/energia_2018.pdf (accessed on 8 October 2019).
26. Kuziemska, B.; Trębicka, J.; Wieremej, W.; Klej, P.; Pieniak-Lendzion, K. Benefits and risks in the production of biogas. *Zesz. Nauk. Uniw. Przyr. Humanist. Siedlcach Ser. Adm. Zarządzanie* **2014**, *103*, 99–113. (In Polish)
27. Maj, G. Emission factors and energy properties of agro and forest biomass in aspect of sustainability of energy sector. *Energies* **2018**, *11*, 1516. [CrossRef]
28. Sikora, J. The research on efficiency of biogas production from organic fraction of municipal solid waste mixed with agricultural biomass. *Infrastruct. Ecol. Rural Areas* **2012**, *2*, 89–98.
29. Sikora, J.; Tomal, A. Determination of the energy potential of biogas in selected farm household. *Infrastruct. Ecol. Rural Areas* **2016**, *3*, 971–982.
30. Czekała, W. Agricultural Biogas Plants as a Chance for the Development of the Agri-Food Sector. *J. Ecol. Eng.* **2018**, *19*, 179–183. [CrossRef]
31. Jędrejek, A.; Jarosz, Z. Regionalne możliwości produkcji biogazu rolniczego (Regional possibilities of agricultural biogas production). *Rocz. Nauk. Stowarzyszenia Ekon. Rol. Agrobiz.* **2016**, *18*, 61–65. (In Polish)
32. Cukrowski, A.; Oniszk-Popławska, A.; Mroczkowski, P.; Zowsik, M.; Wiśniewski, G.; The Institute of

Renewable Energy, Warszawa. A Guide for Investors Interested in Building the Agricultural Biogas Plants. 2011. Available online: http://www.mg.gov.pl/node/13229 (accessed on 28 September 2018).

33. Piwowar, A.; Dzikuć, A.; Adamczyk, J. Agricultural biogas plants in Poland—Selected technological, market and environmental aspects. *Renew. Sustain. Energy Rev.* **2016**, *58*, 69–74. [CrossRef]
34. Obrycka, E. Korzyści społeczne i ekonomiczne budowy biogazowni rolniczych (Social and economic benefits of agricultural biogas plants). *Zesz. Nauk. Szk. Gl. Gospod. Wiej. Ekon. Organ. Gospod. Zywnosciowej* **2014**, *107*, 163–176. (In Polish)
35. Act of 10 July 2007 on Fertilizers and Fertilization Law of Poland. Available online: http://prawo.sejm.gov.pl/isap.nsf/download.xsp/WDU20071471033/T/D20071033L.pdf (accessed on 12 November 2019).
36. Marszałek, M.; Banach, M.; Kowalski, Z. Utylizacja gnojowicy na drodze fermentacji metanowej i tlenowej—Produkcja biogazu i kompostu (Utilization of liquid manure by methane and oxygen fermentation—Biogas and compost production). *Czas. Tech.* **2011**, *10*, 143–158. (In Polish)
37. Banaszkiewicz, T.; Wysmyk, J. Ecological aspect of utilization of agricultural feedstock. *Eur. Reg.* **2015**, *23*, 21–34. (In Polish)
38. Pawlak, J. Biogaz z rolnictwa—Korzyści i bariery (Biogas from agriculture—Benefits and limitations). *Probl. Agric. Eng.* **2013**, *3*, 99–108. (In Polish)
39. Borowski, S.; Domański, J.; Weatherley, L. Anaerobic co-digestion of swine and poultry manure with municipal sewage sludge. *Waste Manag.* **2014**, *34*, 513–521. [CrossRef] [PubMed]
40. Bujoczek, G.; Oleszkiewicz, J.; Sparling, R.; Cenkowski, S. High Solid Anaerobic Digestion of Chicken Manure. *J. Agric. Eng. Res.* **2000**, *76*, 51–60. [CrossRef]
41. Hryniewicz, M.; Grzybek, A. Available straw surplus for use for energy purposes in 2016. *Probl. Inz. Rol.* **2017**, *25*, 15–31. (In Polish)
42. Jarosz, Z. Energy potential of agricultural crops biomass and their use for energy purposes. *Sci. Noteb. Wars. Univ. Life Sci. SGGW Wars. Probl. World Agric.* **2017**, *17*, 81–92. (In Polish)
43. Jasiulewicz, M. Possibility of Liquid Biofuels, Electric and Heat Energy Production from Biomass in Polish Agriculture. *Polish J. Environ. Stud.* **2010**, *19*, 479–483.
44. Michalski, T. Biogazownia w każdej gminie—Czy wystarczy surowca (Biogas plant in every municipality—Is there enough raw material). *Wieś Jutra* **2009**, *3*, 12–14. (In Polish)
45. Romaniuk, W.; Domasiewicz, T.; Borek, K.; Borusiewicz, A.; Marczuk, T. *Analiza Potrzeb Techniczno-Technologicznych oraz Propozycje Rozwiązań w Produkcji Biogazu w Gospodarstwach Rodzinnych i Farmerskich (The Analysis of Technical Demands and Possible Solutions for Biogas Production in Family Agriculture and on Farms)*; Wydawnictwo Wyższej Szkoły Agrobiznesu w Łomży: Lomży, Poland, 2015. (In Polish)
46. Kowalczyk-Juśko, A.; Kościk, B.; Jóźwiakowski, K.; Marczuk, A.; Zarajczyk, J.; Kowalczuk, J.; Szmigielski, M.; Sagan, A. Effects of biochemical and thermochemical conversion of sorghum biomass to usable energy. *Przem. Chem.* **2015**, *94*, 1838–1840.
47. Król, A. Kiszonki—Cenny substrat do produkcji biogazu (Silage—A valuable substrate for biogas production). *Autobusy Tech. Eksploat. Syst. Transp.* **2011**, *10*, 249–254. (In Polish)
48. Księżak, J. Produkcja kukurydzy w różnych rejonach Polski (Maize production in various regions of Poland). *Wieś Jutra* **2009**, *3*, 16. (In Polish)
49. Daniel, Z.; Juliszewski, T.; Kowalczyk, Z.; Malinowski, M.; Sobol, Z.; Wrona, P. The method of solid waste classification from the agriculture and food industry. *Infrastruct. Ecol. Rural Areas* **2012**, *2*, 141–152.
50. Czyżyk, F.; Strzelczyk, M. Rational utilization of production residues generated in agri-food. *Arch. Waste Manag. Environ. Prot.* **2015**, *17*, 99–106.
51. Kruczek, M.; Drygaś, B.; Habryka, C. Pomace in fruit industry and their contemporary potential application. *World Sci. News* **2016**, *48*, 259–265.
52. Wesołowska-Trojanowska, M.; Targoński, Z. The whey utilization in biotechnological processes. *Eng. Sci. Technol.* **2014**, *1*, 102–119.
53. Michalska, K.; Pazera, A.; Bizukojć, M.; Wolf, W.; Sibiński, M. Innovative dairies-energy independence and waste-free technologies as a consequence of biogas and photovoltaic investments. *Acta Innov.* **2013**, *9*, 5–16.
54. Adamski, M.; Pilarski, K.; Dach, J. Possibilities of Usage of the Distillery Residue as a Substrate for Agricultural Biogas Plant. *J. Res. Appl. Agric. Eng.* **2009**, *54*, 10–15. (In Polish)
55. Kozłowski, K.; Cieślik, M.; Smurzyńska, A.; Lewicki, A.; Jas, M. The usage of waste from meat processing for energetic purposes. *Eng. Sci. Technol.* **2015**, *1*, 36–46.

56. Czekała, W.; Smurzyńska, A.; Cieślik, M.; Boniecki, P.; Kozłowski, K. Biogas Efficiency of Selected Fresh Fruit Covered by the Russian Embargo. In Proceedings of the 16th International Multidisciplinary Scientific Conference SGEM 2016, Albena, Bulgaria, 30 June–6 July 2016; Volume 3, pp. 227–234.
57. Smurzyńska, A.; Czekała, W.; Lewicki, A.; Cieślik, M.; Kozłowski, K.; Janczak, D. The biogas output of vegetables utilized in the polish market due to introduction of the Russian embargo. *Tech. Rol. Ogrod. Leśna* **2016**, *6*, 24–27.
58. Weiland, P. Biogas production: Current state and perspectives. *Appl. Microbiol. Biotechnol.* **2010**, *85*, 849–860. [CrossRef]
59. Kupryś-Caruk, M. Agri-food industry as a source of substrates for biogas production. *Postępy Nauk. Technol. Przem. Rolno Spoz.* **2017**, *72*, 69–85. (In Polish)
60. Pilarska, A.; Pilarski, K.; Ryniecki, A. The use of methane fermentation in the development of selected waste products of food industry. *Nauk. Inz. Technol.* **2014**, *4*, 100–111. (In Polish)
61. Prask, H.; Fugol, M.; Szlachta, J. Biogaz z wytłoków z białych i czerwonych winogron. *Przem. Ferment. Owocowo Warzywny.* **2012**, *5*, 45–46. (In Polish)
62. Bożym, M.; Florczak, I.; Zdanowska, P.; Wojdalski, J.; Klimkiewicz, M. An analysis of metal concentrations in food wastes for biogas production. *Renew. Energy* **2015**, *77*, 467–472. [CrossRef]
63. Kwaśny, J.; Banach, M.; Kowalski, Z. Technologies of biogas production from different sources—A review. *Chem. Tech. Trans.* **2012**, *17*, 83–102. (In Polish)
64. Maślanka, S.; Siołek, M.; Hamryszak, Ł.; Łopot, D. Zastosowanie odpadów z przemysłu mleczarskiego do produkcji polimerów biodegradowalnych. *Chemik* **2014**, *68*, 703–709. (In Polish)
65. Janczukowicz, W.; Zieliński, M.; Dębowski, M. Biodegradability evaluation of dairy effluents originated in selected sections of dairy production. *Bioresour. Technol.* **2008**, *99*, 4199–4205. [CrossRef] [PubMed]
66. Sobczak, A.; Błyszczek, E. Ways of management of by-products from meat industry. *Czas. Tech. Chem.* **2009**, *106*, 141–151.
67. Adhikari, B.B.; Chae, M.; Bressler, D.C. Utilization of Slaughterhouse Waste in Value-Added Applications: Recent Advances in the Development of Wood Adhesives. *Polymers* **2018**, *10*, 176. [CrossRef]
68. Zakrzewski, P. Technologia utylizacji odpadów poubojowych w instalacjach biogazowych (Technology of post-slaughter waste utilization in biogas installations). *Czysta Ènerg.* **2009**, *10*, 40–41. (In Polish)
69. Dach, J.; Kozłowski, K.; Czekała, W. Odpady poubojowe na biogaz—Czy to sie opłaca? (Post-slaughter waste for biogas—Is it profitable). *Biomasa* **2017**, *8*, 38–40.
70. EUR-Lex. Available online: https://eur-lex.europa.eu/eli/reg/2009/1069/oj (accessed on 8 October 2019).
71. Fuess, L.T.; Garcia, M.L. Anaerobic digestion of stillage to produce bioenergy in the sugarcane-to-ethanol industry. *Environ. Technol.* **2014**, *35*, 333–339. [CrossRef]
72. Dubrovskis, V.; Plume, I. Methane production from stillage. In Proceedings of the 16th International Scientific Conference Engineering for Rural Development, Latvia University of Life Sciences and Technology, Jelgava, 24–26 May 2017. [CrossRef]
73. Czupryński, B.; Kotarska, K. Recyrkulacja i sposoby zagospodarowania wywaru gorzelniczego (Recirculation and methods of managing stillage). *Inz. Apar. Chem.* **2011**, *50*, 21–23. (In Polish)
74. Skowrońska, M.; Filipek, T. Nawozowe wykorzystanie wywaru gorzelnianego (The use of stillage for fertilizing purposes). *Proc. ECOpole* **2012**, *6*, 267–271. (In Polish) [CrossRef]
75. Kotarska, K.; Kłosowski, G.; Czupryński, B. Zagospodarowanie wywaru gorzelniczego na cele paszowe (Managing of stillage for fodder purposes). *Przem. Ferment. Owocowo Warzywny* **1996**, *40*, 27–30. (In Polish)
76. Kotarska, K.; Dziemianowicz, W. Effect of Different Conditions of Alcoholic Fermentation of Molasses on Its Intensification and Quality of Produced Spirit. *Zywnosc Nauka. Technologia. Jakosc* **2015**, *2*, 150–159. (In Polish) [CrossRef]
77. Owczuk, M.; Wardzińska, D.; Zamojska-Jaroszewicz, A.; Matuszewska, A. The use of biodegradable waste to produce biogas as an alternative source of renewable energy. *Studia Ecol. Bioethicae UKSW* **2013**, *11*, 133–144. (In Polish)
78. Jasiulewicz, M. Implementation of the innovative investment in food industry and anaerobic digestion in the field of bioenergy. *Rocz. Nauk. Stowarzyszenia Ekon. Rol. Agrobiz.* **2017**, *19*, 88–94. (In Polish)
79. Romaniuk, W.; Domasiewicz, T. Substraty dla biogazowni rolniczych (Substrates for agricultural biogas plants). *Agrotech. Porad. Rolnika* **2014**, *11*, 74–75. (In Polish)

80. Janczak, D.; Kozłowski, K.; Zbytek, Z.; Cieślik, M.; Bugała, A.; Czekała, W. Energetic Efficiency of the Vegetable Waste Used as Substrate for Biogas Production. *Environment &Chem.* **2016**, *64*, 06002.
81. Kozłowski, K.; Lewicki, A.; Cieślik, M.; Janczak, D.; Czekała, W.; Smurzyńska, A.; Brzoski, M. The possibility of improving the energy and economic balance of agricultural biogas plant. *Tech. Rol. Leśna* **2017**, *3*, 10–13.
82. Kasprzycka, A. Causes of interference methane fermentation. *Autobusy* **2011**, *10*, 224–228. (In Polish)
83. Szymanska, D.; Lewandowska, A. Biogas power plants in Poland—Structure, capacity, and special distribution. *Sustainability* **2015**, *7*, 16801–16819. [CrossRef]
84. Rzeznik, W.; Mielcarek, P. Agricultural biogas plants in Poland. *Eng. Rural Dev.* **2018**, *17*, 1760–1765.
85. Biomass Media Group. Available online: http://rynekbiogazu.pl/2018/03/21/potencjal-rozwoju-sektora-biogazu-w-polsce (accessed on 8 October 2019).
86. Kowalczyk-Juśko, A.; Listosz, A.; Flisiak, M. Spatial and social conditions for the location of agricultural biogas plants in Poland (case study). *E3S Web Conf.* **2019**, *86*, 00036. [CrossRef]
87. Caruso, M.C.; Braghieri, A.; Capece, A.; Napolitano, F.; Romano, P.; Galgano, F.; Altieri, G.; Genovese, F. Recent updates on the use of agro-food waste for biogas production. *Appl. Sci.* **2019**, *9*, 1217. [CrossRef]
88. Bartoli, A.; Hamelin, L.; Rozakis, S.; Borzęcka, M.; Brandão, M. Coupling economic and GHG emission accounting models to evaluate the sustainability of biogas policies. *Renew. Sustain. Energy Rev.* **2019**, *106*, 133–148. [CrossRef]
89. Koszel, M.; Lorencowicz, E. Agricultural use of biogas digestate as a replacement fertilizers. *Agric. Agric. Sci. Procedia* **2015**, *7*, 119–124. [CrossRef]
90. Losak, T.; Hlusek, J.; Zatloukalova, A.; Musilova, L.; Vitezova, M.; Skarpa, P.; Zlamalova, T.; Fryc, J.; Vitez, T.; Marecek, J.; et al. Digestate from biogas plants is an attractive alternative to mineral fertilisation of kohlrabi. *J. Sustain. Dev. Energy Water Environ. Syst.* **2014**, *2*, 309–318. [CrossRef]
91. Šimon, T.; Kunzová, E.; Friedlová, M. The effect of digestate, cattle slurry and mineral fertilization on the winter wheat yield and soil quality parameters. *Plant Soil Environ.* **2015**, *61*, 522–527. [CrossRef]
92. Koszel, M.; Kocira, A.; Lorencowicz, E. The evaluation of the use of biogas plant digestate as a fertilizer in alfalfa and spring wheat cultivation. *Fresenius Environ. Bull.* **2016**, *25*, 3258–3264.
93. Pilarska, A.A.; Piechota, T.; Szymańska, M.; Pilarski, K.; Wolna-Maruwka, A. Ocena wartości nawozowej pofermentów z biogazowni oraz wytworzonych z nich kompostów (Evaluation of fertilizer value of digestate and its composts obtained from biogas plant). *Nauka Przyr. Technol.* **2016**, *10*, 35. (In Polish) [CrossRef]
94. Czekała, W.; Pilarski, K.; Dach, J.; Janczak, D.; Szymańska, M. Analysis of management possibilities for digestate from biogas plant. *Tech. Rol. Ogrod. Leśna* **2012**, *4*, 11–13. (In Polish)
95. Mystkowski, E. Poferment dla rolnictwa (Digestate for agriculture). *Rol. ABC* **2015**, *9*, 1–5. (In Polish). Available online: https://studylibpl.com/doc/1424615/masa-pofermentacyjna-nawozem-dla-rolnictwa (accessed on 8 October 2019).
96. Łagocka, A.; Kamiński, M.; Cholewński, M.; Pospolita, W. Korzyści ekologiczne ze stosowania pofermentu z biogazowni rolniczych jako nawozu organicznego (Ecological benefits from the use of digestate from agricultural biogas plants as organic fertilizer). *Kosmos* **2016**, *65*, 601–607. (In Polish)
97. Pantaleo, A.; De Gennaro, B.; Shah, N. Assessment of optimal size of anaerobic co-digestion plants: An application to cattle farms in the province of Bari (Italy). *Renew. Sustain. Energy Rev.* **2013**, *20*, 57–70. [CrossRef]
98. Konstantinis, A.; Rozakis, S.; Maria, E.A.; Shu, K. A definition of bioeconomy through bibliometric networks of the scientific literature. *AgBioForum* **2018**, *21*, 64–85.
99. Krzywonos, M.; Marciszewska, A.; Domiter, M.; Borowiak, D. Bioeconomy -current status, trends and prospects. The challenge for universities, business and government. *Pol. J. Agron.* **2016**, *27*, 71–79. (In Polish)

7

Modeling and Simulation of Particle Motion in the Operation Area of a Centrifugal Rotary Chopper Machine

Andrzej Marczuk [1], Jacek Caban [1,*], Alexey V. Aleshkin [2], Petr A. Savinykh [2], Alexey Y. Isupov [2] and Ilya I. Ivanov [3]

1 Department of Agricultural, Forestry and Transport Machines, University of Life Sciences in Lublin, 20-612 Lublin, Poland
2 N.V. Rudnitsky North-East Agricultural Research Institute, Kirov 610007, Russia
3 FSBEI HE Vologda State Dairy Farming Academy (DSFA) named after N.V. Vereshchagin, Vologda—Molochnoye 160555, Russia
* Correspondence: jacek.caban@up.lublin.pl

Abstract: The article presents approaches to the formation of a general computational scheme for modeling (simulating) the particle motion on an axisymmetric rotating curved surface with a vertical axis of rotation. To describe the complex particle motion over a given surface, the fundamental equation of particle dynamics in a non-inertial reference frame was used, and by projecting it onto the axes of cylindrical coordinates, the Lagrange's differential equations of the first kind were obtained. According to the proposed algorithm in C#, an application was developed that enables graphical and numerical control of the calculation results. The program interface contains six screen forms with tabular baseline data (input) and a table of a step-by-step calculation of results (output); particle displacement, velocity, and acceleration diagrams constructed along the axes of the system of cylindrical coordinates ρ and z; graphical presentation of the generate of the surface of revolution and the trajectory of the absolute motion of a particle over the axisymmetric rotating surface developed in polar coordinates. Examples of the calculation of the particle motion are presented. The obtained results can be used for the study and design of machines, for example, centrifugal rotary chopper machines.

Keywords: accelerator; axisymmetric surface; general equation of dynamics; non-inertial reference frame

---

1. Introduction

The manufacturing sector has one of the largest impacts on worldwide energy use and natural resource consumption. Traditionally, research on manufacturing processes was mainly conducted to improve efficiency and accuracy and to lower costs [1]. One of the modern challenges in the field of designing manufacturing systems is to determine the optimal level of their flexibility from the point of view of the production tasks being performed. Whether or not a production process to be executed is capable of achieving the assumed performance parameters depends, among others, on the reliability of the machines and technological devices that make up the system under design [2]. As commonly known, the oldest manufacturing techniques used by humans are the grinding, chopper, and milling processes. Research on grinding tries to enhance economic and ecological properties and performance to extend grinding applications in the overall process chain—on the one hand, in the direction of increased material removal rates, avoiding turning and milling, and on the other hand, in the direction of fine finishing, thus making further abrasive finishing processes, such as lapping and polishing, obsolete [3].

Many factors influence the sustainability of a manufacturing process, and a combination of qualitative and quantitative methods are needed to discover them [4]. To resolve these issues, namely, the identification of the type of underlying impact factors, the uncertainty of the influencing factors, and the incompleteness of the evaluation information, the researchers used the evaluation method with analytical hierarchy process [5,6] and simulations [2,7–9] in the manufacturing sector. Sustainability encompasses the three pillars of economic, environmental, and social sustainability [10–12]. In the investigation of Ahmad [10], generally, there were approximately 44% (25/57) of highly applicable indicators. Among them, there were 28% (7/25) environmental, 28% (7/25) economic, and 44% (11/25) social indicators [10]. We can, therefore, see that economic and environmental factors are treated on an equal level. Social indicators in sustainable development are highly rated. Nevertheless, in the case of the two previous ones, they are more adequate for production processes, which have a more indirect nature on social impact. Beekaroo et al. [13] revealed an index with nine environmental, four economic, and two social indicators which were pertinent in sustainability measurement.

One of the most commonly used methods focusing on environmental sustainability is the product life cycle assessment (LCA). This method is used in many sectors of the economy [14–16] such as energy sector [17–19], transport sector [18,19], and food and agricultural industry [20,21]. The limitation of using this method is the need for specific and very detailed databases covering a lot of quantitative data.

Another method uses sustainability indicators (SI). In this method, the indicator can be defined "as a measure or aggregation of measures from which one can infer about the phenomenon of interest" [11]. This method can capture all three dimensions of sustainable development and help with the evaluation on many levels (e.g., enterprises, objects, processes, and products). In particular, for perpetrators with limited means and resources, SI provide a good method of analyzing sustainable development. Enterprises can assess their real situation using indicators, raise awareness, and set targets. The results of a literature review show that energy costs and GHG (Greenhouse Gas) indicators [10,11,13,19,22–25] are becoming the most commonly used indicators in sustainable planning. On this basis [11,26], four main directions of future research on sustainability indicators can also be mentioned: implementation of new optimization methods; adding further sustainable development indicators; extension of the model on a larger scale of the production system; and loosening certain assumptions. Finding optimal indicators in production processes is not an easy task; generally, we focus on minimizing energy consumption, waste production, improving the efficiency of machinery and equipment, as well as production and logistic processes in a given company. Sustainability indicators are based on measured and/or estimated data that have to be normalized, scaled, and aggregated consistently [27].

The evolution of design criteria for grinding and chopper machines is driven by functional requirements, general trends in machine tools, and cost [28]. The primary functional requirements, as named by Möhring et al. [29], are similar for all machine tools: high static and dynamic stiffness, fatigue strength, damping, thermal and long-term stability, and low weight of moving parts. The grinding and chopper process is carried out in many areas of food manufacturing [30–32] and agriculture [33–35], as well as in the industrial sector [36–38]. The mixing process is highly complicated, with a number of affecting parameters, such as the particle properties, the structure and performance of the mixer, the mixing process parameters, and the particle feeding order [38]. The crushing method and the machinery used for this purpose should be adapted to the type of material being ground and, in particular, to its mechanical properties [33,39].

Laboratories and modern industry require fast and effective grinding in small volumes [40]. In the field of agricultural mechanization, centrifugal chopper machines with a horizontal rotor have recently attracted increasing interest. Their principle of action lies in the acceleration of grains due to centrifugal inertia forces with their subsequent grinding (fragmentation) by impact or cutting [41,42]. This is due to the lower energy consumption for the grinding process in comparison with other choppers (crushers), which, in addition to the direct costs of grain destruction, have the energy costs for drifting whole and crushed grain using air. In centrifugal rotary chopper machines, the material can be supplied to crusher/shredder hammers or to a cutting pair by centrifugal inertia forces created by a horizontally rotating rotor with working elements. In this, one of the key roles of a centrifugal rotary material grinding machine is the distribution bowl or accelerator, which functions to provide a uniform and stable supply of material to the chopper bodies and impart to the material (particles) located on its surface the necessary linear velocity and trajectory of motion. Despite the rotary blenders relying upon the action of gravity to cause the powder to cascade and mix within a rotating vessel, the convective blenders employ a paddle, impeller, blade, or screw which stirs the powder inside a static vessel [43].

Most of the centrifugal rotary grinding machines in their design have a central feed of the material into the accelerator made in the form of a flat disc. However, with this material feed, the acceleration of a particle that is closest to the axis of rotation will be difficult because of the lack of initial velocity and centrifugal forces to overcome the friction forces, which provokes the appearance of a stagnant zone and an increase in the cost of overcoming them. To solve this problem, two approaches can be used: feeding the material with an offset from the axis of rotation, leading to design complexity, or using the reflective surface of the splitter, which is a straight conical surface located coaxially with the axis of rotation, ensuring the separation of the particles to be ground from the axis of rotation and giving them the initial velocity. In centrifugal impact grinders using the "stone-crushing-stone" principle with self-lining, the trajectory of the particles to be ground (milled) is equally important.

Figure 1 shows the rotary chopper machine analyzed in this study.

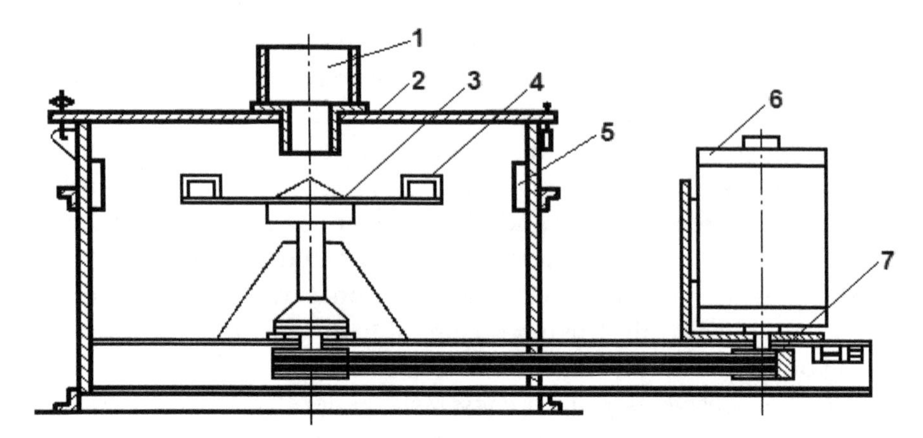

Figure 1. Tornado chopper machine: 1—feed-in tube; 2—case cover; 3—accelerator; 4—accelerating blade; 5—shredding element; 6—electric motor; 7—V-belt drive.

Figure 2 shows a scheme of self-lining formation in the accelerator.

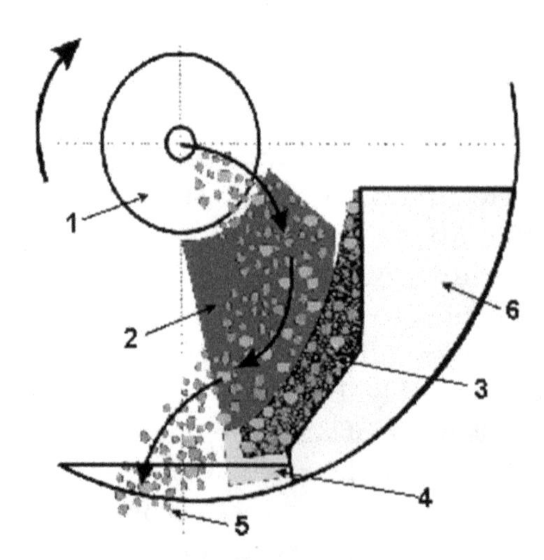

Figure 2. Scheme of self-lining formation in the accelerator: 1—a splitter; 2—wear plate; 3—self-lining pocket; 4—carbide blade; 5—descent of the material from the accelerator; 6—accelerator case.

The main contribution of this study is the modeling (simulating) of the particle motion on an axisymmetric rotating curved surface with a vertical axis of rotation, whose calculation can be used for the study and design of machines, for example, centrifugal rotary chopper machines. A mathematical model of motion was developed on the basis of step-by-step numerical integration of the obtained closed system of differential equations with an equation of constraint describing the surface of revolution. An application based on the algorithm in C# under MS Visual Studio (Microsoft, Redmond, Washington, USA) was developed that enables graphical and numerical control of the calculation results. Moreover, the results of the calculations of the particle motion trajectory and kinematic indicators are presented. The conducted simulation tests show possibilities of improving the process efficiency and shortening the operation times, which results in economic benefits during sustainable manufacturing.

2. Materials and Methods

Hypothesis. One of the methods to achieve this goal takes into account the form of the accelerator, allowing us to set the initial kinematic parameters and particle trajectory, which in turn will affect the geometric dimensions, metal consumption, and technological modes of operation of a centrifugal rotary chopper or its nodes, for example, rotor (crusher) speed and material feed rate, which would then reduce the specific energy consumption per unit of output.

Methods. To solve this problem, we modeled the motion of a particle of the material on a rotating curved surface. To do this, we used the basic law of particle dynamics (dynamics of a material point) and considered the motion of a particle on the surface of an axisymmetric spinning bowl. We used step-by-step numerical integration, which allowed us to obtain data for analyzing the particle.

3. Results and Discussion

3.1. Modeling Results

The curved surface rotates around the vertical axis z (Figure 3). The equation of the surface is described in cylindrical coordinates by the equation

$$\eta(\rho, \varphi, z) = 0 \tag{1}$$

where ρ is the cylindrical radius, φ is the polar (vectorial) angle, and z is the application (z-axis).

The radius vector of a material point moving on a curved surface in the coordinate axes, rotating with it, is a function of these three coordinates, which can transform over time without breaking Equation (1):

$$\vec{r} = \vec{r}(\rho, \varphi, z). \tag{2}$$

The differential equation of the relative motion of a point on a rotating surface is written in vector form [40]

$$m\frac{d^2\vec{r}}{dt^2} = m\vec{g} + \vec{N} + \vec{F_{TR}} + \vec{\Phi_e} + \vec{\Phi_c}, \tag{3}$$

where m is the mass of a particle, $\frac{d^2\vec{r}}{dt^2}$ is its acceleration, $m\vec{g}$ is the gravity force, $\vec{N}$ is the normal reaction of a bowl surface, $\vec{F_{TR}}$ is the friction force from the surface directed opposite to the relative velocity of a particle, $\vec{\Phi_e}$ is the centrifugal inertial force, and $\vec{\Phi_c}$ is the Coriolis inertial force.

Forces and accelerations on the direction of the cylindrical axes of coordinates are projected. The gravity force is opposite to the z axis,

$$m\vec{g} = m\vec{g}(0,\ 0,\ -mg), \tag{4}$$

where $\vec{g}$ is free-fall acceleration. The normal reaction of a rotating curved surface is

$$\vec{N} = \lambda \cdot grad(\eta), \tag{5}$$

where $\lambda = \lambda(t)$ is Lagrange's indeterminate multiplier [44], $grad(\eta)$ is the vector gradient to the equation of surface (1), which has projections on the axial cylindrical coordinate system with the φ axis directed perpendicular to the ρ, z, axes and passing through the moving point, so that the axes ρ, φ, z form the right-hand system of vectors:

$$\begin{cases} (grad(\eta))_\rho = \frac{\partial \eta}{\partial \rho} \\ (grad(\eta))_\varphi = \frac{\partial \eta}{\partial \varphi} \cdot \frac{1}{\rho} \\ (grad(\eta))_z = \frac{\partial \eta}{\partial z} \end{cases} . \tag{6}$$

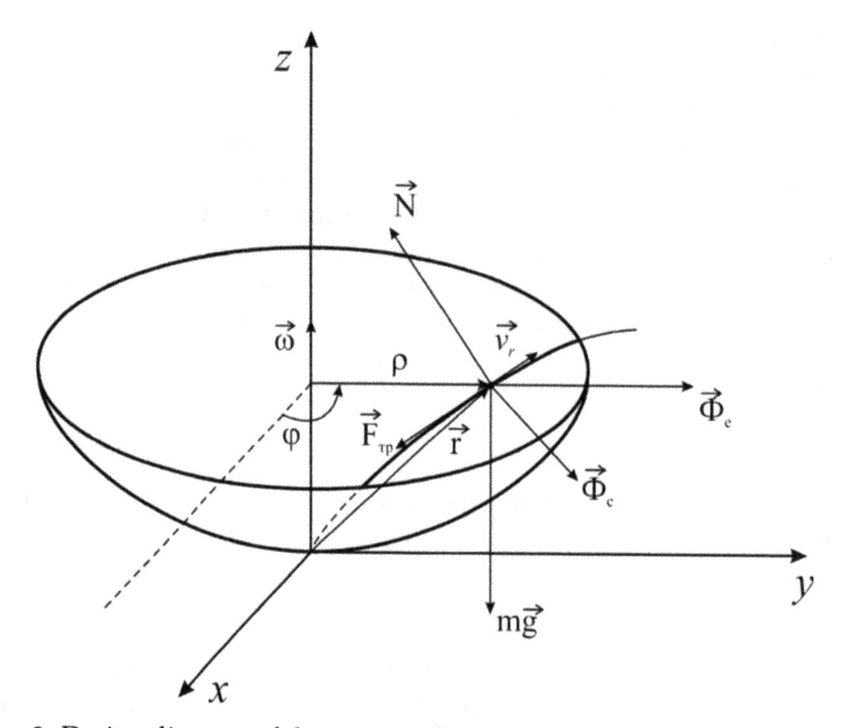

Figure 3. Design diagram of the motion of a particle on a rotating curved surface.

The projections of the normal reaction $\vec{N}$ are:

$$\begin{cases} N_\rho = \lambda \frac{\partial \eta}{\partial \rho} \\ N_\varphi = \lambda \frac{\partial \eta}{\partial \varphi} \cdot \frac{1}{\rho} \\ N_z = \lambda \frac{\partial \eta}{\partial z} \end{cases} . \tag{7}$$

The modulus of the normal reaction is found by

$$N = \left|\vec{N}\right| = |\lambda| \sqrt{\left(\frac{\partial \eta}{\partial \rho}\right)^2 + \left(\frac{\partial \eta}{\partial \varphi}\frac{1}{\rho}\right)^2 + \left(\frac{\partial \eta}{\partial z}\right)^2}. \tag{8}$$

At a constant angular velocity of rotation (spin rate) of the curved surface $\vec{\omega}$, the centrifugal inertial force $\vec{\Phi_e}$ is directed along the radius ρ in accordance with the result of vector products in its definition

$$\vec{\Phi_e} = -m\vec{\omega} \times (\vec{\omega} \times \vec{r}), \tag{9}$$

where $\vec{r}$ is determined by Equation (2), then

$$\begin{cases} \Phi_{e\rho} = m\omega^2\rho \\ \Phi_{e\varphi} = 0 \\ \Phi_{ez} = 0 \end{cases}. \quad (10)$$

Coriolis inertial force $\vec{\Phi}_c$ is by definition equal to

$$\vec{\Phi}_c = -2m\vec{\omega} \times \vec{v}, \quad (11)$$

where $\vec{v}$ is the relative velocity of a material point in the moving axes of coordinates, and the modulus of relative velocity is associated with the cylindrical coordinates by

$$v = \sqrt{\dot{\rho}^2 + (\rho\dot{\varphi})^2 + \dot{z}^2}. \quad (12)$$

Then, the projections of the Coriolis force equal

$$\begin{cases} \Phi_{c\rho} = 2m\omega\dot{\varphi}\rho \\ \Phi_{c\varphi} = -2m\dot{\rho}\omega \\ \Phi_{cz} = 0 \end{cases}. \quad (13)$$

The friction force of a particle on the bowl surface $\vec{F}_{TR}$ is determined by Coulomb's law through a normal reaction and it is opposite in direction to the relative velocity:

$$\vec{F}_{TR} = -\left|\vec{N}\right| f \frac{\vec{v}}{v} \quad (14)$$

$$\begin{cases} F_{TR\rho} = -\left|\vec{N}\right| f \frac{\dot{\rho}}{v} \\ F_{TR\varphi} = -\left|\vec{N}\right| f \frac{\dot{\varphi}\rho}{v} \\ F_{TRz} = -\left|\vec{N}\right| f \frac{\dot{z}}{v} \end{cases}. \quad (15)$$

Let us project Equation (3) on the axis of the moving system of cylindrical coordinates rotating together with the curved surface by using the found projections of all forces:

$$\begin{cases} m\left(\ddot{\rho} - \rho\dot{\varphi}^2\right) = \lambda\frac{\partial\eta}{\partial\rho} - Nf\frac{\dot{\rho}}{v} + m\omega^2\rho + 2m\omega\dot{\varphi}\rho \\ m\frac{1}{\rho}\frac{d}{dt}\left(\rho^2\dot{\varphi}\right) = -Nf\frac{\dot{\varphi}\rho}{v} - 2m\dot{\rho}\omega \\ m\ddot{z} = -mg + \lambda\frac{\partial\eta}{\partial z} - Nf\frac{\dot{z}}{v} \end{cases}. \quad (16)$$

Let us transform the second equation of the system, taking into account the projection of acceleration on the φ axis:

$$\frac{1}{\rho}\frac{d}{dt}\left(\rho^2\dot{\varphi}\right) = 2\dot{\rho}\dot{\varphi} + \rho\ddot{\varphi}. \quad (17)$$

By bringing all forces to a unit mass of a particle $m = 1$, we obtain

$$\begin{cases} \ddot{\rho} = \rho\dot{\varphi}^2 + \lambda\frac{\partial\eta}{\partial\rho} - Nf\frac{\dot{\rho}}{v} + \omega^2\rho + 2\omega\dot{\varphi}\rho \\ \ddot{\varphi} = \frac{1}{\rho}\left(-2\dot{\rho}\dot{\varphi} - Nf\frac{\dot{\varphi}\rho}{v} - 2\dot{\rho}\omega\right) \\ \ddot{z} = -g + \lambda\frac{\partial\eta}{\partial z} - Nf\frac{\dot{z}}{v} \end{cases}. \quad (18)$$

To obtain a closed system of equations, we supplement the system (18) with a coupling equation (of constraints) (1) (see Supplementary Material). Then, four unknown functions $\rho(t)$, $\varphi(t)$, $z(t)$, $\lambda(t)$ can be found by solving a system of three differential (18) and one algebraic (1) equations.

In the equation, various forms of execution of the surface of revolution are possible (1). Let us define the coupling equation in the form of an axisymmetric surface described by a power function passing through the given starting and final points of motion, written as

$$\frac{z}{z_K} = a_0 + \left(\frac{\rho}{\rho_K}\right)^n, \tag{19}$$

where a_0 is a summand determined from the condition of belonging to a given surface of revolution of the initial point of the trajectory with the coordinates

$$\rho(0) = \rho_0, \ \varphi(0) = 0, \ z(0) = 0 \tag{20}$$

ρ_K, z_K are coordinates of the final point of the trajectory, where the surface described by Equation (19) ends for all values of the polar (vectorial) angle φ; n is an exponent in the equation of surface, which changes the degree of the bowl concavity and which can be varied to achieve the required parameters of the particle motion.

Thus, the origin of coordinates of the moving frame of reference, in which the particle motion is described, always corresponds along the z axis to the starting point of the motion trajectory, and the parabola vertex in the axial section of the surface of revolution lies on this axis below zero by the value a_0:

$$a_0 = -\left(\frac{\rho_0}{\rho_K}\right)^n. \tag{21}$$

Let us reduce Equation (19) to the form (1)

$$\eta(\rho, \varphi, z) = \frac{z}{z_K} - a_0 - \left(\frac{\rho}{\rho_K}\right)^n = 0, \tag{22}$$

and calculate the partial derivatives in the gradient expression to the constraint surface

$$\begin{cases} \frac{\partial \eta}{\partial \rho} = -n\left(\frac{\rho}{\rho_K}\right)^{n-1} \frac{1}{\rho_K} \\ \frac{\partial \eta}{\partial \varphi} = 0 \\ \frac{\partial \eta}{\partial z} = \frac{1}{z_K} \end{cases}. \tag{23}$$

These expressions in Equation (23) should be substituted in the projection of force $\vec{N}$, a normal reaction to the constraint surface in Equation (18).

From the last equation of the system (18) we find the Lagrange multiplier

$$\lambda = \frac{1}{\left(\frac{\partial \eta}{\partial z}\right)}\left(\ddot{z} + g + Nf\frac{\dot{z}}{v}\right). \tag{24}$$

Since, when moving along the surface, only two cylindrical coordinates are independent, then for step-by-step numerical integration, we take the first two differential Equations (18) as the basis and calculate the coordinate z and its time derivatives through coupling Equation (22):

$$z = z_K a_0 + z_K\left(\frac{\rho}{\rho_K}\right)^n \tag{25}$$

$$\dot{z} = nz_K \frac{1}{\rho_K}\left(\frac{\rho}{\rho_K}\right)^{n-1} \dot{\rho} \tag{26}$$

$$\ddot{z} = nz_K \frac{1}{\rho_K}(n-1)\left(\frac{\rho}{\rho_K}\right)^{n-2} \frac{\dot{\rho}^2}{\rho_K} + nz_K \frac{1}{\rho_K}\left(\frac{\rho}{\rho_K}\right)^{n-1} \ddot{\rho}. \tag{27}$$

In Equations (24)–(27), the coordinate values ρ and φ, as well as their time derivatives $\dot{\rho}$, $\dot{\varphi}$ at the next integration step, are assumed to be equal to their value at the end of the previous integration step. The procedure of numerical integration of the system of Equation (18) was carried out by the method of averaged acceleration [45], using the original algorithmic program for "Windows" coded in C# under MS Visual Studio.

3.2. Results and Discussion

The results of the analysis of the kinematic parameters of the motion and trajectory of the particle, which initiate motion, with no initial linear velocity v_0 and with an angular velocity ω_0 equal in magnitude to the angular velocity of rotation ω_e, for different curvilinear surfaces (Figure 4) of the accelerator, are presented in Figure 5. It shows that the use of a concave curvilinear surface (n_a = 2) allowed to double the value of the acquired velocity v_{abs} in comparison with the flat surface (n_a = 0) and to increase it 1.7 times compared to a conical surface (n_a = −0.1) at a descent from a rotating surface, which is a consequence of the time spent by the particle on the surface, for example, when n_a = 2, the time was t = 2.05 s, when n_a = 0, the time was t = 0.84 s, and when n_a = −0.1, the time was t = 0.38 s. When the shape of the rotating surface changed from concave n_a = 2 to convex n_a = −0.1, the total length of the path traveled by the particle—that is, its trajectory—decreased.

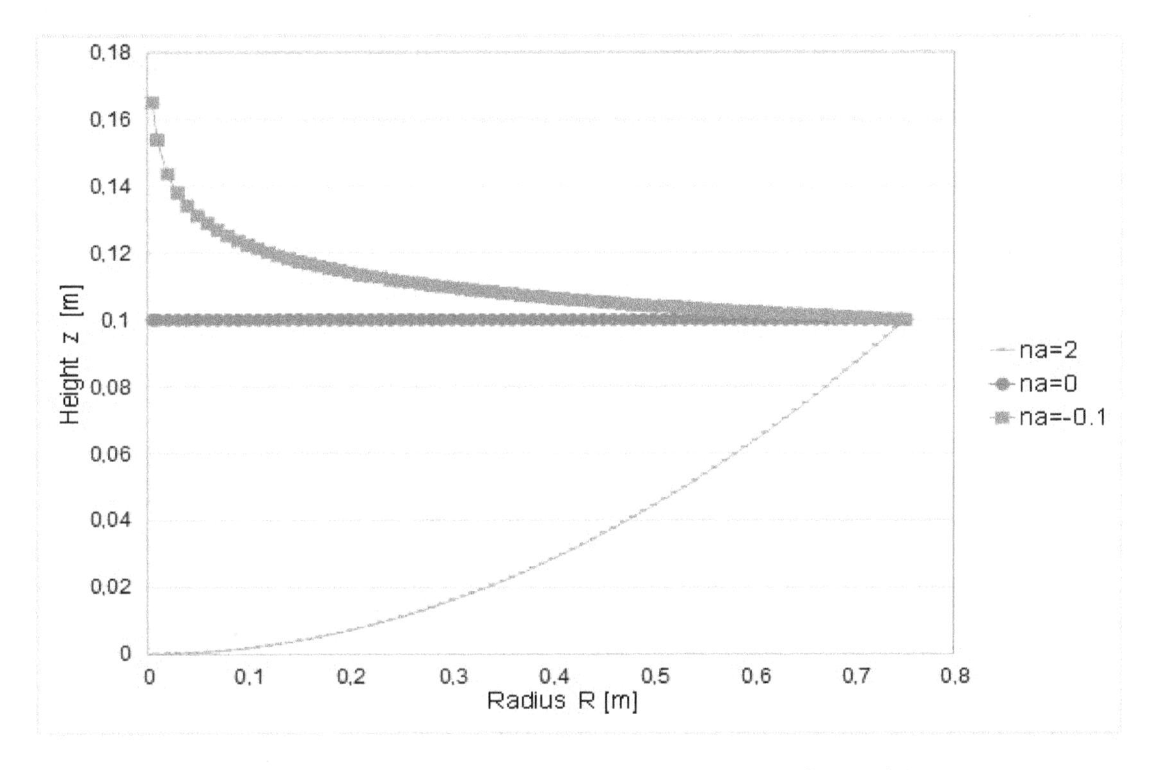

Figure 4. Variants of generates of the studied curvilinear surfaces of the accelerator.

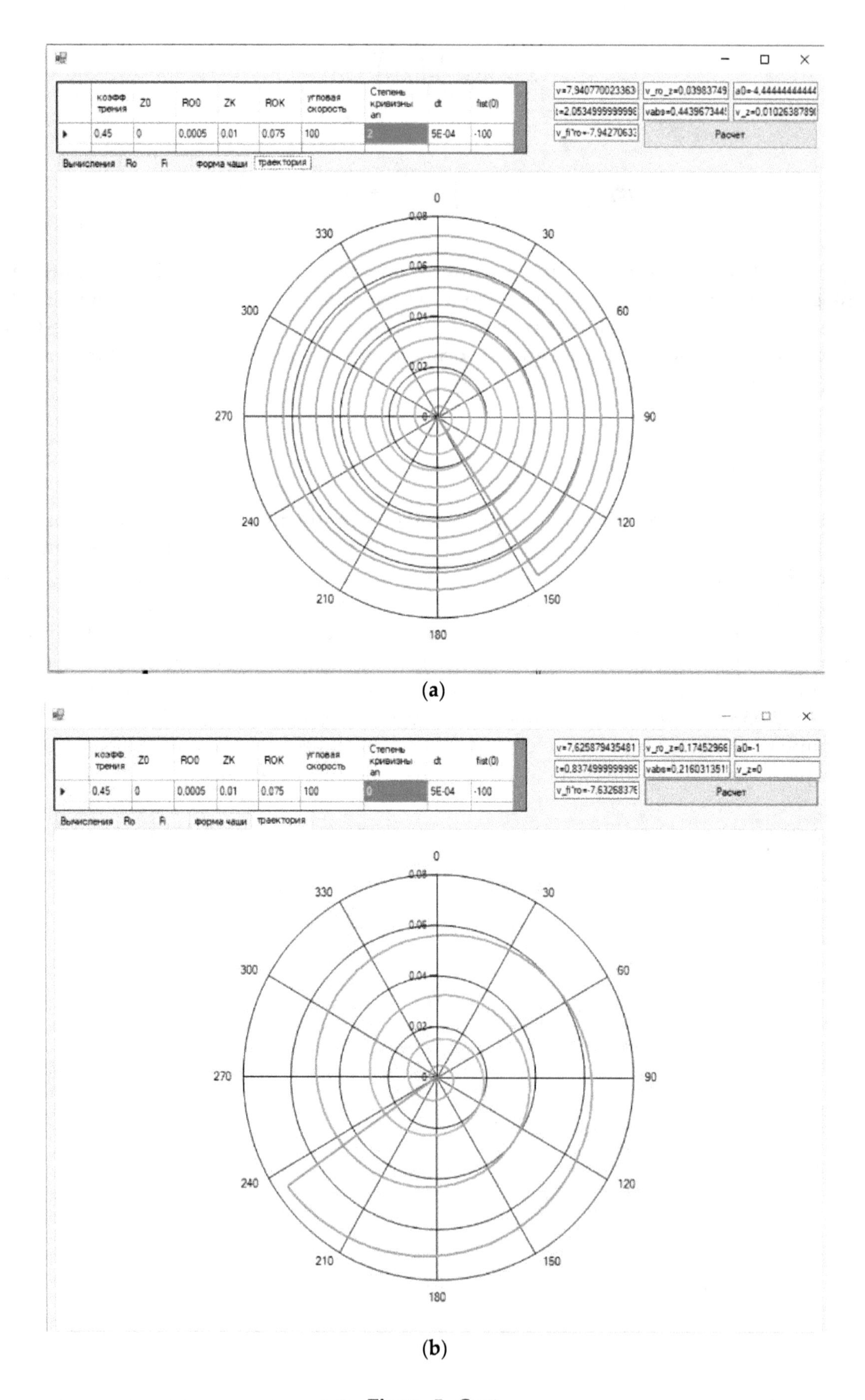

Figure 5. *Cont.*

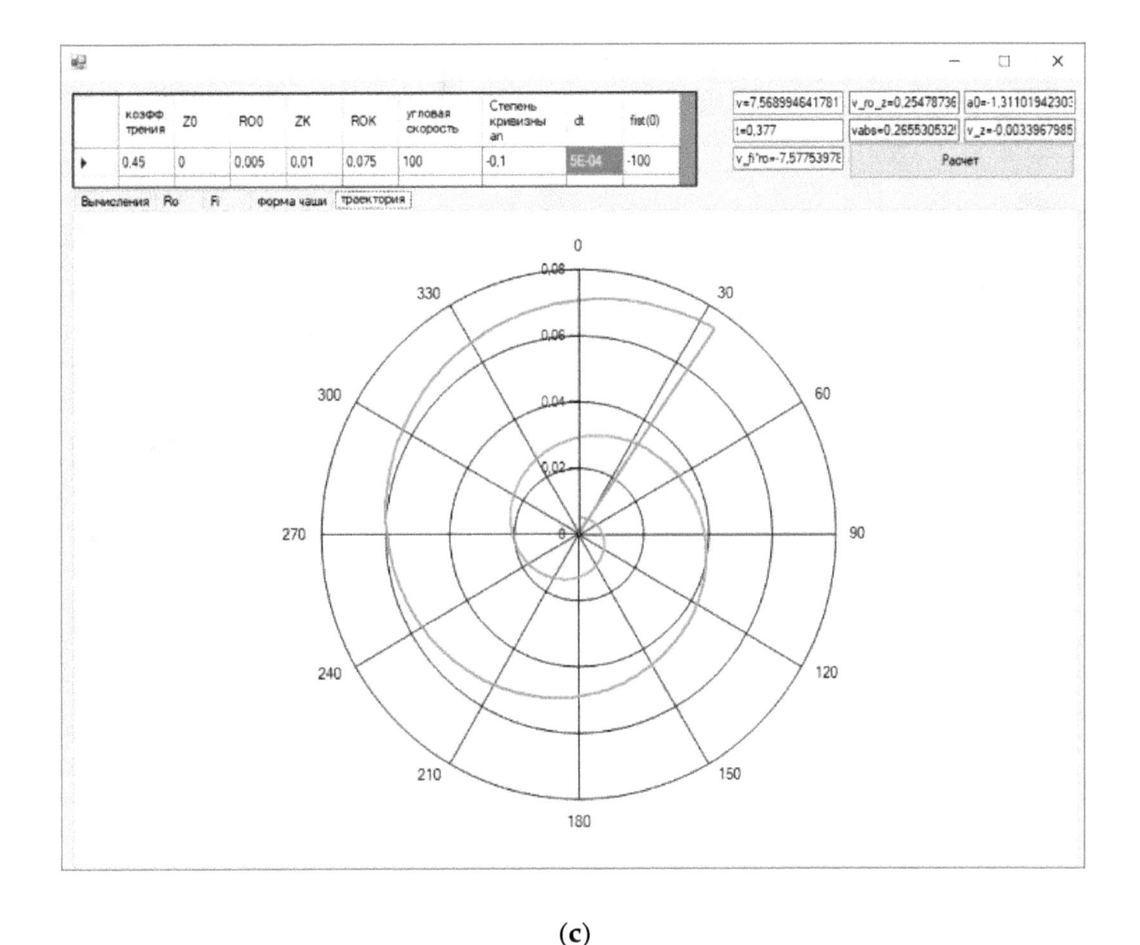

(c)

Figure 5. Data for analyzing the velocity and trajectory of the particle motion, which is introduced with no initial linear velocity and with angular velocity equal in magnitude to the angular velocity of rotation for different curvilinear surfaces of the accelerator: (**a**) $n_a = 2$ (concave); (**b**) $n_a = 0$ (straight line); (**c**) $n_a = -0.1$ (conical curvilinear).

The calculations were carried out for different variable characteristics of the surface, initial indicators of kinematic parameters, and friction coefficient. Figure 6a presents an example of tabular data with the results of calculations and diagrams of changes in the motion trajectory, velocity, and acceleration of a particle along the cylindrical axes ρ and z.

Analysis of the diagrams presented in Figure 6b, c made it possible to note that the particle reached its maximum velocity in 0.313 s, being at a distance of $\rho = 0.0092$ m from the axis of rotation. At the same time, the subsequent changes in velocity, down to its descent from the concave surface, had a damped harmonic nature.

Thus, curvilinear surfaces of concave nature combining conical and concave surfaces are of great interest for further research. The trajectories of the particles and their kinematic indicators can be used to determine the relationship between the design factors and the process parameters of centrifugal choppers, such as the time when the particle is on a dispensing bowl, the descent velocity of the particle from the bowl, and the shape of accelerating blades.

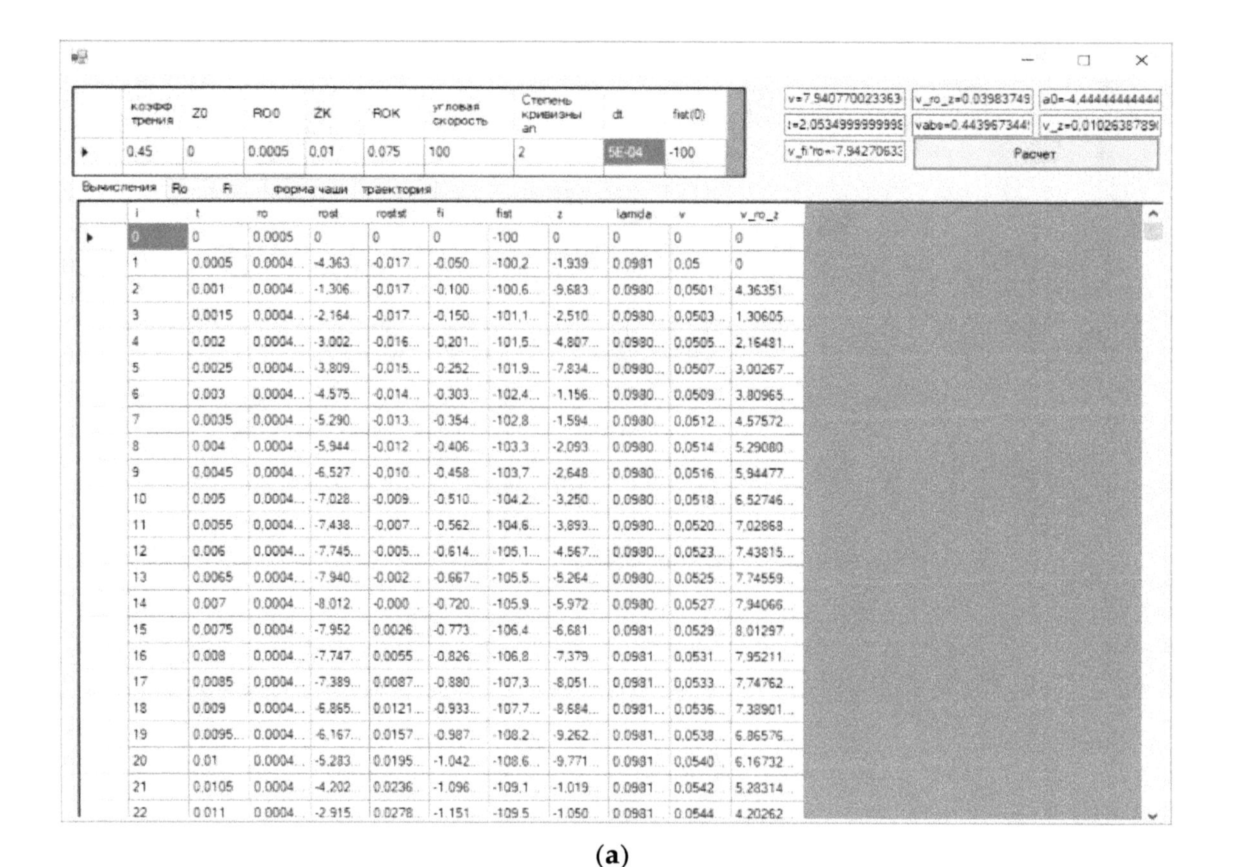

коэфф трения	Z0	RO0	ZK	ROK	угловая скорость	Степень кривизны an	dt	fist(0)
0.45	0	0.0005	0.01	0.075	100	2	5E-04	-100

i	t	ro	rost	rostst	fi	fist	z	lamda	v	v_ro_z
0	0	0.0005	0	0	0	-100	0	0	0	0
1	0.0005	0.0004...	-4.363...	-0.017...	-0.050...	-100.2...	-1.939...	0.0981	0.05	0
2	0.001	0.0004...	-1.306...	-0.017...	-0.100...	-100.6...	-9.683...	0.0980...	0.0501...	4.36351...
3	0.0015	0.0004...	-2.164...	-0.017...	-0.150...	-101.1...	-2.510...	0.0980...	0.0503...	1.30605...
4	0.002	0.0004...	-3.002...	-0.016...	-0.201...	-101.5...	-4.807...	0.0980...	0.0505...	2.16481...
5	0.0025	0.0004...	-3.809...	-0.015...	-0.252...	-101.9...	-7.834...	0.0980...	0.0507...	3.00267...
6	0.003	0.0004...	-4.575...	-0.014...	-0.303...	-102.4...	-1.156...	0.0980...	0.0509...	3.80965...
7	0.0035	0.0004...	-5.290...	-0.013...	-0.354...	-102.8...	-1.594...	0.0980...	0.0512...	4.57572...
8	0.004	0.0004...	-5.944...	-0.012...	-0.406...	-103.3...	-2.093...	0.0980...	0.0514...	5.29080...
9	0.0045	0.0004...	-6.527...	-0.010...	-0.458...	-103.7...	-2.648...	0.0980...	0.0516...	5.94477...
10	0.005	0.0004...	-7.028...	-0.009...	-0.510...	-104.2...	-3.250...	0.0980...	0.0518...	6.52746...
11	0.0055	0.0004...	-7.438...	-0.007...	-0.562...	-104.6...	-3.893...	0.0980...	0.0520...	7.02868...
12	0.006	0.0004...	-7.745...	-0.005...	-0.614...	-105.1...	-4.567...	0.0980...	0.0523...	7.43815...
13	0.0065	0.0004...	-7.940...	-0.002...	-0.667...	-105.5...	-5.264...	0.0980...	0.0525...	7.74559...
14	0.007	0.0004...	-8.012...	-0.000...	-0.720...	-105.9...	-5.972...	0.0980...	0.0527...	7.94066...
15	0.0075	0.0004...	-7.952...	0.0026...	-0.773...	-106.4...	-6.681...	0.0981...	0.0529...	8.01297...
16	0.008	0.0004...	-7.747...	0.0055...	-0.826...	-106.8...	-7.379...	0.0981...	0.0531...	7.95211...
17	0.0085	0.0004...	-7.389...	0.0087...	-0.880...	-107.3...	-8.051...	0.0981...	0.0533...	7.74762...
18	0.009	0.0004...	-6.865...	0.0121...	-0.933...	-107.7...	-8.684...	0.0981...	0.0536...	7.38901...
19	0.0095...	0.0004...	-6.167...	0.0157...	-0.987...	-108.2...	-9.262...	0.0981...	0.0538...	6.86576...
20	0.01	0.0004...	-5.283...	0.0195...	-1.042...	-108.6...	-9.771...	0.0981...	0.0540...	6.16732...
21	0.0105	0.0004...	-4.202...	0.0236...	-1.096...	-109.1...	-1.019...	0.0981...	0.0542...	5.28314...
22	0.011	0.0004...	-2.915...	0.0278...	-1.151...	-109.5...	-1.050...	0.0981...	0.0544...	4.20262...

(**a**)

(**b**)

Figure 6. *Cont.*

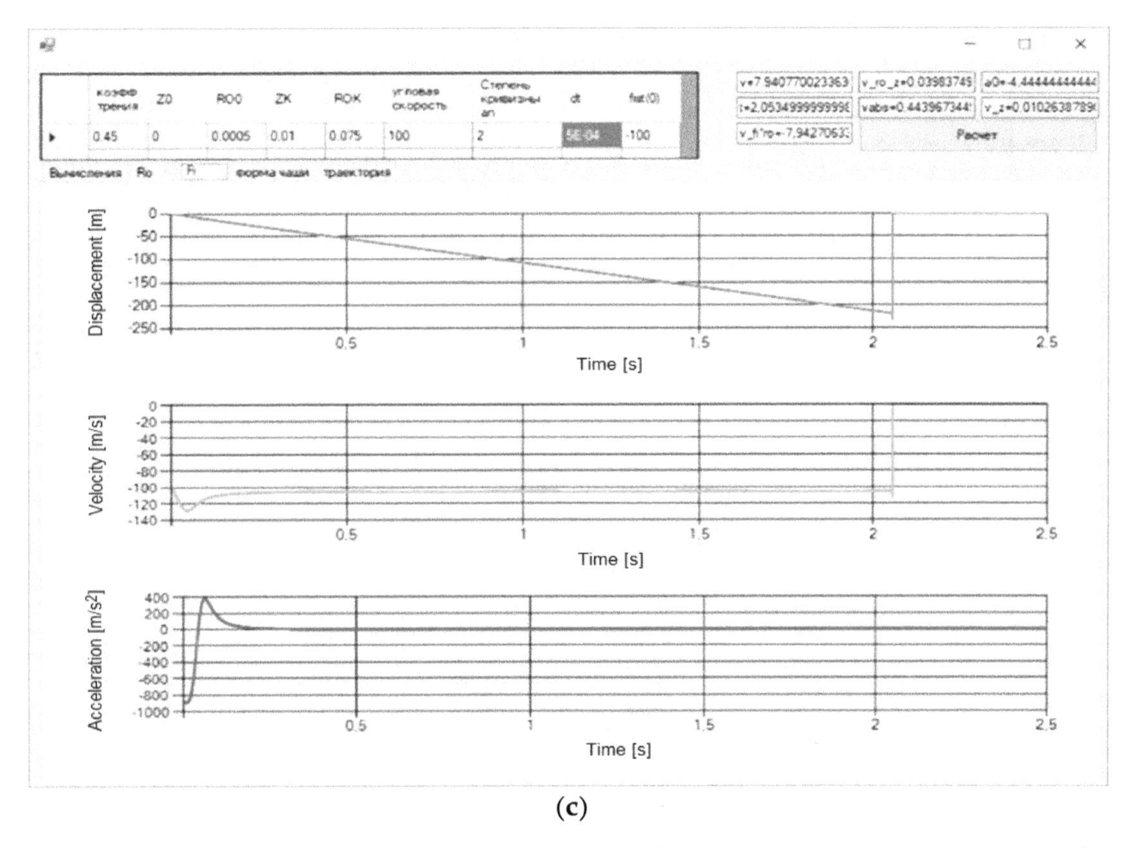

(c)

Figure 6. Example of the results of calculations of the particle motion trajectory and kinematic indicators for $n_a = 2$; (**a**) tabular data; (**b**) diagrams of particle displacement, velocity, and acceleration along the axis ρ; (**c**) diagrams of particle displacement, velocity, and acceleration along the axis z.

The ability to achieve the assumed performance parameters of the production process planned for implementation depends on the degree of reliability of the machines and technological devices included in the designed system [2]. Based on previous research [34], it can be confirmed that with the optimal values of the technological process (efficiency, time of a given operation—mixing or mining, coefficient of friction, and particle movement), it is possible to reduce energy consumption. Data on the possibility of improving the operation of production processes of various enterprises in Germany are presented by Steinhofel and others [46]. As we know, many factors influence the production process, for example, technical, economic, and social factors. Due to the above reasons, obtaining such large benefits in the whole production process, where various technological machines operate (technological operations), is difficult; nevertheless, it enables the introduction of sustainable production processes. The conducted simulation tests show possibilities of improving process efficiency and shortening the operation times, which results in economic benefits. The lower energy consumption and the increase in the efficiency of technological processes will also lead to the generation of less post-production waste (ecological benefits), which is in line with the principles of sustainable development.

4. Conclusions

In this study, a closed system of equations was obtained with a coupling equation (of constraints) that allows modeling a particle motion on a rotating surface; when numerically integrating them, it is possible to evaluate the use of a particular accelerator surface of a centrifugal rotary chopper machine to achieve the desired results in terms of downsizing, increasing productivity, or improving the quality of shredding. The use of numerical integration methods allows to consider the influence of physical and geometrical processes and kinematic parameters on the output result, for example, traveling time and velocity in cylindrical coordinates. Thus, the simulation of the particle motion along an accelerator allows choosing rational geometrical dimensions of the chopper machine and its optimal operating practice.

Author Contributions: Conceptualization, J.C., P.A.S. and I.I.I.; Data curation, A.V.A.; Formal analysis, A.M., A.V.A., A.Y.I. and I.I.I.; Methodology, A.V.A., P.A.S. and A.Y.I.; Resources, J.C.; Software, P.A.S.; Writing—original draft, J.C., P.A.S. and A.Y.I.; Writing—review & editing, J.C. and I.I.I.

Acknowledgments: We are thankful to the Editor and the reviewers for their valuable comments and detailed suggestions to improve the paper.

References

1. Linke, B.S.; Corman, G.J.; Dornfeld, D.A.; Tonissen, S. Sustainability indicators for discrete manufacturing processes applied to grinding technology. *J. Manuf. Syst.* **2013**, *32*, 556–563. [CrossRef]
2. Gola, A. Reliability analysis of reconfigurable manufacturing system structures using computer simulation methods. *Eksploat. I Niezawodn. Maint. Reliab.* **2019**, *21*, 90–102. [CrossRef]
3. Wegener, K.; Bleicher, F.; Krajnik, P.; Hoffmeister, H.W.; Brecher, C. Recent developments in grinding machines. *Cirp Ann. Manuf. Technol.* **2017**, *66*, 779–802. [CrossRef]
4. Zhou, Z.; Dou, Y.; Sun, J.; Jiang, J.; Tan, Y. Sustainable Production Line Evaluation Based on Evidential Reasoning. *Sustainability* **2017**, *9*, 1811. [CrossRef]
5. Shankar, K.M.; Kumar, P.U.; Kannan, D. Analyzing the Drivers of Advanced Sustainable Manufacturing System Using AHP Approach. *Sustainability* **2016**, *8*, 824. [CrossRef]
6. Wu, G.; Duan, K.; Zuo, J.; Zhao, X.; Tang, D. Integrated Sustainability Assessment of Public Rental Housing Community Based on a Hybrid Method of AHP-Entropy Weight and Cloud Model Sustainability. *Sustainability* **2017**, *9*, 603.
7. Bardelli, M.; Cravero, C.; Marini, M.; Marsano, D.; Milingi, O. Numerical Investigation of Impeller-Vaned Diffuser Interaction in a Centrifugal Compressor. *Appl. Sci.* **2019**, *9*, 1619. [CrossRef]
8. Faria, P.M.C.; Rajamain, R.K.; Tavares, L.M. Optimization of Solids Concentration in Iron Ore Ball Milling through Modeling and Simulation. *Minerals* **2019**, *9*, 366. [CrossRef]
9. Koltai, T.; Lozano, S.; Uzonyi-Kecskés, J.; Moreno, P. Evaluation of the results of a production simulation game using a dynamic DEA approach. *Comput. Ind. Eng.* **2017**, *105*, 1–11. [CrossRef]
10. Ahmad, S.; Wong, K.Y. Development of weighted triple-bottom line sustainability indicators for the Malaysian food manufacturing industry using the Delphi method. *J. Clean. Prod.* **2019**, *229*, 1167–1182. [CrossRef]
11. Joung, C.B.; Carrell, J.; Sarkar, P.; Feng, S.C. Categorization of indicators for sustainable manufacturing. *Ecol. Indic.* **2013**, *24*, 148–157. [CrossRef]
12. Krajn, D.; Glavic, P. Indicators of sustainable production. *Clean Technol. Environ. Policy* **2003**, *5*, 279–288. [CrossRef]
13. Beekaroo, D.; Callychurn, D.S.; Hurreeram, D.K. Developing a sustainability index for Mauritian manufacturing companies. *Ecol. Indic.* **2019**, *96*, 250–257. [CrossRef]
14. Filleti, R.A.P.; Silva, D.A.L.; da Silva, E.J.; Ometto, A.R. Productive and environmental performance indicators analysis by a combined LCA hybrid model and real-time manufacturing process monitoring: A grinding unit process application. *J. Clean. Prod.* **2017**, *161*, 510–523. [CrossRef]
15. Machado, C.G.; Despeisse, M.; Winroth, M.; Ribeiro da Silva, E.H.D. Additive manufacturing from the sustainability perspective: Proposal for a self-assessment tool. In Proceedings of the 52nd CIRP Conference on Manufacturing Systems, Procedia CIRP, Ljubljana, Slovenia, 12–14 June 2019; Volume 81, pp. 482–487.
16. Zarte, M.; Pechmann, A.; Nunes, I.L. Decision support systems for sustainable manufacturing surrounding the product and production life cycle—A literature review. *J. Clean. Prod.* **2019**, *219*, 336–349. [CrossRef]
17. Piasecka, I.; Tomporowski, A.; Flizikowski, J.; Kruszelnicka, W.; Kasner, R.; Mroziński, A. Life Cycle Analysis of Ecological Impacts of an Offshore and a Land-Based Wind Power Plant. *Appl. Sci.* **2019**, *9*, 231. [CrossRef]
18. Wang, R.; Wu, Y.; Ke, W.; Zhang, S.; Zhou, B.; Hao, J. Can propulsion and fuel diversity for the bus fleet achieve the win–win strategy of energy conservation and environmental protection? *Appl. Energy* **2015**, *147*, 92–103. [CrossRef]
19. Ventura, J.A.; Kweon, S.J.; Hwang, S.W.; Tormay, M.; Li, C. Energy policy considerations in the design of an

alternative-fuel refueling infrastructure to reduce GHG emissions on a transportation network. *Energy Policy* **2017**, *111*, 427–439. [CrossRef]
20. Arzoumanidis, I.; Salomone, R.; Petti, L.; Mondello, G.; Raggi, A. Is there a simplified LCA tool suitable for the agri-food industry? An assessment of selected tools. *J. Clean. Prod.* **2017**, *149*, 406–425. [CrossRef]
21. Tassielli, G.; Renzulli, P.A.; Mousavi-Avval, S.H.; Notarnicola, B. Quantifying life cycle inventories of agricultural field operations by considering different operational parameters. *Int. J. Life Cycle Assess.* **2019**, *24*, 1075–1092. [CrossRef]
22. Helleno, A.L.; Isaias de Moraes, A.J.; Simon, A.T. Integrating sustainability indicators and Lean Manufacturing to assess manufacturing processes: Application case studies in Brazilian industry. *J. Clean. Prod.* **2017**, *153*, 405–416. [CrossRef]
23. Skrucany, T.; Ponicky, J.; Kendra, M.; Gnap, J. Comparison of railway and road passenger transport in energy consumption and GHG production. In Proceedings of the Third International Conference on Traffic and Transport Engineering (ICTTE), Lucerne, Switzerland, 6–10 July 2016; pp. 744–749.
24. Skrucany, T.; Semanova, S.; Figlus, T.; Sarkan, B.; Gnap, J. Energy intensity and GHG production of chosen propulsions used in road transport. *Commun. Sci. Lett. Univ. Zilina* **2017**, *19*, 3–9.
25. Tucki, K.; Mruk, R.; Orynych, O.; Wasiak, A.; Botwinska, K.; Gola, A. Simulation of the Operation of a Spark Ignition Engine Fueled with Various Biofuels and Its Contribution to Technology Management. *Sustainability* **2019**, *11*, 2799. [CrossRef]
26. Akbar, M.; Irohara, T. Scheduling for sustainable manufacturing: A review. *J. Clean. Prod.* **2018**, *205*, 866–883. [CrossRef]
27. Singh, R.K.; Murty, H.R.; Gupta, S.K.; Dikshit, A.K. An overview of sustainability assessment methodologies. *Ecol. Indic.* **2012**, *15*, 281–299. [CrossRef]
28. Leonesio, M.; Bianchi, G.; Cau, N. Design criteria for grinding machine dynamic stability. *Procedia Cirp* **2018**, *78*, 382–387. [CrossRef]
29. Möhring, H.C.; Brecher, C.; Abele, E.; Fleischer, J.; Bleicher, F. Materials in machine tool structures. *Cirp Ann. Manuf. Technol.* **2015**, *64*, 725–748. [CrossRef]
30. Madej, O.; Kruszelnicka, W.; Tomporowski, A. Wyznaczanie procesowych charakterystyk wielokrawędziowego rozdrabniania ziaren kukurydzy. *Inżynieria I Apar. Chem.* **2016**, *4*, 144–145.
31. Stadnyk, I.; Vitenko, T.; Droździel, P.; Derkach, A. Simulation of components mixing in order to determine rational parameters of working bodies. *Adv. Sci. Technol. Res. J.* **2016**, *10*, 30–138. [CrossRef]
32. Tong, J.; Xu, S.; Chen, D.; Li, M. Design of a Bionic Blade for Vegetable Chopper. *J. Bionic Eng.* **2017**, *14*, 163–171. [CrossRef]
33. Flizikowski, J.; Topolinski, T.; Opielak, M.; Tomporowski, A.; Mrozinski, A. Research and analysis of operating characteristics of energetic biomass micronizer. *Eksploat. I Niezawodn. Maintanance Reliab.* **2015**, *17*, 19–26. [CrossRef]
34. Marczuk, A.; Misztal, W.; Savinykh, P.; Turubanov, N.; Isupov, A.; Zyryanov, D. Improving efficiency of horizontal ribbon mixer by optimizing its constructional and operational parameters. *Eksploat. I Niezawodn. –Maint. Reliab.* **2019**, *21*, 220–225. [CrossRef]
35. Rocha, A.G.; Montanhini, R.N.; Dilkin, P.; Tamiosso, C.D.; Mallmann, C.A. Comparison of different indicators for the evaluation of feed mixing efficiency. *Anim. Feed Sci. Technol.* **2015**, *209*, 249–256. [CrossRef]
36. Tropp, M.; Tomasikova, M.; Bastovansky, R.; Krzywonos, L.; Brumercik, F. Concept of deep drawing mechatronic system working in extreme conditions. *Procedia Eng.* **2017**, *192*, 893–898. [CrossRef]
37. Vitenko, T.; Droździel, P.; Rudawska, A. Industrial usage of hydrodynamic cavitation device. *Adv. Sci. Technol. Res. J.* **2018**, *12*, 158–167. [CrossRef]
38. Xiao, X.; Tan, Y.; Zhang, H.; Jiang, S.; Wang, J.; Deng, R.; Cao, G.; Wu, B. Numerical investigation on the effect of the particle feeding order on the degree of mixing using DEM. *Procedia Eng.* **2015**, *102*, 1850–1856. [CrossRef]
39. Tomporowski, A.; Flizikowski, J.; Al-Zubledy, A. An active monitoring of biomaterials grinding. *Przem. Chem.* **2018**, *97*, 250–257.
40. Camargo, I.L.; Erbereli, R.; Lovo, J.F.P.; Fortulan, C.A. Planetary Mill with Friction Wheels Transmission Aided by an Additional Degree of Freedom. *Machines* **2019**, *7*, 33. [CrossRef]
41. Druzhynin, R.A. Improving the Working Process of an Impact Centrifugal Grinding Machine. Ph.D. Thesis,

Voronezh State University, Voronezh, Russia, 2014.
42. Stepanovich, S.N. Centrifugal Rotary Grain Grinders. Ph.D. Thesis, Chelyabinsk State Agroengineering University, Chelyabinsk, Russia, 2008.
43. Cho, J.; Zhu, Y.; Lewkowicz, K.; Lee, S.H.; Bergman, T.; Chaudhuri, B. Solving granular segregation problems using a biaxial rotary mixer. *Chem. Eng. Process.* **2012**, *57–58*, 42–50. [CrossRef]
44. Nikitin, N.N. *The Course of Theoretical Mechanics*; Higher School: Moscow, Russia, 1990; p. 607.
45. Timoshenko, S.P.; Young, D.H.; Weaver, W. *Vibration Problems in Engineering*; Trans. from English; D. Van Nostrand Company INC.: New York, NY, USA, 1985; p. 472.
46. Steinhöfel, E.; Galeitzke, M.; Kohl, H.; Orth, R. Sustainability Reporting in German Manufacturing SMEs. 16th Global Conference on Sustainable Manufacturing—Sustainable Manufacturing for Global Circular Economy. *Proc. Manuf.* **2019**, *33*, 610–617.

8

The Effect of Wheat Moisture and Hardness on the Parameters of the Peleg and Normand Model during Relaxation of Single Kernels at Variable Initial Loading

Jawad Kadhim Al Aridhee [1], Grzegorz Łysiak [2,*], Ryszard Kulig [2], Monika Wójcik [2] and Marian Panasiewicz [2]

[1] College of Agriculture, Al Muthanna University, Al Muthanna 66001, Iraq; jawadaridhee@gmail.com
[2] Department of Food Engineering and Machines, University of Life Sciences in Lublin, 20-033 Lublin, Poland; ryszard.kulig@up.lublin.pl (R.K.); monika.wojcik@up.lublin.pl (M.W.); marian.panasiewicz@up.lublin.pl (M.P.)
* Correspondence: grzegorz.lysiak@up.lublin.pl

Abstract: The aim of the study was to examine the Peleg and Normand model to characterize the overall stress relaxation behavior of wheat kernel at varying load conditions. The relaxation experiments were made with the help of the universal testing machine, Zwick Z020, by subjecting the samples to compression at four distinct initial load levels, i.e., 20 N, 30 N, 40 N, and 50 N. The measurements were made for four wheat varieties (two soft and two hard-type endosperms) and seven levels of moisture content. Relaxation characteristics were approximated with the help of the Peleg and Normand equation. An interactive influence of the load level, moisture, and wheat hardness on the Peleg and Normand constants has been confirmed. For moist kernels, a higher amount of absorbed compression energy was released, since less energy was required to keep the deformation at a constant level. The constants differed depending on wheat hardness. Higher values of k_1 revealed that the initial force decay was slower for hard varieties. This is more characteristic of elastic behavior. Similarly, higher values of k_2 pointed to a larger amount of elastic (recoverable) energy at the end of the relaxation. The initial loading level had no or only a slight effect on the model coefficients ($Y(t)$, k_1, and k_2). The parameters of the Peleg and Normand model decreased with an increase in the water content in the kernels.

Keywords: wheat; stress relaxation; Initial load; Peleg and Normand; compression

---

1. Introduction

Agricultural and food products most often have a complex structure and exhibit complex behavior during processing. These determine the suitability for processing and constitute the basic quality criterion for acceptance by contractors and customers. This also applies to wheat grains. Wheat grain has a multi-layered and complex structure that depends on the genetic characteristics, the environment, and the specific cultivation conditions. The individual components of its structure exhibit different mechanical properties, including resistance to external loads and resistance to the formation and propagation of cracks. It results, on the one hand, in different susceptibilities of the varieties to the production of flour and, consequently of bread, biscuits, or pasta, and on the other hand, in the complexity and diversity of numerous grain processing technologies. Hence, the optimization of wheat breeding programs as well as the appropriate wheat selection for technology for many years required proper assessment of the technological quality of grain. However, the quality of wheat cannot be simply defined, since it changes depending on the farmer, grain dealer, seed company, milling industry, pasta industry, and consumer. For the farmers it can be yield or resistance to disease, for miller's protein content, hardness, or many others. By knowing the quality, processors can avoid purchasing grain that does not meet their needs. There are a lot of methods used to determine the technological quality of kernels. A number of these are time consuming and expensive (farinograph, alveograph), and hence are often impractical to use as the way to pay producers premiums based on kernel quality expectations. Finding measures of wheat which can be conducted quickly and less expensively gives the opportunity to develop new approaches to predict kernel or dough behavior. Between them, the research on mechanical properties of kernels continues to expand.

Mechanical properties of wheat kernels are naturally associated with the conditions of mechanical separation of the endosperm and the outer bran layer, the breaking resistance of the bran or the breaking susceptibility of the endosperm itself, or starch and proteins. Khazaei and Mann [1] stated that the relaxation data could be useful in estimating the susceptibility of materials to damage. Ponce-García et al. [2] successfully applied rheological measurements to distinguish among wheat classes, varieties, and different moisture contents. The test on intact kernels was used by Figueroa et al. [3,4] to establish relationships among protein composition, viscoelasticity of dough, and baking outcomes. However, there are many different conclusions and opinions regarding the scope of application of many studies. The reason for this is undoubtedly the differences arising from the use of different materials, equipment, procedures, detailed measurement conditions, and different methods of interpretation of results.

The interpretation of stress relaxation test results is most often based on rheological models (mechanical analogs) known from the rheology theory. According to these, food materials can be assumed to be composed of springs (representing ideal solids) and dashpots (considered ideal fluids) arranged in different ways. The most frequently used mechanical analogs are the Maxwell model, the Kelvin–Voigt model, and the standard linear model, SLM [5]. The spring constants in the Maxwell model give information on the stiffness, that is, the solid nature of the sample. The dashpots constants represent the relaxation times of individual Newtonian elements. The Maxwell model is suitable for representing stress relaxation data, but it does not consider the equilibrium stress. For this reason, a generalized Maxwell–Wiechert model consisting of a few elements in parallel with a spring is better for describing the viscoelastic behavior of food [5]. It was reported by Sozer and Dalgic [6] that most viscoelastic foods do not follow the simple Maxwell model; therefore, it is necessary to use more complex models to describe their stress relaxation behavior.

The Peleg and Normand model has fewer constants than the Maxwell model, however it is a simple, quick, and effective method to handle stress relaxation data [7]. According to Sozer et al. [8], the Peleg and Normand equation can be a good alternative to the Maxwell model. The authors observed that at large deformations, the Maxwell model did not fit their data very well, and the Peleg and Normand model helped overcome this problem. Apart from its debated mathematical convenience, the Peleg and Normand simulation are easy to perform and analyze the results. Stress relaxation parameters were sensitive enough to account for the textural characteristics of raw, dried, and cooked noodles [9–11]. Lewicki and Łukaszuk [12] concluded that the rheological properties, based on the Peleg model, strongly correlated with the moisture content and reflected the changes in the distribution of components and structure of apple tissue during convective drying. On the other hand, Buňka et al. [13] stated that only the parameter expressing the extent of material relaxation provided an adequate description of the actual changes occurring during the ripening of Edam cheese. The initial rate at which the stress relaxed provided no relevant description. Similarly, Singh et al. [14] and Filipčev [15] support the opinion of the inadequacy of parameter k_1 to follow textural changes or in differentiating various food products. Other authors found, however, that this parameter differed between various samples, such as cooked spaghetti [8], cooked Asian noodles [9], or wheat breads [16].

More studies have reported rheological properties evaluated under large strains (e.g., 5% strain or greater) [17]. This results from the fact that during real processing conditions, large deformations occur, resulting in the expected irreversible deformation, flow, or fracture of the material [18]. Creep and relaxation tests for large deformations were carried out for various food products, but their results indicate different opinions on the adequacy and variability of parameters of various rheological models. Higher applied stress can result in faster relaxation, according to Guo et al. [19], and larger proportion of the unrelaxed stress, according to [20]. The constants of the Peleg and Normand model for steamed bread [7] and bulk relaxation test of Jatropha curcas seeds [21] significantly decreased with larger strains. Karaman et al. [22] indicated a slightly more elastic behavior of cheese at higher compression levels. According to Khazaei and Mann [1], the deformation level had a definite effect on the decay forces for Maxwell elements for buckthorn berries, whereas the effect on relaxation times was not always ambiguous and different for each Maxwell element. Also, Bargale and Irudayaraj [23] stated that the effect of deformation level on relaxation time was stable and did not show a clear tendency. Moreover, Faridi and Faubion [24], following Shelef and Bousso [25], pointed to the independence of the relaxation parameters on the initial stress.

In summary, there are still different opinions on specific rheological measurements, and many particular cases must be treated individually. The goal of this study is the determination of the influence of the initial loading level on relaxation characteristics described by the model of Peleg and Normand. The effect was examined for different wheat types and varying moisture content.

2. Materials and Methods

2.1. Material Selection

Four wheat varieties were used in the experiments, two soft-type endosperms and two hard-type endosperms. Of these, two soft Polish cultivars (Zawisza and Wydma) and one hard SMH87 originated from the plant breeding station HR Smolice (harvested in 2015). The second hard variety was obtained from the pasta production company, Lubella. The names of the wheat varieties are denoted here as: Zawisza—SOFT1, Wydma—SOFT2, SMH87—HARD1, and wheat from Lubella—HARD2. The selection of both the soft-type and hard-type endosperm varieties was motivated by their confirmed significant differences in stress–strain behavior and fragmentation susceptibility.

2.2. Chemical Analyses

Chemical analyses were performed in the Central Agroecological Laboratory of the University of Life Sciences in Lublin, Poland. The Infratec™ 1241 (Foss, Denmark), a whole grain analyzer that uses near-infrared transmittance technology, was used to estimate the amounts of protein, wet gluten, starch, and Zeleny index. The apparatus uses a wavelength in the range of 570–1100 nm. From the batch sample about 200 g, 10 subsamples were scanned. The result of the analysis was the average of 10 measurements.

2.3. Moisture Determination

The initial moisture content of a batch sample of wheat kernels was determined by applying the air oven method. Three 5 g samples of grains were dried for over 2 h at a temperature of 130 °C, in accordance with the Polish Standard, PN-EN ISO 712:2012. The initial moisture content (m_i) was calculated according to the formula (wet basis):

$$m_i = \frac{M_w - M_d}{M_w} \cdot 100\%, \tag{1}$$

where M_w is mass of the sample before drying and, M_d is mass of the sample after drying.

Different moisture levels, i.e., 8%, 10%, 12%, 14%, 16%, 18%, and 20% (wet basis), were assigned in the experimental plan. To obtain the respective levels, each batch sample with known weight and moisture was dried at 40 °C until its weight corresponding to the driest sample (8%) was obtained. The batch sample was then divided into seven smaller samples, and to each of them, the required amount of water was added. The amount of water was calculated using simple mass balance equations.

2.4. Stress Relaxation Procedures

The measurements were performed in the laboratory of the Department of Food Engineering and Machines. A Zwick Z020 universal testing machine was used for relaxation measurements. It was equipped with 0.1 kN capture. Application of the 0.1 N capture with an accuracy grade of 0.5% resulted in a 0.5 N accuracy of force measurement. The machine was operated using the *testXpert* firmware supplied by Zwick (version 7.1). Measurements were carried out for all four wheat varieties with the seven specified levels of moisture content. A single kernel of wheat was placed with its ventral side on the bottom of the machine's plate and then loaded axially until a required load level was obtained. The constant compression rate during the loading at 10 mm/min was adjusted. The strain (deformation) of the kernel was kept constant during next 300 s. The decreasing value of the loading force as a function of time was measured with the help of the *testXpert* software at a frequency of 100 Hz. Twenty replications were performed.

The experiments were conducted for four distinct initial load levels, i.e., 20 N, 30 N, 40 N, and 50 N. The maximum applied value (50 N) corresponds to about half of the value of the force causing kernel rupture (at ambient conditions) (Figure 1).

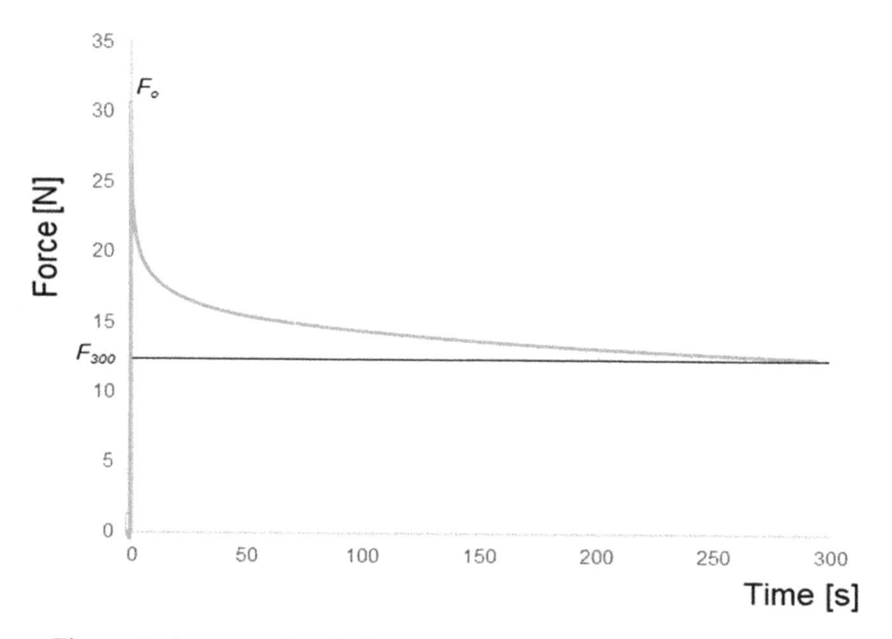

Figure 1. An example of relaxation characteristics of wheat kernel.

2.5. Calculations and Statistical Analysis

Peleg and Normand [26,27] suggested that stress relaxation data could be calculated as normalized stress and fitted to formula (2). The formula has been used by many authors, and depending on the assumptions made, it is based on decay in stress, modulus, or force. Force, which was used in the study, is also a valid criterion [28]. Hence, the relaxation of wheat kernel can be expressed in terms of force decay, as:

$$\frac{F_o\, t}{F_o - F_t} = k_1 + k_2 t, \tag{2}$$

where F_o is the initial loading force in Newtons at time $t = 0$; F_t is the force in Newtons at relaxation time t; t is the time of relaxation in seconds; k_1 and k_2 are constants; the reciprocal of k_1 determines the initial decay rate; and k_2 is the hypothetical value of the asymptotic normalized force.

The decay parameter *Y(t)* was calculated according to:

$$Y(t) = \frac{F_o}{F_o - F_{(300)}}, \tag{3}$$

where $F_{(300)}$ – is the force established at 300 s.

k_1 indicates how fast the material in question is relaxing in energy (at least initially). If the reciprocal of the k_1 represents the initial decay rate, then its value can be associated with a low rate of decay, indicating a pronounced elastic behavior. A higher value of k_1 suggests a harder, more solid-like material that dissipates less energy, thus needing more force to be compressed [29]. Parameter k_2 represents the degree of solidity, and it varies between 1, for a material that is an ideal liquid, to infinity, for an ideal elastic solid where the stress does not relax at all [27].

The constants of the Peleg and Normand model were estimated by fitting the experimental data to the above formulas using Excel software. Experimental data were analyzed using Statistica, Dell Inc. [30] version 13. In all the analyses, a significance level at 0.05 was acquired.

3. Results

The chemical composition of wheat varieties is presented in Table 1. Hard varieties had higher gluten and protein content. The amount of wet gluten was close to 34–35% for hard wheats and 25–26% for soft types. The protein amounts were close to 15% and 12%, respectively. Higher starch content was characteristic for soft wheats (near 71%) in comparison to hard ones (about 64%). Finally, the ability of the flour proteins to swell in an acid medium expressed by Zeleny index was about two times higher for hard wheats.

The amount of protein in soft, hard, and durum can vary according to the differences in grading systems in different countries. However, within a species, wheat cultivars are further classified in terms of protein content as soft wheats (protein content about 10%), hard (protein content higher than in soft), and durum (are generally high in gluten-producing proteins, 15%). Generally, soft wheat has been bred to yield flour containing less protein than hard wheats, about 8% to 11% versus 10% to 14% protein, respectively [31]. Although the hardness is postulated to be strongly correlated with protein content [32], soft wheats can have a higher protein content than the hard cultivars, or wheat lines with different protein content can have very close hardness scores [33]. This all means that the cause of starch–protein adhesion is not straightforward, and is still being discussed [32–35].

Table 1. Chemical composition of wheat.

Wheat	Parameter	Unit	Result
HARD1	Protein content (d.b.)	%	15.2 ± 0.6
	Starch content (d.b.)	%	63.7 ± 1.9
	Wet gluten content	%	35.2 ± 3.6
	Zeleny index	mL	61.8 ± 10.7
HARD2	Protein content (d.b.)	%	15.0 ± 0.7
	Starch content (d.b.)	%	64.6 ± 2.0
	Wet gluten content	%	33.7 ± 3.4
	Zeleny index	mL	64.6 ± 11.7
SOFT1	Protein content (d.b.)	%	11.8 ± 0.5
	Starch content (d.b.)	%	70.3 ± 2.2
	Wet gluten content	%	26.2 ± 2.6
	Zeleny index	mL	34.0 ± 5.9
SOFT2	Protein content (d.b.)	%	11.9 ± 0.5
	Starch content (d.b.)	%	71.1 ± 2.2
	Wet gluten content	%	25.4 ± 2.6
	Zeleny index	mL	34.5 ± 6.0

d.b.—dry basis.

3.1. The Force Decay

Figure 2 shows the effect of wheat variety and moisture on the force decay rate. The decay $Y(t)$ represents a rate of decline in the force as a result of relaxation. The Tukey test was used for the testing of hypotheses of means equality. Distinct letters denote groups for which the means differed significantly at $p < 0.05$. Four groups were distinguished for studied cultivars. The highest values of the force decay rate were characteristic of hard wheats. The differences between the average values were particularly noticeable for lower grain moistures (up to 14%). For higher water amounts, the average values for hard and soft varieties were close. However, significant differences between soft and hard kernels were confirmed.

A considerable decline in the force decay was caused by increasing the moisture of kernels. Only for 18% and 20% moisture levels were the means were not statistically different, though the slower decline rate was observed already above 16%. The interactive effect between the two factors was also significant and confirmed by ANOVA analysis.

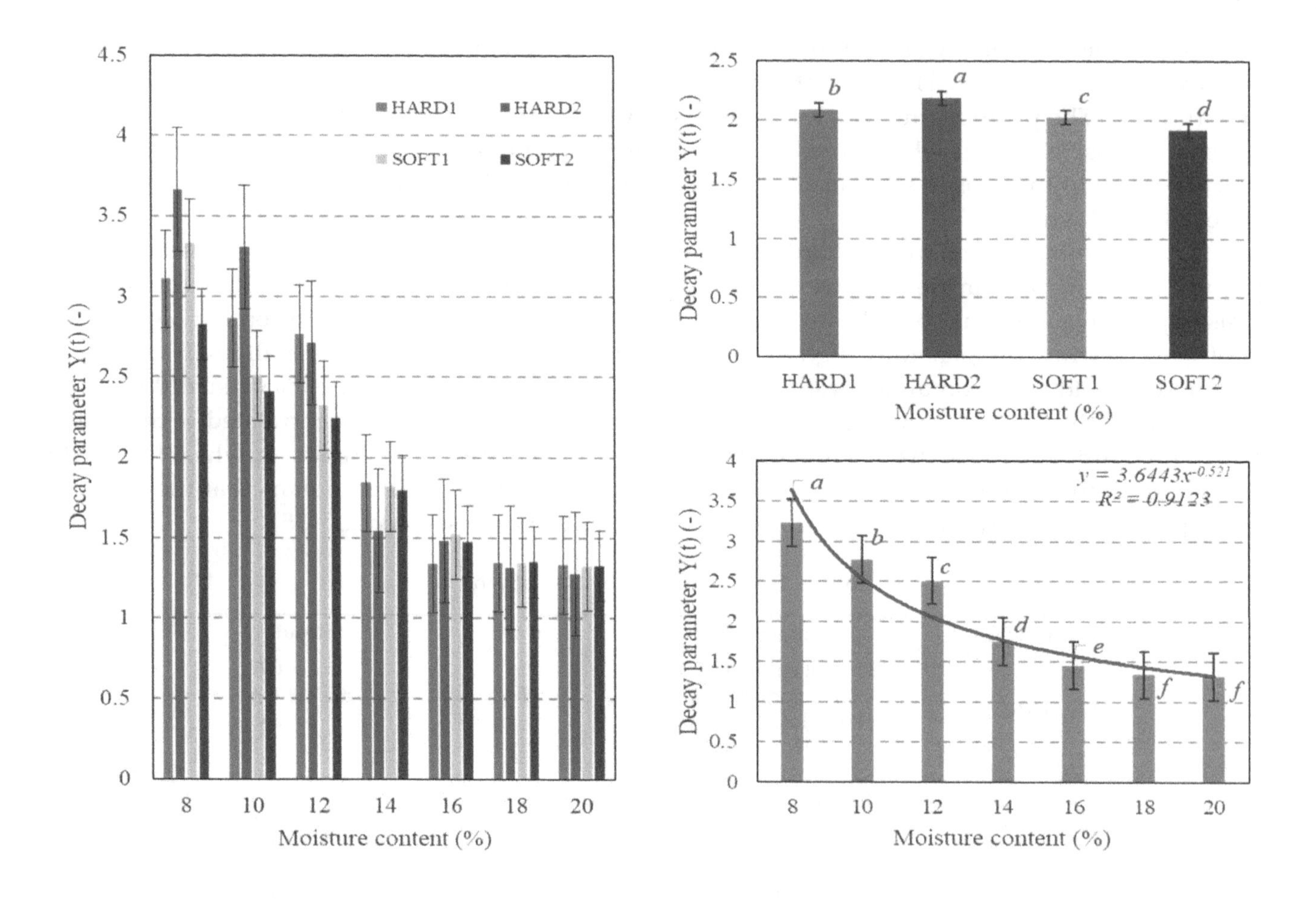

Figure 2. Influence of moisture content and wheat type on the force decay based on the Peleg and Normand equation: *a–d* and *a–f*—homogeneous groups; whiskers denote the standard error of the mean.

Figure 3 shows the effect of initial load level on the decay parameter for varied moisture and wheat type. For individual samples differing in moisture content, the effect of loading was not confirmed, although the significant effects of moisture, initial loads, as well as interaction of the two factors (moisture content × initial load) on the parameter were statistically evidenced by variance analysis. However, the effect of load was unclear both for hard and soft-type endosperm. In this case, two-way ANOVA proved no influence of initial load nor wheat type versus initial load interaction on the force decay rate $Y(t)$.

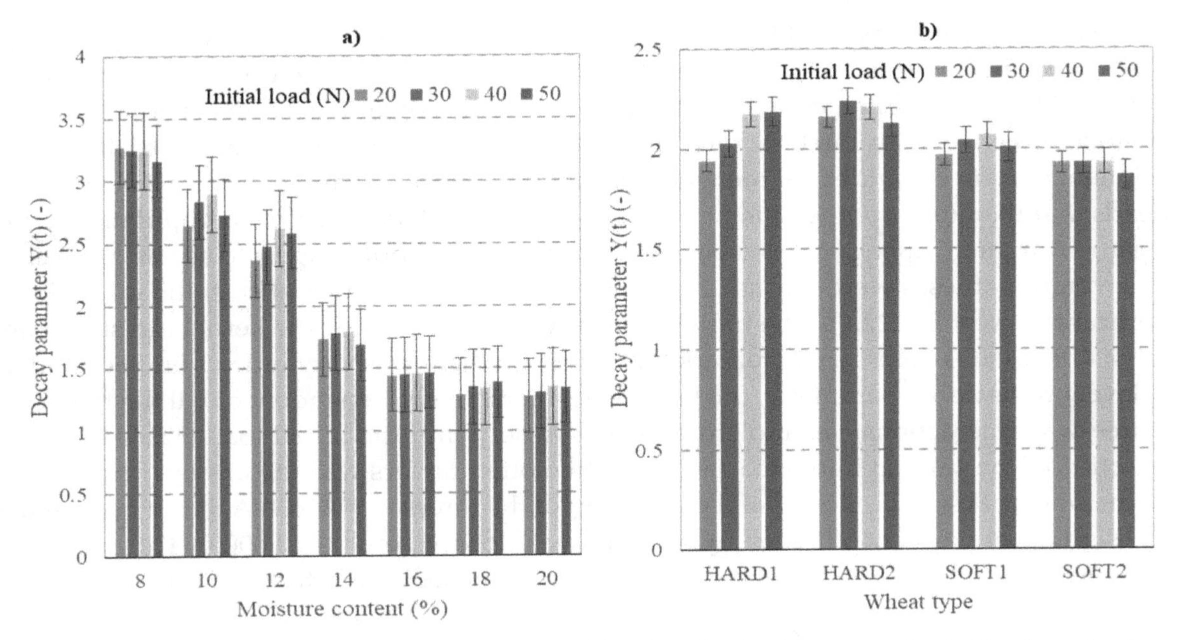

Figure 3. Plot of interactions: (**a**) moisture–initial load and (**b**) wheat type–initial load on the force decay based on the Peleg and Normand equation; whiskers denote the standard error of the mean.

Next, Figure 4 shows the general effect of the initial loading level on the average values and the distribution of the decay parameter. Only slight variations of the parameter were observed for the initial loads applied. The averages ranged from 2.0 to 2.1, although some means were statistically different. The observed small decrease in decay for 50 N loading might be caused by higher plastic deformation for moist kernels. Apart from the averages, the variability of individual results represented by their histograms (Figure 4b) was also similar. The scale parameter of the distributions was found to be also similar and not statistically dependent on the initial loading. The two-fold higher number of observations in the range of 1.0–1.5 was a reason for wheat softness at higher moistures (at 18–20%) and fact that the decay tended to unity for softer (liquid) materials.

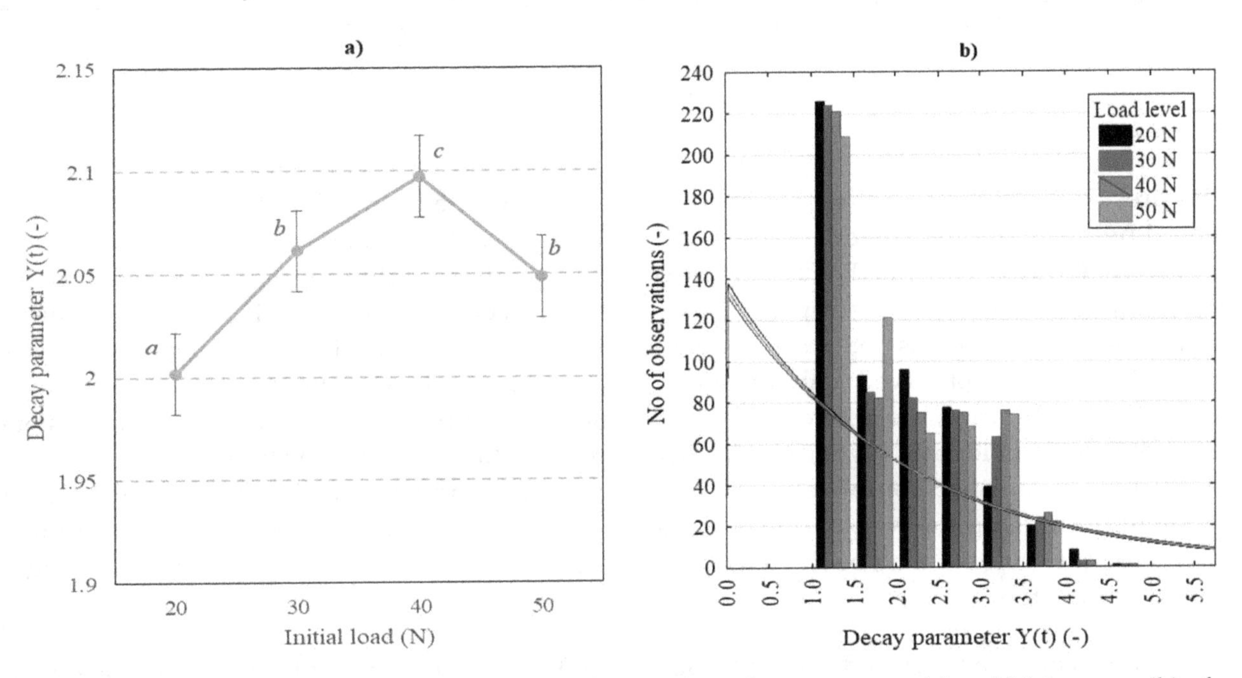

Figure 4. Influence of the initial loading level of wheat kernels on averages (**a**) and histograms (**b**) of $Y(t)$;/ *a*–*c*/—homogeneous groups; whiskers denote the standard error of the mean.

The results of three-way ANOVA analysis are presented in Table 2. There were seven null hypotheses with the analysis: (1–3) the population means of each of the three independent factors are equal; (4–6) there is no interaction between all possible combinations of two factors; and (7) there is no interaction between all three factors. The F-statistic is the mean square for each main effect, and the interaction effect divided by the within variance is simply a ratio of two variances. A large F ratio means that the null hypothesis is wrong (the data are not sampled from populations with the same mean)—variation among group means is more than you would expect to see by chance. In our case, all seven hypotheses were wrong. From the three analyzed main effects, the initial load was less significant (lowest F = 22.4). The highest effect was evidenced for the kernel moisture contents (largest F = 4783.6). It is necessary to remark that although there was very little or limited influence of loading level on the decay parameter, the latter was interactively dependent on all the two-way and all the three-way interactions, i.e., moisture, wheat variety, and the initial load (Table 2). However, the interaction of the studied initial load with the two other factors was the lowest. The "p" column presents the statistical significance level (i.e., p-value) of the three-way ANOVA. It can be seen that the statistical significance level of all seven hypotheses was close to zero ($p = 0.0000$). This value is <0.05 (i.e., it satisfies $p < 0.05$), which means that there was a statistically significant effect.

Table 2. Analysis of variance of the force decay parameter $Y(t)$.

Effect	**Univariate Tests of Significance for $Y(t)$**				
	SS	**DF**	**MS**	**F**	p
Intercept	9431.540	1	9431.540	233,734.0	0.0000
Wheat type	21.051	3	7.017	173.9	0.0000
Moisture	1158.090	6	193.015	4783.3	0.0000
Load	2.713	3	0.904	22.4	0.0000
Wheat type × Moisture	72.133	18	4.007	99.3	0.0000
Wheat type × Load	5.558	9	0.618	15.3	0.0000
Moisture × Load	5.316	18	0.295	7.3	0.0000
Wheat type × Moisture × Load	21.858	54	0.405	10.0	0.0000
Error	85.788	2126	0.040		

SS—sum of squares, DF—degrees of freedom, MS—mean square, F—F ratio, p—p-value

3.2. Coefficients k_1 and k_2

Figures 5 and 6 show the effect of moisture content on the coefficients k_1 and k_2 of the model presented by Peleg and Normand. Values of k_1 decreased with an increase in the kernel moisture content from about 17.0 to 5.0. The means of k_1 were significantly different for each moisture level, although the relationship was not proportional. A relatively fast decrease in the parameter with moisture rising up to 14% was noticeably slower for higher water levels, with practically no difference at 18% and 20%. This was observed both for soft and hard type wheats. On the contrary, for low moisture levels, in the 8–12% range, the differences between the kernel types were very clear. The values of the k_1 parameter were evidently higher for the two hard varieties. In turn, for high moisture contents, these differences vanished completely. This is an interesting and noteworthy observation regarding the possibility of using the relaxation test to distinguish varietal differences. According to this, a too high moisture content may be the reason for disappearing differences in the values of Peleg and Normand parameters.

The values of parameter k_2 decreased from about 3.3 to 1.4 together with the rise in the moisture contents used. The impact of both moisture and wheat type was very similar to the parameter k_1 described above.

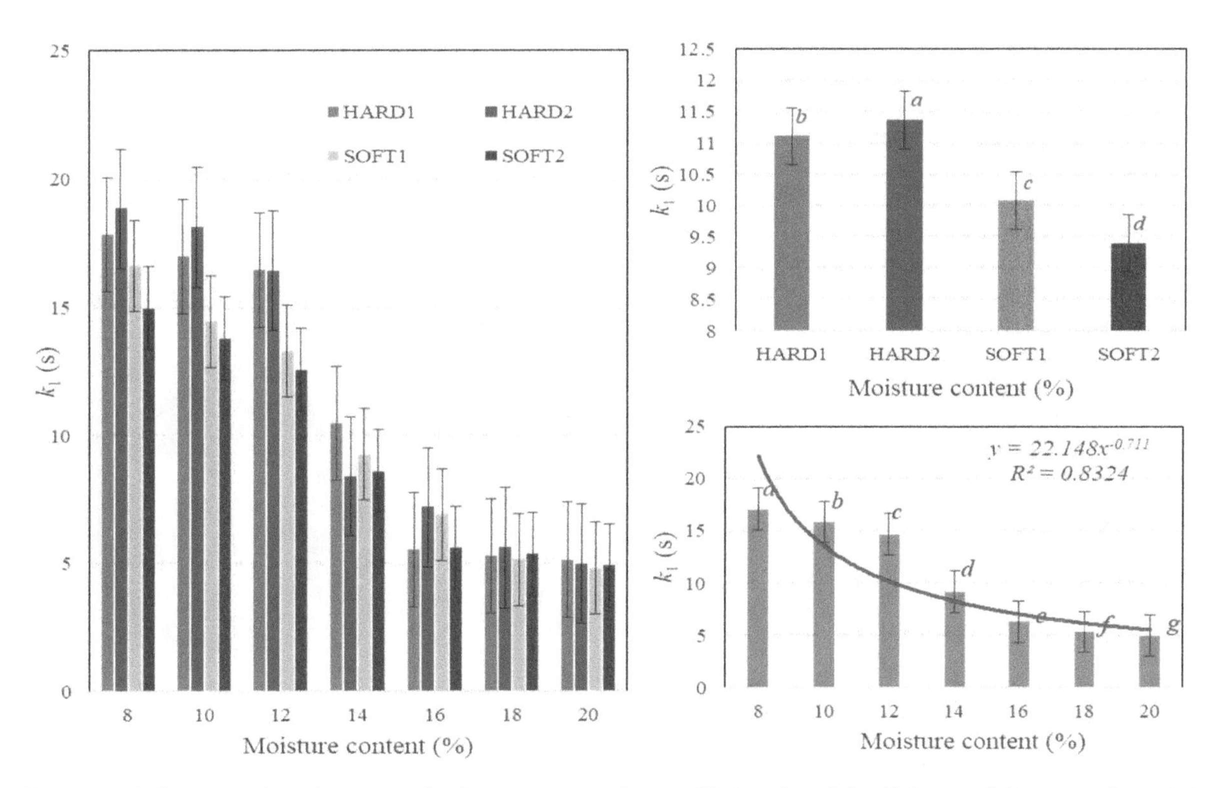

Figure 5. Influence of moisture and wheat type on the coefficient k_1 of the Peleg and Normand model; *a–d* and *a–g*—homogeneous groups, whiskers denote the standard error of the mean.

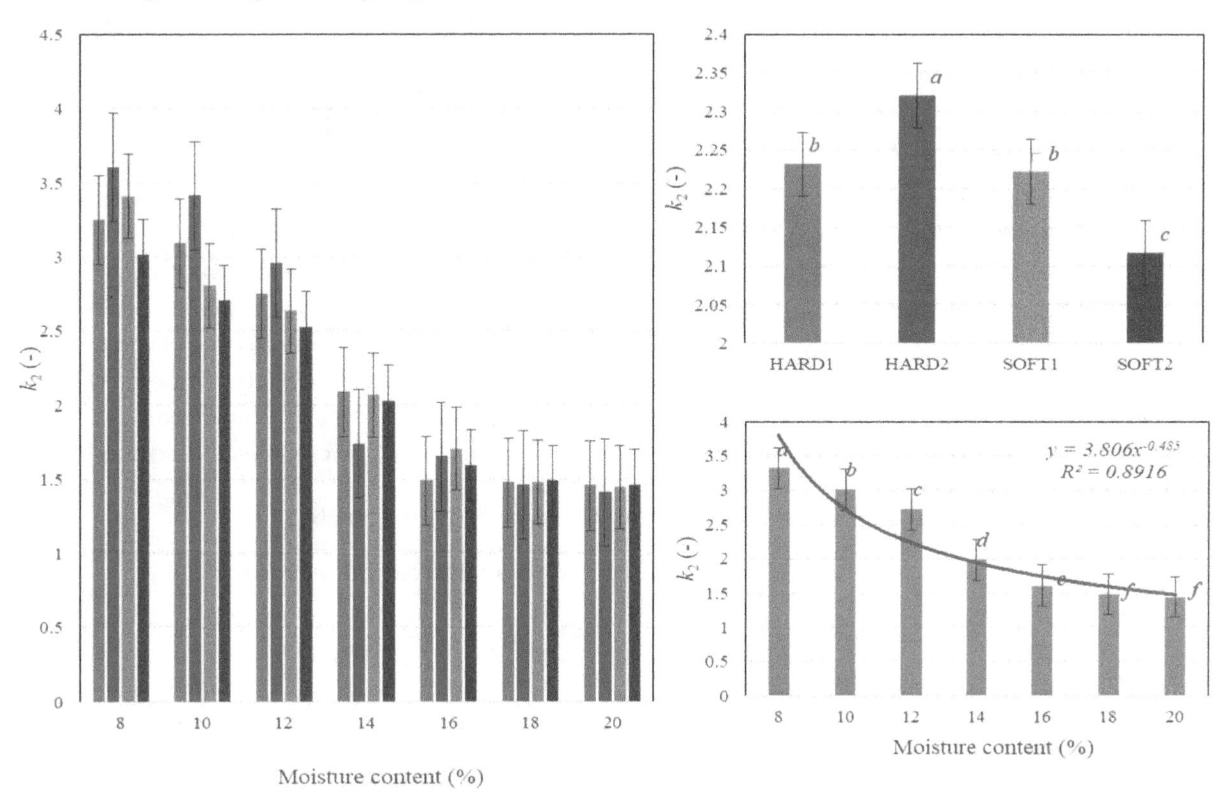

Figure 6. Influence of moisture and wheat type on the coefficient k_2 of the Peleg and Normand model; *a–c* and *a–g*—homogeneous groups, whiskers denote the standard error of the mean.

The effect of the loading level on the coefficients k_1 and k_2 is presented in Figures 7 and 8.

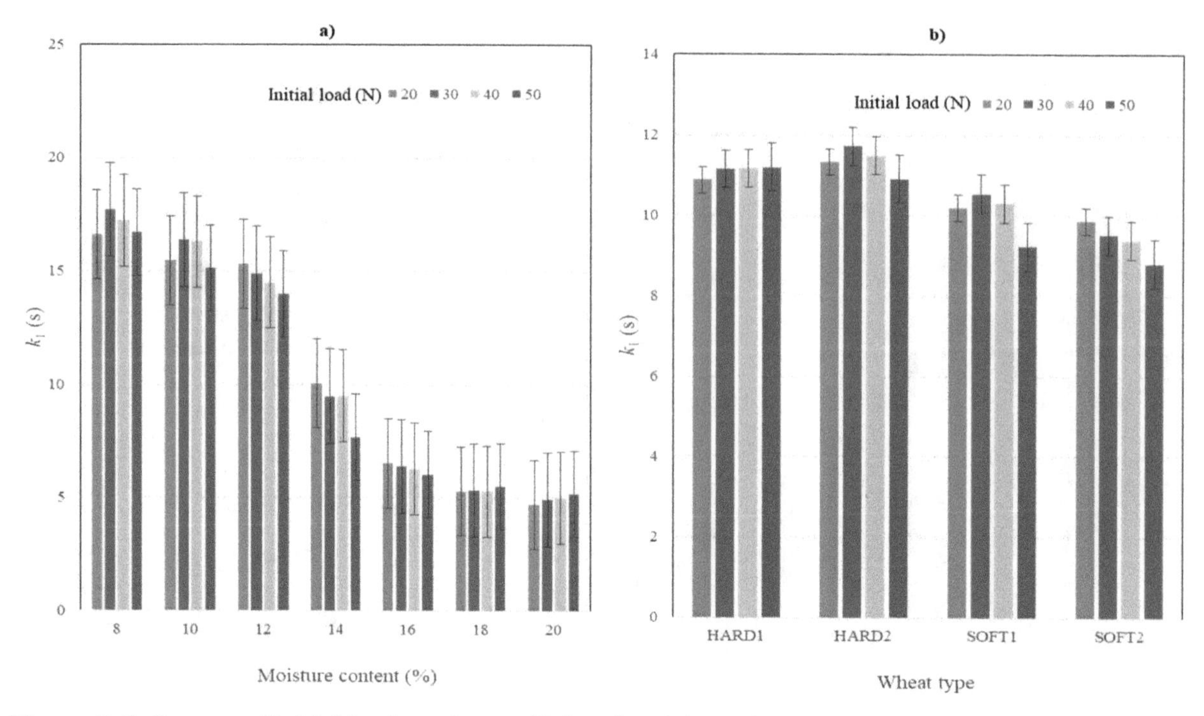

Figure 7. Influence of initial load on the coefficient k_1 of the Peleg and Normand model for different moisture (**a**) and wheat type (**b**), whiskers denote the standard error of the mean.

The analysis for the individual moisture levels showed that the mean values of k_1 did not differ significantly. Some differences were noted for 14% and 20% moisture levels. One-way ANOVA, in which the data were completely randomized, did not confirm any significant effect of the initial load on k_1. However, two-way ANOVA, including both effects of the loading level and moisture level, confirmed their interaction on the values of the coefficient. The interactive effect was also confirmed for load and wheat type interaction. Similarly, three-way ANOVA analysis, which was the most appropriate for our three-factorial experiment, proved the interactive influence of the three analyzed factors, i.e., moisture, wheat variety, and load, on k_1 (Table 3). Similarly, to the above analyzed decay parameter, from three factors, the initial load was less significant (F = 16.66). The highest effect was confirmed for wheat moisture (largest F = 2340.4). As for the previous analyses, the influence of wheat type was better evidenced than that of initial loading. K_1 was interactively influenced by all the two-way and all the three-way interactions, i.e., moisture, wheat variety, and the initial load (Table 2). However, the interaction of the studied initial load with the two other factors was also the lowest.

Table 3. Analysis of variance of the force decay parameter k_1.

Effect	Univariate Tests of Significance for k_1				
	SS	DF	MS	F	p
Intercept	1,215,041	1	1,215,041	69,015	0.0000
Wheat type	5419	3	1806	102.61	0.0000
Moisture	247,224	6	41,204	2340.4	0.0000
Load	880	3	293	16.66	0.0000
Wheat type × Moisture	6622	18	368	20.90	0.0000
Wheat type × Load	680	9	76	4.29	0.0000
Moisture × Load	1883	18	105	5.94	0.0000
Wheat type × Moisture × Load	3690	54	68	3.88	0.0000
Error	37,464	2128	18		

SS—sum of squares, DF—degrees of freedom, MS—mean square, F—F ratio, p—p-value.

In the case of k_2, the changing initial load had a more significant impact. We were able to distinguish two, three, and even four homogeneous groups, depending on the moisture level; however, no straightforward relationship was found. For k_2, main effects, like wheat type and load, were significant, but their interaction was not. Neither the load nor the interactive effect was significant ($p > 0.05$). Finally, no effect of load was observed on the relaxation coefficients for either soft or hard endosperms. Like for k_1, the three-way ANOVA analysis proved the interactive influence of the three analyzed factors on k_2 (Table 4). All the individual factors and interactive effects were confirmed to be statistically significant at very low p-values ($p = 0.0000$). All the observations for k_1 were similar, with two exceptions. The F-statistics for k_2 were higher (stronger effect), and the effect of load in this case was slightly higher (F = 70.6) than that of wheat type (67.8).

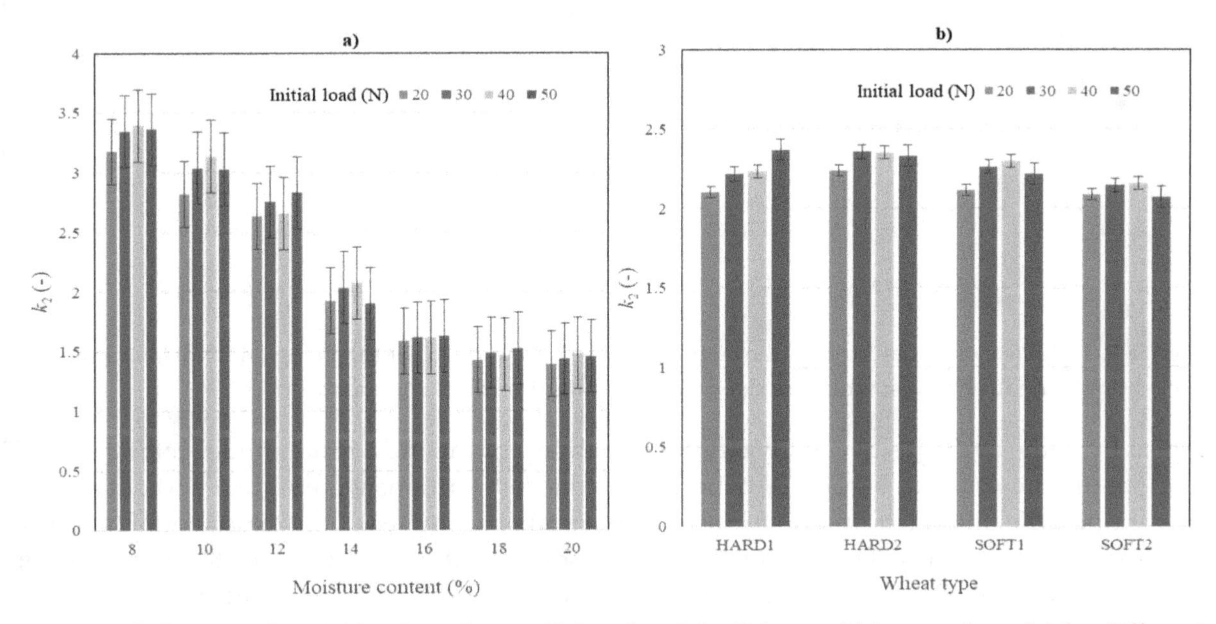

Figure 8. Influence of initial load on the coefficient k_2 of the Peleg and Normand model for different moisture (**a**) and wheat type (**b**), whiskers denote the standard error of the mean.

Table 4. Analysis of variance of the force decay parameter k_2.

Effect	Univariate Tests of Significance for k_2				
	SS	DF	MS	F	p
Intercept	7925.104	1	7925.104	249,207.8	0.0000
Wheat type	6.472	3	2.157	67.8	0.0000
Moisture	732.497	6	122.083	3838.9	0.0000
Load	6.734	3	2.245	70.6	0.0000
Wheat type × Moisture	31.723	18	1.762	55.4	0.0000
Wheat type × Load	3.305	9	0.367	11.5	0.0000
Moisture × Load	4.171	18	0.232	7.3	0.0000
Wheat type × Moisture × Load	13.557	54	0.251	7.9	0.0000
Error	67.673	2128	0.032		

SS—sum of squares, DF—degrees of freedom, MS—mean square, F—F ratio, p—p-value.

The overall effect of the initial load no the coefficients k_1 and k_2 is presented in Figure 9. The values of the two constants changed only very little. k_1 ranged from 10.04 to 10.73 and k_2 from 2.14 to 2.27. The lowest and most statistically different value of k_1 was noted at 50 N, whereas in the range from 20 to 30 N, the means were very close. In turn, the lowest values of k_2 were obtained for the 20 N load level. The homogeneous groups marked in the figure were established by the tree-way ANOVA analysis.

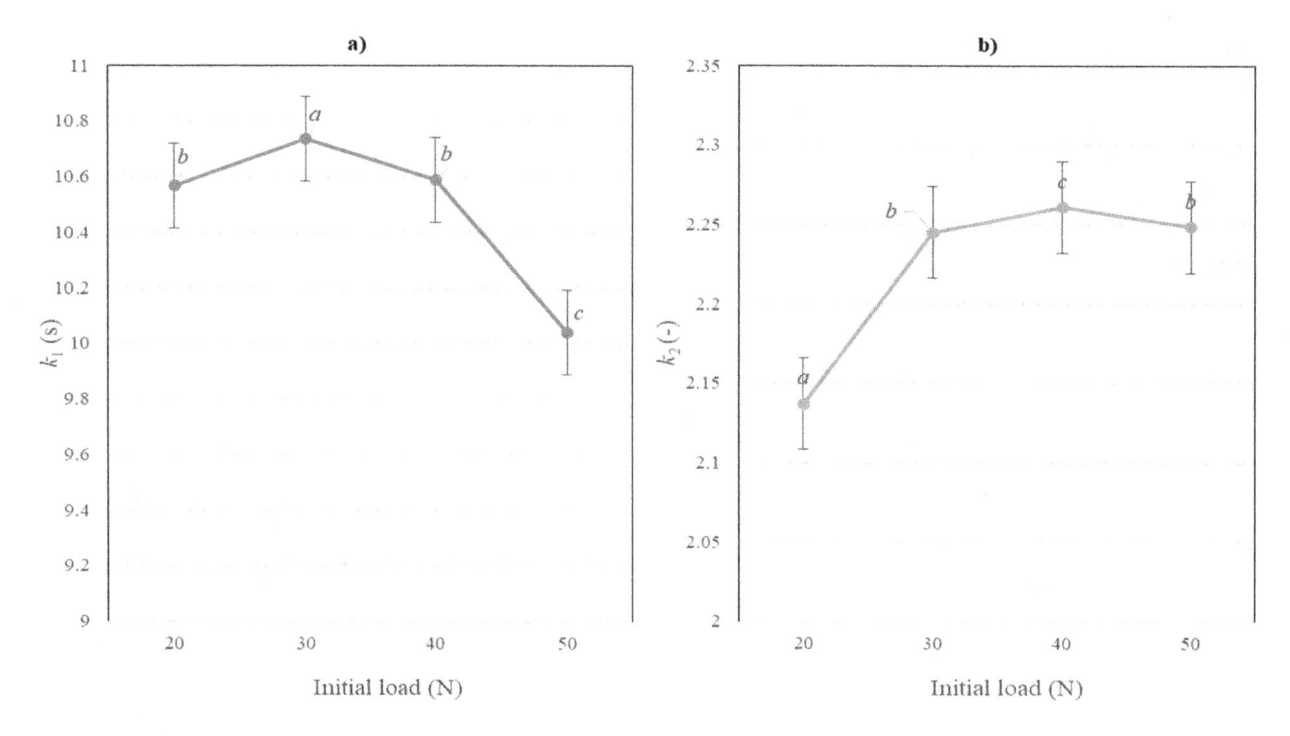

Figure 9. Influence of initial load on the coefficients k_1 (**a**) and k_2 (**b**) of the Peleg and Normand model; *a–c*—homogeneous groups; whiskers denote the standard error of the mean.

Apart from the averages, we analyzed the variability of the individual model coefficients. Their distributions are presented in Figure 10. In both cases the distributions were similar and not statistically dependent upon the initial loading. For example, the scale parameter of the exponential distributions was 0.093–0.099 for k_1 and 0.44–0.47 for k_2. Independent of the distribution fitting model applied (normal, lognormal, Weibull, or others) their parameters were very close and independent of the loading level.

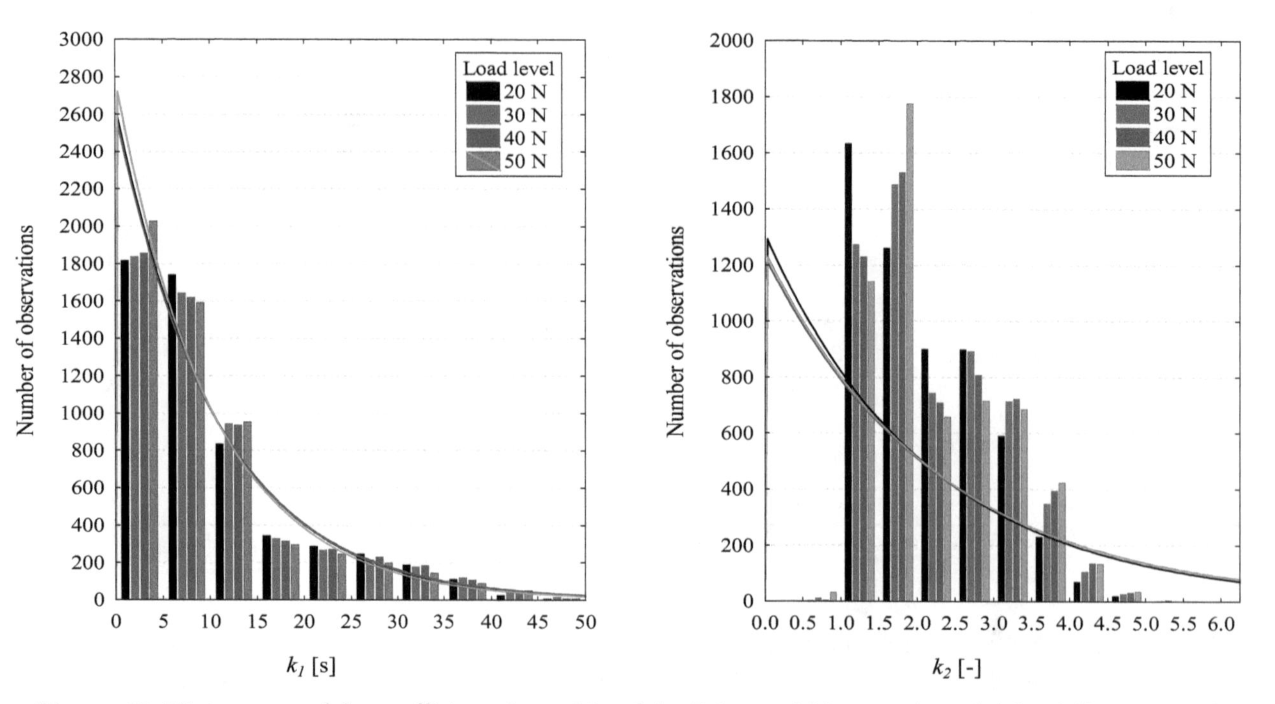

Figure 10. Histograms of the coefficients k_1 and k_2 of the Peleg and Normand model for different initial loading levels.

Finally, the analyzed impact of the initial load on the model parameters can be compared with the wheat type, which, as observed, was very small or insignificant for higher moisture contents. The histograms of the parameters k_1 and k_2 are shown in Figure 11. In the case of k_1, the distributions (medians) were clearly shifted towards higher values for hard wheats, with a slightly larger spreading (scale factor). However, in the case of k_2, the median values for the examined wheats were similar, with a slightly different scale factor.

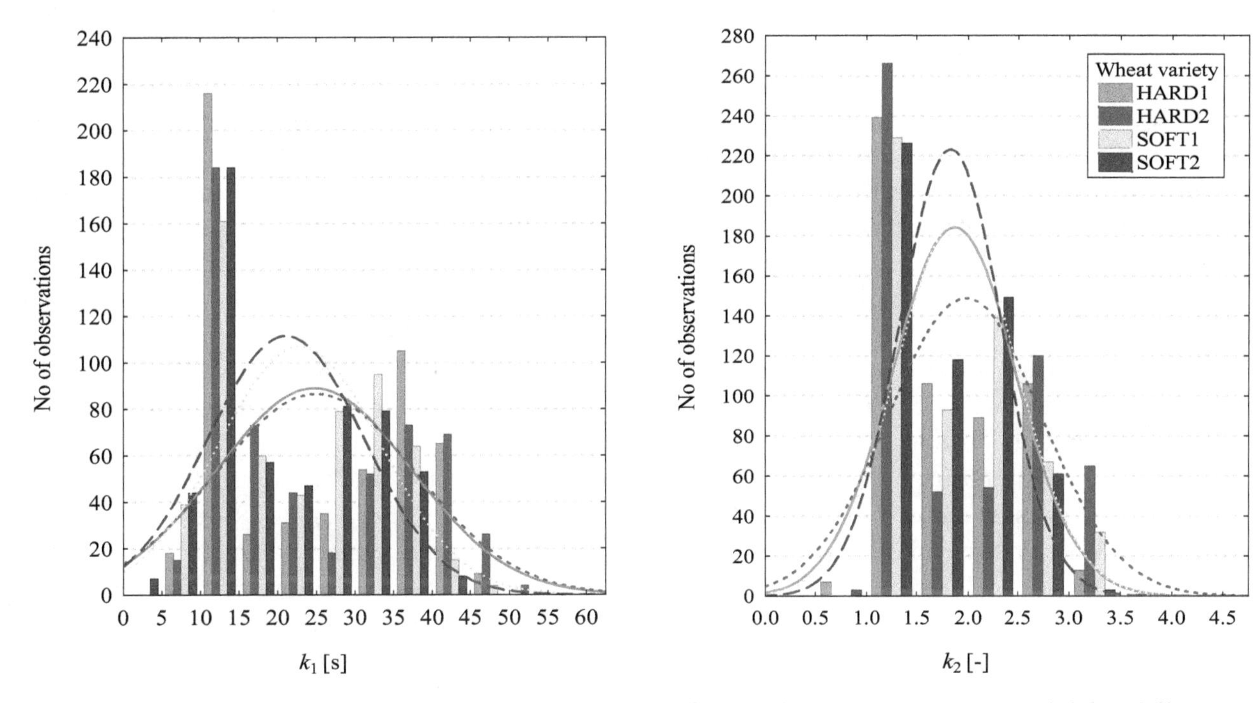

Figure 11. Histograms of the coefficients k_1 and k_2 of the Peleg and Normand model for different wheat hardness.

4. Discussion

The observed decrease in $Y(t)$ with moisture means that for dry kernels, only a small amount of the initial force was relaxed, and the residual force established at 300 s was relatively high. In fact, as moisture content increases, the kernel becomes softer, and consequently, a lower force is required to maintain a certain deformation level [1]. Water influences the rheological properties of both liquid and solid foods. It affects the response of solid foods to force [36]. Ozturk and Takhar [37] explained a decreasing trend in the values of relaxation constants as moisture level increased, with higher resistance to the relaxation of stresses at lower moisture levels. Samples with higher moisture content relaxed faster than the ones with lower moisture, as a result of the softening process. In the study, it was observed that an increase in the moisture content of kernels caused the force to decrease more quickly. This is due to the increased plasticity of the kernel and energy losses in non-recoverable deformations [28,38]. In the model of Peleg and Normand, the coefficient k_1 represents the reciprocal of the initial decay rate. This means that a high value of k_1 corresponds to a slower rate of force relaxation in the first few seconds. The coefficient k_2 represents the hypothetical value of the asymptotic normalized force (not relaxed). With an increase in moisture content, the residual force at the end of relaxation decreased, as did the values of the coefficient. This demonstrates that for moist samples, much more energy was relaxed and less energy was necessary to keep the deformation at a constant level. Similar results have also been reported in other research works [28,37].

Generally, well noticeable differences between hard and soft cultivars were observed and statistically confirmed. All three parameters, Y(t), k_1, and k_2, were statistically higher for the hard type endosperm, but the differences could be confirmed only at lower moisture levels and they diminished with increasing water content in kernels (especially above 16%). The changes in all obtained coefficients of $Y(t)$, k_1, and k_2, for the studied wheat variety were very slight, and no clear tendency was evidenced. According to Edwards [39], the moisture sensitivity of starch strength, storage protein strength, and/or the strength of starch/protein adhesion may be responsible for differences in the strength of wheats with varying moisture contents. Such responses to water may be due to the structure of the endosperm and strength of the cell wall architecture. Soft and hard wheats exhibit the same trend with moisture content; however, they may do so at different response rates. A similar trend for soft and hard wheats was observed in our study, with little difference in rates, mainly for lower moisture contents. Glenn and Johnston [35] pointed to no significant differences in any of the mechanical properties of the starch and protein components, either within a variety or among soft, hard, and durum wheat types, though mean values for each parameter were highest for durum and the lowest soft caryopses. Many studies have underlined the meaning of the overall porosity of the endosperm structure and indicated that endosperm strength is unrelated to the protein content [40,41]. Accordingly, endosperm can be very resistant, either at low protein or high protein content, depending on the adherence between the protein and starch granules. Also, Haddad et al. [42] stated that relaxation tests make it possible to relate endosperm behavior to different types listed in materials science. However, their proposal for wheat classification was based only on the relationship between the rupture energy and the Young modulus. Following Ponce-Garcia [43] the elastic, viscoelastic, and flow properties allow the indirect measurement of wheat characteristics related to the chemical composition, including nongluten components. The elastic behavior of intact kernels was related to the sedimentation volume and composition of glutenin. He demonstrated that hard wheats showed higher plastic deformation work than durum wheat kernels, and therefore had higher elastic properties [44]. Some relationships between wheat hardness and Zeleny index and protein quality were observed in [45]. In our study, we determined the basic chemical composition of wheats, however any conclusions on their relationship to measured rheological parameters were not direct or simply not justified. Some generalization based on wheat hardness may be briefly discussed here; nevertheless, the link between the biochemical studies and mechanical studies is not established.

Irrespective of the wheat softness, no obvious or very little effect (some interactions were observed) of the initial load on the model of Peleg and Normand was demonstrated. Strain level is postulated to be meaningful in relaxation experiments; however, no unique opinion in research literature exists. Some reports showed that the initial stress can influence the amount of unrelaxed energy [20] and the relaxation speed [19], which directly relates to the constants of the relaxation models [21–23]. Lewicki and Spiess [20] showed that the proportion of the unrelaxed stress is larger if the developed stress is higher. Guo et al. [19] stated that higher applied stress can result in higher relaxation speed. The obtained results confirm those of Bargale and Irudayaraj [23] on the uni-axial compression of barley kernels, in which the authors reported that the effect of the deformation level on relaxation time was consistent and did not show a clear trend. Also, according to Faridi and Faubion [24] and Shelef and Bousso [25], stress relaxation can be fairly independent of the initial stress. An analysis of the available research results allows one to state that many conclusions are true for specific circumstances only. These results demonstrate that wheat properties, which may differ very slightly, influence the relaxation characteristics more significantly than the level of loading applied. This allows a statement on its little meaning for wheat compression in the range of loads applied in the study. The results showed that this method could be a useful tool to distinguish among wheat classes, cultivars, and different moisture levels in the kernels.

To the above-mentioned different conclusions, the general opinion about the limited applicability of the Peleg and Normand model must be emphasized. To date, relaxation tests have had few practical applications, however the reasons for this seem to be not in the method itself but in the dependence of the outcomes on a number of experimental conditions (material properties, load, speed, rate of deformation, time) and research works, which cannot be easily compared. Moreover, the presented results underline the meaning of the interactive effect between various measurement conditions. Hence, the obtained parameters of the Peleg and Normand model must be examined or interpreted together with other, often very important, factors. One of these is the time of relaxation used for the model development. Only a few studies have reported the effect of relaxation time on the model parameters [13,46]. The dependence of the model parameters on the relaxation time and varied applicability to differentiate cheese samples was also reported in studies of Bunka et al. [13]. Morales et al. [46] observed that both k_1 and k_2 changed with relaxation time. The time significantly influenced the accuracy of both the rate the force decreased at the initial relaxation phase and the value of residual force. Depending on the relaxation time, the parameters were over or underestimated. Similar conclusions can be derived from the work of Al Aridhee [47]. He demonstrated very strong dependences of the both model parameters on the relaxation holding time—k_1 significantly and linearly increased with time, whereas for longer relaxation periods k_2 decreased exponentially. For times approaching zero, k_2 increased to infinity. It must be added that the effect of time was quantitatively and qualitatively comparable to that of moisture contents.

5. Conclusions

The study led to a clear identification of the effect of wheat moisture and confirmed the influence of the wheat type on the Peleg and Normand model coefficients. The parameters k_1 and k_2, and the force decay $Y(t)$, decreased with the increase in water content in the kernels. The relatively fast decrease in the parameter with moisture rising up to 14% was noticeably slower for higher water levels, practically no difference at 18% and 20%. This was observed both for soft and hard type wheats.

The constants differed depending on the wheat hardness. The highest values of the force decay rate were characteristic of hard cultivars. For higher water amounts, the average values for hard and soft varieties were similar. On the contrary, for low moisture levels, in the 8–12% range, the differences between the grain types were very clear. This is an interesting and important observation regarding the possibility of using the relaxation test to distinguish varietal differences. According to this, too high moisture levels may be the reason for disappearing differences in the values of Peleg and Normand parameters.

The initial loading level had no effect or only a slight effect on the parameters of the Peleg and Normand model. The values of all three constants changed only very little. The effect was similar for all parameters, and none of the three analyzed parameters were influenced more than others. Apart from the averages, the distributions were very close and independent of the loading level. It must be added, however, that an interactive influence of moisture, wheat variety, and load level on the Peleg and Normand constants was statistically confirmed.

The research showed that in comparison to wheat moisture and wheat type, the influence of initial loading was very weak, ambiguous, or simply not present. Very small differences between the parameters of the Peleg and Norman model show the very weak influence of the initial load on wheat relaxation experiments, and the practical application of different loads will probably lead to similar results. It is reasonable to state that experiments on larger loads than those applied here may lead to the development and propagation of cracks, making the relaxation tests more difficult to control and interpret.

The grain protein content and protein quality aspects together with grain hardness explain most of the variation in wheat behavior. However, endosperm can be very resistant, either at low or high protein content. Thus, any relationships between wheat chemical composition and measured rheological parameters are not straightforward, and the link between the biochemical and mechanical studies has not been established yet.

Attempts to apply the Peleg and Normand model into practice need further study, including the influence of all relevant factors.

Author Contributions: Conceptualization: J.K.A.A., G.Ł.; methodology: J.K.A.A., G.Ł.; software: J.K.A.A.; validation: J.K.A.A., G.Ł.; formal analysis: G.Ł.; investigation: J.K.A.A.; resources: J.K.A.A., G.Ł., M.W., M.P.; data curation: J.K.A.A., G.Ł.; writing—original draft preparation: J.K.A.A.; writing:— J.K.A.A., Gr.Ł.; writing - review

& editing: G.Ł., M.W., M.P.; visualization: J.K.A.A.; supervision: G.Ł., R.K.; project administration: J.K.A.A., G.Ł., R.K.; funding acquisition: J.K.A.A., G.Ł., M.W., M.P.

Acknowledgments: I deeply appreciate the scholarship from the Iraqi Ministry of the Higher Education and Scientific Research.

References

1. Khazaei, J.; Mann, D. Effects of moisture content and number of loadings on force relaxation behavior of chickpea kernels. *Int. Agrophys.* **2005**, *19*, 305–313.
2. Ponce-García, N.; Ramírez-Wong, B.; Torres-Chávez, P.I.; Figueroa-Cárdenas, J.D.; Serna Saldívar, S.O.; Cortez-Rocha, M.O. Effect of moisture content on the viscoelastic properties of individual wheat kernels evaluated by the uniaxial compression test under small strain. *Cereal Chem.* **2013**, *90*, 558–563. [CrossRef]
3. Figueroa, J.D.C.; Hernández, Z.J.E.; Véles, J.J.; Rayas-Duarte, P.; Martínez-Flores, H.E.; Ponce García, N. Evaluation of degree of elasticity and other mechanical properties of wheat kernels. *Cereal Chem.* **2011**, *88*, 12–18. [CrossRef]
4. Figueroa, J.D.C.; Ramírez-Wong, B.; Peña, J.; Khan, K.; Rayas-Duarte, P. Potential use of the elastic properties of intact wheat kernels to estimate millings, rheological and bread—making quality of wheat. In *Wheat Science Dynamics: Challenges and Opportunities*; Chibbar, R.N., Dexter, J., Eds.; Agrobios: Jodhpur, India, 2011; pp. 317–325.
5. Steffe, J.F. *Rheological Methods in Food Process Engineering*, 2nd ed.; Freeman press: East Lansing, MI, USA, 1996.
6. Sozer, N.; Dalgic, A.C. Modeling of rheological characteristics of various spaghetti types. *Eur. Food Res. Technol.* **2007**, *225*, 183–190. [CrossRef]
7. Wu, M.Y.; Chang, Y.H.; Shiau, S.Y.; Chen, C.C. Rheology of fiber—Enriched steamed bread: Stress relaxation and texture profile analysis. *J. Food Drug Anal.* **2012**, *20*, 133–142.
8. Sozer, N.; Kaya, A.; Dalgic, A.C. The effect of resistant starch addition on viscoelastic properties of cooked spaghetti. *J. Texture Stud.* **2008**, *39*, 1–16. [CrossRef]
9. Hatcher, D.W.; Bellido, G.G.; Dexter, J.E.; Anderson, M.J.; Fu, B.X. Investigation of uniaxial stress relaxation parameters to characterize the texture of yellow alkaline noodles made from durum and common wheats. *J. Texture Stud.* **2008**, *39*, 695–708. [CrossRef]
10. Shiau, S.Y.; Wu, T.T.; Liu, Y.L. Effect of the amount and particle size of wheat fiber on textural and rheological properties of raw, dried and cooked noodles. *J. Food Qual.* **2012**, *35*, 207–216. [CrossRef]
11. Shiau, S.Y.; Chang, Y.H. Instrumental textural and rheological properties of raw, dried, and cooked noodles with transglutaminase. *Int. J. Food Prop.* **2013**, *16*, 1429–1441. [CrossRef]
12. Lewicki, P.; Łukaszuk, A. Changes of rheological properties of apple tissue undergoing convective drying. *Dry. Technol.* **2000**, *18*, 707–722. [CrossRef]
13. Buňka, F.; Pachlová, V.; Pernická, L.; Burešová, I.; Kráčmar, S.; Lošák, T. The dependence of Peleg's coefficients on selected conditions of a relaxation test in model samples of edam cheese. *J. Texture Stud.* **2013**, *44*, 187–195. [CrossRef]
14. Singh, H.; Rockall, A.; Martin, C.R.; Chung, O.K.; Lookhart, G.L. The analysis of stress relaxation data of some viscoelastic foods using a texture analyzer. *J. Texture Stud.* **2006**, *37*, 383–392. [CrossRef]
15. Filipčev, B.V. Texture and stress relaxation of spelt—amaranth composite breads. *Food Feed Res.* **2014**, *41*, 1–9. [CrossRef]
16. Mandala, I.; Karabela, D.; Kostaropoulos, A. Physical properties of breads containing hydrocolloids stored at low temperature. I. Effect of chilling. *Food Hydrocoll.* **2007**, *21*, 1397–1406. [CrossRef]
17. ASAE S368.4. *Compression Test of Food Materials of Convex Shape*; American Society of Agricultural and Biological Engineers: Saint Joseph, MI, USA, 2006; pp. 554–556.
18. Karim, A.A.; Norziah, M.H.; Seow, C.C. Methods for the study of starch retrogradation. *Food Chem.* **2000**, *71*, 9–36. [CrossRef]
19. Guo, L.; Wang, D.; Tabil, L.G.; Wang, G. Compression and relaxation properties of selected biomass for briquetting. *Biosyst. Eng.* **2016**, *148*, 101–110. [CrossRef]
20. Lewicki, P.; Spiess, W.E. Rheological properties of raisins: Part, I. Compression test. *J. Food Eng.* **1995**, *24*, 321–338. [CrossRef]
21. Herak, D.; Kabutey, A.; Choteborsky, R.; Petru, M.; Sigalingging, R. Mathematical models describing the

relaxation behaviour of Jatropha curcas L. bulk seeds under axial compression. *Biosyst. Eng.* **2015**, *131*, 77–83. [CrossRef]

22. Karaman, S.; Yilmaz, M.T.; Toker, O.S.; Dogan, M. Stress relaxation/creep compliance behaviour of kashar cheese: Scanning electron microscopy observations. *Int. J. Dairy Technol.* **2016**, *69*, 254–261. [CrossRef]
23. Bargale, P.C.; Irudayaraj, J. Mechanical strength and rheological behavior of barley kernels. *Int. J. Food Sci. Technol.* **1995**, *30*, 609–623. [CrossRef]
24. Faridi, H.; Faubion, J.M. *Dough Rheology and Baked Product Texture*; Springer: Berlin/Heidelberg, Germany, 2012.
25. Shelef, L.; Bousso, D. A new instrument for measuring relaxation in flour dough. *Rheol. Acta* **1964**, *3*, 168–172. [CrossRef]
26. Peleg, M. Linearization of relaxation and creep curves of solid biological materials. *J. Rheol.* **1980**, *24*, 451–463. [CrossRef]
27. Peleg, M.; Normand, M.D. Comparison of two methods for stress relaxation data presentation of solid foods. *Rheol. Acta* **1983**, *22*, 108–113. [CrossRef]
28. Łysiak, G. Influence in of moisture on rheological characteristics of the kernel of wheat. *Acta Agrophys* **2007**, *9*, 91–97.
29. Guo, Z.; Castell-Perez, M.E.; Moreira, R.G. Characterization of masa and low-moisture corn tortilla using stress relaxation methods. *J. Texture Stud.* **1999**, *30*, 197–215. [CrossRef]
30. Dell Inc. Dell Statistica (data analysis software system), version 13. 2016. Available online: software.dell.com (accessed on 6 December 2019).
31. Delcour, J.A.; Joye, I.J.; Pareyt, B.; Wilderjans, E.; Brijs, K.; Lagrain, B. Wheat gluten functionality as a quality determinant in cereal-based food products. *Annu. Rev. Food. Sci. Technol.* **2012**, *3*, 1–24. [CrossRef]
32. Mikulíková, D. The effect of friabilin on wheat grain hardness: A review. *Czech J. Gen. Plant Breed.* **2018**, *43*, 35–43. [CrossRef]
33. Turnbull, K.M.; Rahman, S. Endosperm texture in wheat. *J. Cereal Sci.* **2002**, *36*, 327–337. [CrossRef]
34. Pasha, I.; Anjum, F.M.; Morris, C.F. Grain Hardness: A Major Determinant of Wheat Quality. *Food Sci. Tech. Int.* **2010**, *16*, 511–522. [CrossRef]
35. Glenn, G.M.; Johnston, R.K. Mechanical properties of starch, protein and endosperm and their relationship to hardness in wheat. *Food Struct.* **1992**, *11*, 187–199.
36. Lewicki, P. Water as the determinant of food engineering properties. A review. *J. Food Eng.* **2004**, *61*, 483–495. [CrossRef]
37. Ozturk, O.K.; Takhar, P.S. Stress relaxation behavior of oat flakes. *J. Cereal Sci.* **2017**, *77*, 84–89. [CrossRef]
38. Al Aridhee, J.; Łysiak, G. Stress relaxation characteristics of wheat kernels at different moisture. *Acta Sci. Pol. Tech. Agrar.* **2015**, *14*, 3–10.
39. Edwards, M.A. Morphological Features of Wheat Grain and Genotype Affecting Flour Yield. Ph.D. Thesis, Southern Cross University, Lismore, NSW, Australia, 2010.
40. Dexter, J.E.; Marchylo, B.A.; MacGregor, A.W.; Tkachuk, R. 1989. The structure and protein composition of vitreous, piebald and starchy durum wheat kernels. *J. Cereal Sci.* **1989**, *10*, 19–32. [CrossRef]
41. Dobraszczyk, B.J.; Schofield, J.D. Rapid Assessment and Prediction of Wheat and Gluten Baking Quality With the 2-g Direct Drive Mixograph Using Multivariate Statistical Analysis. *Cereal Chem.* **2002**, *79*, 607–612. [CrossRef]
42. Haddad, Y.; Benet, J.C.; Delenne, J.Y.; Mermet, A.; Abecassis, J. Rheological Behaviour of Wheat Endosperm—Proposal for Classification Based on the Rheological Characteristics of Endosperm Test Samples. *J. Cereal Sci.* **2001**, *34*, 105–113. [CrossRef]
43. Ponce-García, N.; Ramírez-Wong, B.; Escalante-Aburto, A.; Torres-Chávez, P.I.; Figueroa, J.D.C. Mechanical Properties in Wheat (*Triticum aestivum*) Kernels Evaluated by Compression Tests: A Review. In *Viscoelastic and Viscoplastic Materials*; El-Amin, M., Ed.; IntechOpen: London, UK, 2016.
44. Ponce-García, N.; Figueroa, J.D.C.; López-Huape, G.A.; Martínez, H.E.; Martínez-Peniche, R. Study of viscoelastic properties of wheat kernels using compression load method. *Cereal Chem.* **2008**, *85*, 667–672. [CrossRef]
45. Hrušková, M.; Švec, I. Wheat hardness in relation to other quality factors. *Czech J. Food Sci.* **2009**, *27*, 240–248. [CrossRef]

46. Morales, R.; Arnau, J.; Serra, X.; Guerrero, L.; Gou, P. Texture changes in dry-cured ham pieces by mild thermal treatments at the end of the drying process. *Meat Sci.* **2008**, *80*, 231–238. [CrossRef]
47. Al Aridhee, J. Investigation and Modelling of Stress Relaxation of Wheat Kernels in View of Grinding Prediction. Ph.D. Thesis, University of Life Sciences, Lublin, Poland, 2018.

9

Modelling Water Absorption in Micronized Lentil Seeds with the use of Peleg's Equation

Izabela Kuna-Broniowska [1], Agata Blicharz-Kania [2,*], Dariusz Andrejko [2], Agnieszka Kubik-Komar [1], Zbigniew Kobus [3], Anna Pecyna [3], Monika Stoma [4], Beata Ślaska-Grzywna [2] and Leszek Rydzak [2]

1 Department of Applied Mathematics and Computer Science, University of Life Sciences in Lublin, 20-612 Lublin, Poland; izabela.kuna@up.lublin.pl (I.K.-B.); agnieszka.kubik@up.lublin.pl (A.K.-K.)
2 Department of Biological Bases of Food and Feed Technologies, University of Life Sciences in Lublin, 20-612 Lublin, Poland; dariusz.andrejko@up.lublin.pl (D.A.); beata.grzywna@up.lublin.pl (B.Ś.-G.); leszek.rydzak@up.lublin.pl (L.R.)
3 Department of Technology Fundamentals, University of Life Sciences in Lublin, 20-612 Lublin, Poland; zbigniew.kobus@up.lublin.pl (Z.K.); anna.pecyna@up.lublin.pl (A.P.)
4 Department of Power Engineering and Transportation, University of Life Sciences in Lublin, 20-612 Lublin, Poland; monika.stoma@up.lublin.pl
* Correspondence: agata.kania@up.lublin.pl

Abstract: The aim of the paper was to investigate the effect of infrared pre-treatment on the process of water absorption by lentil seeds. The paper presents the effects of micronization on the process of water absorption by lentil seeds. As a source of infrared emission, 400-W ceramic infrared radiators ECS-1 were used. The seeds were soaked at three temperature values (in the range from 25 to 75 °C) for 8 h, that is, until the equilibrium moisture content was achieved. Peleg's equation was used to describe the kinetics of water absorption by lentil seeds. The results were compared with those obtained in the process of soaking crude seeds. On the basis of the conducted research, it was found that the infrared pre-treatment contributed to a substantial increase in the water absorption rate in the initial period of soaking lentil seeds (especially at 25 °C). Infrared irradiation can be an effective method for intensification of lentil seed hydration at an ambient temperature. It should be assumed that, in accordance with the principles of sustainable development, shortening the heating time will significantly reduce the energy consumption and cost of processing lentil seeds.

Keywords: legumes; infrared processing; acceleration of the process of hydration; Peleg's equation

---

1. Introduction

Water absorption by plant raw materials is influenced by both the process parameters (e.g., temperature, pressure) and the physical and chemical properties of the raw materials. The impact of the temperature of the soaking process on the rate and amount of absorbed water was investigated by, for example, Maskan [1] and Turhan et al. [2]. Their results clearly demonstrate the role of temperature (in the range from 6 to 100 °C) in the water absorption process. An elevated temperature was found to accelerate the water absorption rate in the case of peas, wheat, lupine, soybean, and faba bean.

Research is being carried out to develop a mathematical description of water absorption by granular plant materials. The process of water absorption in some cereal caryopses and legume seeds, that is, wheat, peas, or rice, has been investigated, and a model based mainly on the Fick diffusion equation has been developed [3–5]. However, prediction of the water absorption time based on Fick's law requires a very complex mathematical apparatus, which makes it very inconvenient for practical calculations in many cases. Hence, Peleg [6] proposed a two-parameter, non-exponential, empirical equation for modelling the water absorption process in raw materials and food products, as follows:

$$M_t = M_0 \pm \frac{t}{K_1 + K_2 \cdot t}, \quad (1)$$

where

M_t—Water content after time t, % of d.w.;
M_o—Initial water content, % of d.w.;
K_1—Constant, $h \cdot \%^{-1}$;
K_2—Constant, $\%^{-1}$.

Equation (1) contains the "±" sign. The "+" sign is used to model the water absorption process, and the "–" sign is used to model the drying process.

The main attribute of this equation, compared with other equations, is its simplicity [1]. This formula has been positively verified for several cereal and legume species [2,7–9].

The sorption rate R can be calculated from the first derivative of Peleg's Equation.

$$R = \frac{dM}{dt} = \pm \frac{K_1}{(K_1 + K_2 t)^2} \quad (2)$$

Peleg's constant K_1 refers to the initial sorption rate (R_0, i.e., the value of R at $t = t_0$).

$$R_0 = \frac{dM}{dt}\Big|_{t_0} = \pm \frac{1}{K_1} \quad (3)$$

At $t \to \infty$, Equation (4) describes the relationship between the equilibrium moisture content (M_e) and constant K_2.

$$M\big|_{t_\infty} = Me = M_0 \pm \frac{1}{K_2} \quad (4)$$

Linearization of Equation (1) gives the following formula:

$$\frac{t}{M_t - M_0} = K_1 + K_2 t. \quad (5)$$

Adjustment of the equation of such a line allows estimation of parameters K_1 and K_2.

The value of constant K_1 provides information about the mass transfer rate during the water absorption process, that is, the lower the K_1 value, the higher the water absorption rate [2,7,10, 11]. While the K_1 coefficient has been relatively precisely defined in many publications, there is a large discrepancy as far as the significance of constant K_2 is concerned. In investigations of water absorption by wheat-based products, Maskan [1] has found that constant K_2 refers to the maximum water absorption capacity, that is, the water absorption capacity increases at lower K_2 values. Similar conclusions were formulated by Abu-Ghannam and McKenna [7], as well as Sayar et al. [11]. However, other authors [10,12–14] did not confirm these findings.

Reduction of energy consumption in economy is the most efficient and best-recognized way of implementing the principle of sustainable development. In addition, from the consumer's point of view, the use of legume seeds is determined by the time of preparation of ready-to-eat products. Therefore, seed pre-treatment should ensure the shortest time of the subsequent hydrothermal treatment thereof (e.g., soaking, cooking). As suggested by Cenkowski and Sosulski [15], thermal infrared pre-treatment of seeds applied prior to soaking or cooking can shorten these treatments and provide instant-type products.

Infrared radiation is very widely used in various industries. In the food industry, it is used primarily for drying and pre-treatment of food. The micronization process allows, for example, reducing the content of anti-nutritional substances in legume seeds or increasing the efficiency of pressing oilseeds. Owing to the specific mechanism of heat transport during micronization, the heat and mass exchange conditions are definitely more favourable than when using traditional heat treatment methods. In practice, it results in more effective operation, and thus reduction of the duration of the process [16–18].

However, the available literature provides no reports on the possibility of using Peleg's equation to determine the rate of water absorption by micronized legumes.

Therefore, the paper presents an attempt to investigate the effect of infrared pre-treatment on the process of water absorption by lentil seeds.

2. Materials and Methods

2.1. Research Material

The lentil seeds (variety Anita) used in the experiment were purchased at a local supermarket in Lublin. Prior to the determinations, the raw material was purified by the removal of cracked and damaged seeds. It was stored in closed plastic containers at a temperature of 5 °C. Before further measurements, the lentil seeds were placed at room temperature (approximately 23 °C) for 3 h.

2.2. Measurement of Initial Moisture Content

The first stage of the research involved determination of the initial moisture content of the raw material (W) using the drying method. The material was dried for 3 h in a laboratory dryer SLN 15 STD at a temperature of 103 °C. The total protein content in the material was determined with the Kjeldahl method using an automated Foss Kjeltec 8400 distillation unit, and the total fat content was determined using a Soxtec 8000 apparatus with AN 310 software (Table 1).

Table 1. Effect of selected components in lentil seeds calculated per dry weight.

Nutrients		Content [%]
Protein		22.01 ± 0.53
Fat		0.90 ± 0.02
Moisture	Crude seeds	8.40 ± 0.17
	Micronized seeds	6.00 ± 0.20

2.3. Micronization Process

A portion of seeds was subjected to the micronization process. As a source of infrared radiation emission, 400-W ceramic infrared radiators ECS-1 were used. These temperature radiators supplied by electricity (230 V) have a small fraction of visible radiation (dark radiators) in the spectrum and heat the entire plane uniformly (plane radiators) (Figure 1). The average temperature of the filament is approximately 500 °C and the emission wavelength is λ = 2.5–3.0 μm. The seeds were heated for 120 s at a temperature on the seed surface of 150 °C. Next, the seeds were left to dry in open containers at room temperature for 3 h.

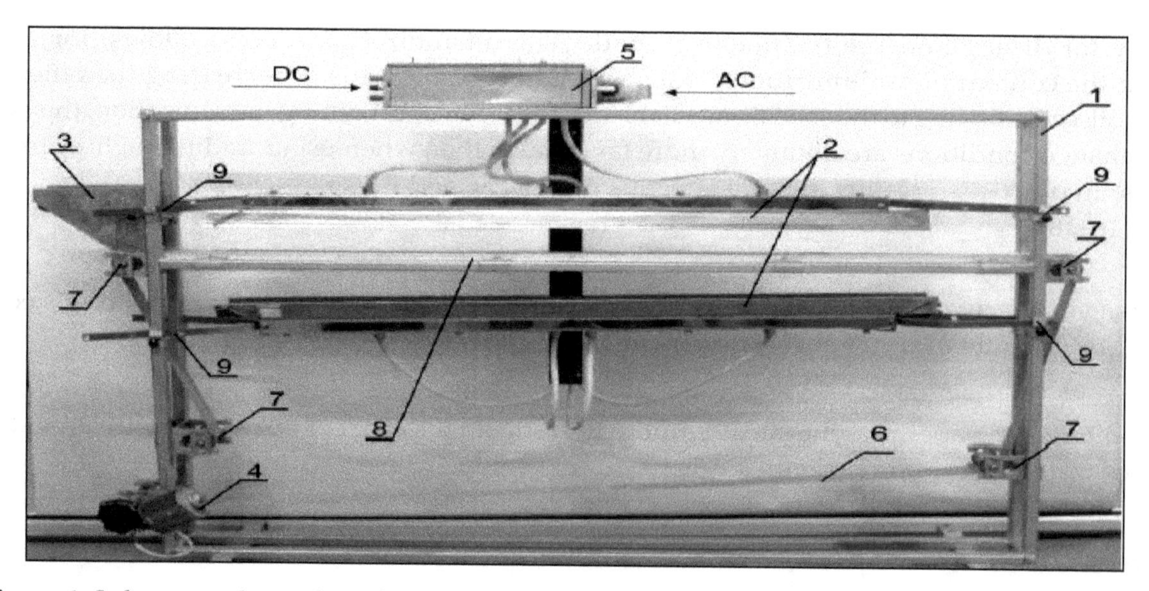

Figure 1. Laboratory device for infrared treatment of granular plant raw materials: DC - direct current, AC - alternating current, 1—frame, 2—infrared radiators, 3—feeding tank, 4—DC motor, 5—control unit, 6—conveyor belt, 7—rollers, 8—heating zone, 9—adjustment of the position of the heads.

2.4. *Measurement of the Water Content*

At 24 h after the micronization process, the moisture content in the micronized seeds was determined and the crude and micronized seeds were soaked. Samples of randomly selected seeds (excluding cracked and damaged seeds) weighing approximately 0.5 g (M_m) were placed in special baskets with a perforated wall and bottom. The material was soaked in a water bath (using distilled water) within a time range from 5 min to 8 h at a temperature of 25 °C, 50 °C, and 75 °C.

After the time intervals adopted in the experiment, the samples were removed, the excess water was dried on filter paper, and the water content in the seeds was determined.

The water content relative to dry weight was calculated using Equation (6):

$$M_t = \frac{M_1 + M_{H_2O}}{M_1} 100\%, \tag{6}$$

where

M_t—Water content after time t, % relative to dry weight [%];
M_{H2O}—Water weight after time t [g];
M_1—Dry weight [g].

The dry weight was calculated based on Equation (7):

$$M_1 = M_m \cdot (100\ \% - W), \tag{7}$$

where

M_m—Initial weight of the material [g];
W—Moisture [%].

The determinations were carried out in triplicate.

2.5. *Statistical Analysis*

The data were analysed statistically. A significance level of $\alpha = 0.05$ was assumed for inference. The analysis was carried out using analysis of variance (ANOVA) (StatSoft Polska, Poland, Cracow).

3. Results and Discussion

Table 1 shows the effect of selected components in lentil seeds calculated per dry weight. The lentil seeds were characterised by typical levels of essential nutrients, for example, protein and fat. For comparison, the investigations of the chemical composition of lentils conducted by Hefnawy [19] revealed a protein content of 26.6 ± 0.5 (g 100 g^{-1} dry weight basis) and fat content of 1.0 ± 0.08 (g 100 g^{-1} dry weight basis). Similarly, the water content in the crude seeds corresponded to the values demonstrated by other authors [19,20].

Figure 2 shows the effect of temperature and time on changes in the water content in the crude lentil seeds. Analysis of variance revealed a significant effect of both of these factors on the water content in the soaked seeds. The water absorption rate decreased with the time of soaking. The highest water content, that is, 142%, was observed after 5 h of soaking of lentil seeds at a temperature of 50 °C. Further, soaking did not change the water content in the treated lentil seeds significantly. The values obtained in this study are in agreement with the results reported by other authors. The mean water content in seeds was estimated at 129% by Chopra and Prasad [21], 140% by Pan and Tangratanavalee [22], and 147% by Quicazán et al. [8].

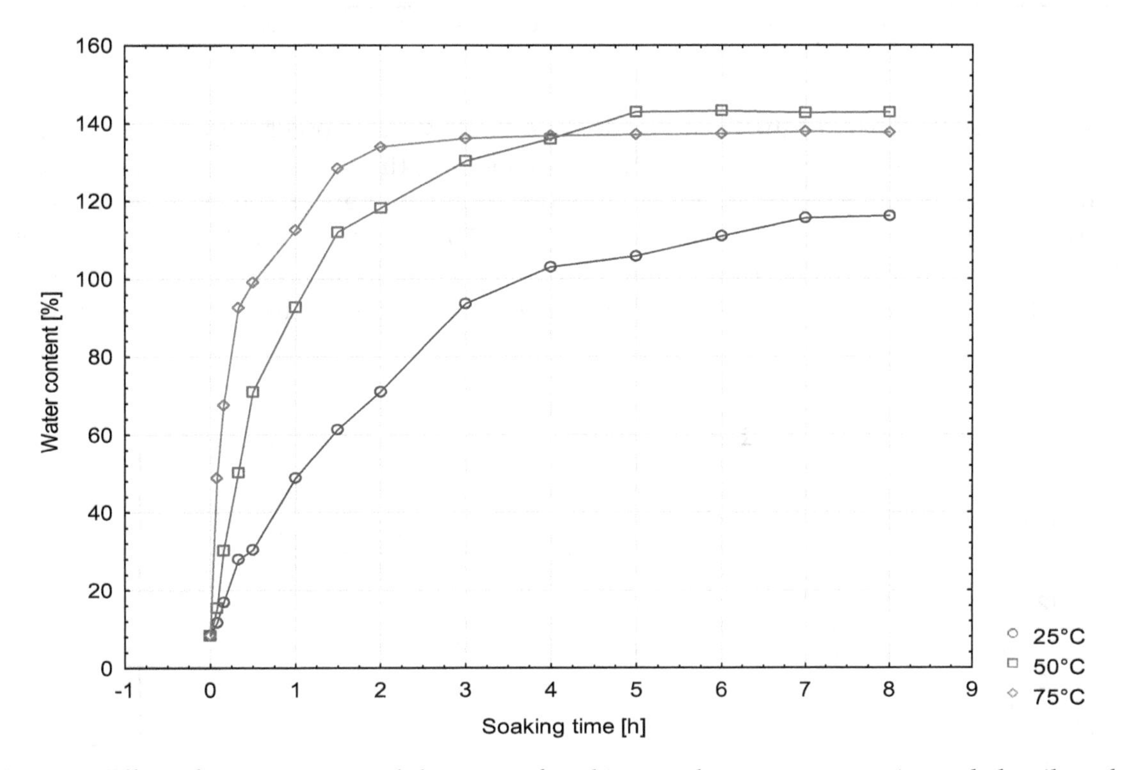

Figure 2. Effect of temperature and duration of soaking on the water content in crude lentil seeds.

In the initial soaking period, the lentil seeds exhibited the lowest water absorption rate at the temperature of 25 °C. An increase in the temperature to 50 °C and then 75 °C resulted in an increase in this parameter. After 3 h of soaking at 75 °C, the lentil seeds reached a maximum moisture level of 138%. Seeds soaked at the temperature of 50 °C absorbed water at a slower rate and reached the maximum level later, but this was by 4% higher than the value obtained at 75 °C. These results are in agreement with the findings presented by other authors. Quicazán et al. [8] reported a slightly higher level of water in seeds soaked at a temperature of 40 °C than the water content in seeds soaked at 80 °C. This phenomenon may be associated with the more intensive denaturation of proteins at a higher temperature, which in turn may reduce the final capacity of water absorption in lentil seeds.

Table 2 shows Peleg's coefficients and the goodness of fit of the model to the experimental data. It was shown that the K_2 parameter value changed slightly with the soaking temperature. This is in line with the results obtained by Sopade et al. [14], in experiments of cereal grain hydration, and Abu-Ghannam and McKenna [7], who investigated the process of hydration of bean seeds. The values of the K_1 parameter declined with an increase in the temperature of soaking. This trend is in good agreement with the results reported by other authors. In experiments involving soybean seed soaking, Quicazán et al. [8] observed a negative correlation between the soaking temperature and the value of the K_1 parameter. The dependence of K_1 on T (absolute temperature) is expressed by Equation (8).

$$K_1 = 0.0006T^2 - 0.4452T + 77.358 \tag{8}$$

Table 2. Parameters of Peleg's model for hydration of crude lentil seeds at three temperatures.

Parameter	T	Soaking Time	K_1	$1/K_1$	K_2	R^2	E
Unit	°C	h					%
Result	25	0–8	1.6845	0.59365	0.6893	0.9960	3.93
	50	0–8	0.5183	1.92938	0.6567	0.9957	6.08
	75	0–8	0.1543	6.48088	0.7461	0.9952	1.94

Notes: R^2—fit of the model; E—goodness of fit of the model.

Figure 3 shows the effect of the temperature and soaking time on changes in the water content in the micronized lentil seeds. The temperature and soaking time contributed to an increase in the water content in the micronized lentil seeds. In the initial soaking period, the lentil seeds exhibited the lowest rate of water absorption at a temperature of 25 °C. The increase in the temperature to 50 °C and then 75 °C increased the water absorption rate. After 4 h of soaking at both 50 °C and 75 °C, the lentil seeds achieved the maximum moisture level. The highest water content, that is, 138%, was observed at the temperature of 75 °C.

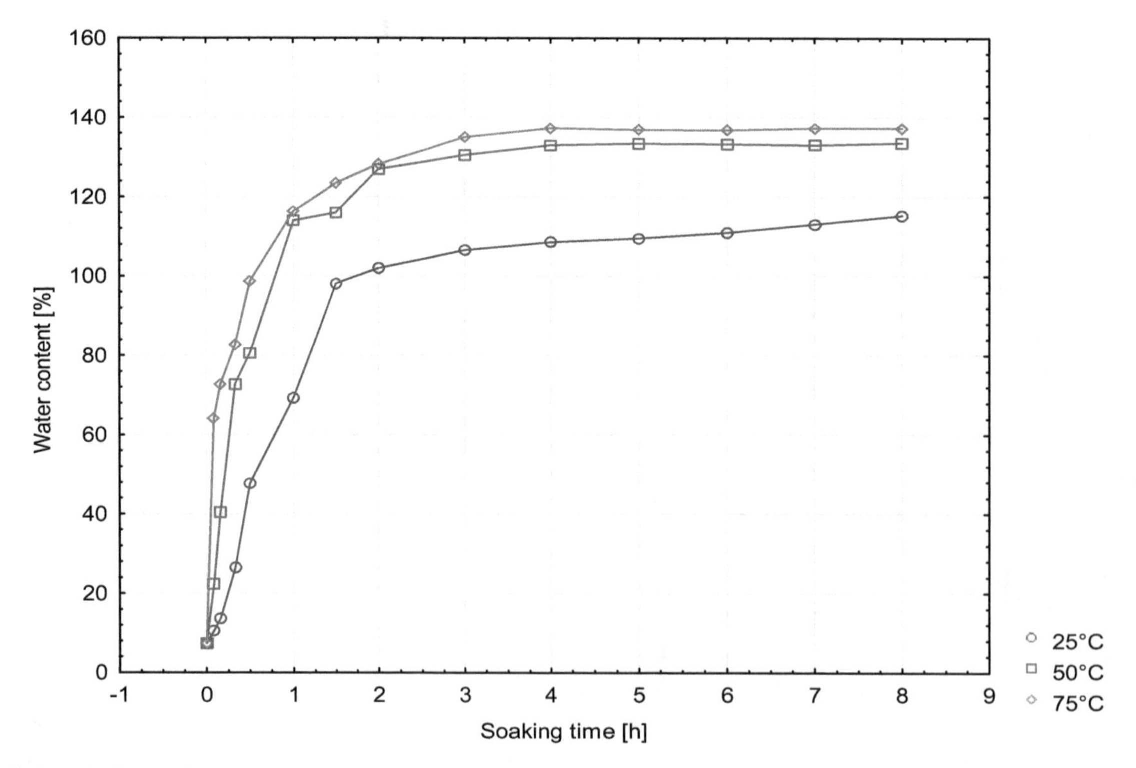

Figure 3. Effect of temperature and soaking time on changes in the water content in the micronized lentil seeds.

Table 3 shows Peleg's model coefficients and the goodness of fit of the model to the experimental data for the micronized seeds. The values of the K_1 parameter declined with the increase in the soaking temperature. In comparison with the crude lentil seeds, the values of the K_1 coefficients were significantly lower at all tested temperatures. This indicates a higher rate of water absorption in the micronized seeds. The dependence of K_1 on T (absolute temperature) is expressed by Equation (9):

$$K_1 = 0.0003T^2 - 0.1838T + 32.056 \quad (9)$$

Table 3. Parameters of Peleg's model for micronized lentil hydration at three temperatures.

Parameter	T	Soaking Time	K_1	$1/K_1$	K_2	R^2	E
Unit	°C	h					%
Result	25	0–8	0.7801	1.28189	0.7977	0.9698	17.38
	50	0–8	0.2920	3.42466	0.7299	0.9895	6.06
	75	0–8	0.1345	7.43494	0.7565	0.9791	4.18

Notes: R^2—fit of the model; E—goodness of fit of the model.

The values of the K_2 coefficient in the case of the micronized seeds changed as in the case of the crude seeds. The highest K_2 coefficient was noted at the temperature of 25 °C; next, it declined to the lowest value at 50 °C and reached an intermediate value at 75 °C. In general, all values of the K_2 coefficient for the micronized seeds were slightly lower than the K_2 values for the crude seeds. This was reflected in the final water content in the micronized seeds, which was slightly lower than that observed in the crude seeds.

4. Conclusions

The paper presents the effect of the infrared radiation treatment on the kinetics of water absorption by lentil seeds. The results were compared with the process of soaking crude seeds. In both cases, the changes in the water content were described using Peleg's model. An impact of the infrared pre-treatment on the water absorption rate and content in lentil seeds was demonstrated. The infrared pre-treatment largely contributes to an increase in the water absorption rate in the initial period of soaking lentil seeds. This is particularly evident at a temperature of 25 °C, at which micronization reduces the time required for achievement of 100% water content by more than half in comparison with crude seeds. At higher temperatures, these differences are smaller, that is, 40% and 17%, for the temperatures of 50 °C and 75 °C, respectively. The impact of micronization on the final water content in lentil seeds depends on the soaking temperature. A statistically significant effect of micronization on the water content was found at 25 °C (reduction of constant K_1 from 1.6845 to 0.7801). Slightly smaller changes in the rate of water absorption were observed at 50 °C and 75 °C.

Infrared treatment can be successfully used to accelerate the process of hydration of lentil seeds at an ambient temperature. As a consequence, limiting the damaging impact of production on the environment will ensure sustainable consumption and processing of legumes. The results obtained in the present study can be regarded as preliminary research on the possibility of using micronization to accelerate the absorption of water by seeds of other plants, including cereals.

Author Contributions: Conceptualization, I.K.-B., D.A. and M.S.; Data curation, A.P.; Formal analysis, I.K.-B. and Z.K.; Investigation, A.B.-K. and A.P.; Methodology, A.B.-K., D.A., A.K.-K. and B.Ś.-G.; Project administration, M.S.; Software, I.K.-B. and A.K.-K.; Supervision, I.K.-B. and D.A.; Validation, A.K.-K.; Visualization, A.B.-K., Z.K. and B.Ś.-G.; Writing–original draft, A.B.-K., Z.K. and A.P.; Writing–review & editing, D.A., A.K.-K. and L.R. All authors have read and agreed to the published version of the manuscript.

Abbreviations

M_t	water content after time t, [% of dry weight];
M_o	initial water content, [% of dry weight];
K_1	constant, [$h \cdot \%^{-1}$];
K_2	constant, [$\%^{-1}$];
R	sorption rate, [-];
M_{H2O}	water weight after time t, [g];
M_1	dry weight, [g];
M_m	initial weight of the material, [g];
W	moisture, [%];
R^2	fit of the model, [-];
E	goodness of fit of the model, [%];

References

1. Maskan, M. Effect of processing on hydration kinetics of three wheat products of the same variety. *J. Food Eng.* **2002**, *52*, 337–341. [CrossRef]
2. Turhan, M.; Sayar, S.; Gunasekaran, S. Application of Peleg model to study water absorption in chickpea during soaking. *J. Food Eng.* **2002**, *53*, 153–159. [CrossRef]
3. Cunningham, S.E.; McMinn, W.A.M.; Magee, T.R.A.; Richardson, P.S. Modelling water absorption of pasta during soaking. *J. Food Eng.* **2007**, *82*, 600–607. [CrossRef]
4. Montanuci, F.D.; Perussello, C.A.; de Matos Jorge, L.M.; Jorge, R.M.M. Experimental analysis and finite element simulation of the hydration process of barley grains. *J. Food Eng.* **2014**, *131*, 44–49. [CrossRef]
5. Munson-McGee, S.H.; Mannarswamy, A.; Andersen, P.K. D-optimal designs for sorption kinetic experiments: Cylinders. *J. Food Eng.* **2011**, *104*, 202–207. [CrossRef]
6. Peleg, M. An empirical model for the description of moisture sorption curves. *J. Food Sci.* **1988**, *53*, 1216–1217, 1219. [CrossRef]
7. Abu-Ghannam, N.; McKenna, B. The application of Peleg's equation to model water absorption during the soaking of red kidney beans (*Phasedus vulgaris* L.). *J. Food Eng.* **1997**, *32*, 391–401. [CrossRef]
8. Quicazán, M.C.; Caicedo, L.A.; Cuenca, M. Applying Peleg's equation to modelling the kinetics of solid hydration and migration during soybean soaking. *Ing. Investig.* **2012**, *32*, 53–57.
9. Sopade, P.A.; Xun, P.Y.; Halley, P.J.; Hardin, M. Equivalence of the Peleg, Pilosof and Singh–Kulshrestha models for water absorption in food. *J. Food Eng.* **2007**, *78*, 730–734. [CrossRef]
10. Hung, T.V.; Liu, L.H.; Black, R.G.; Trewhella, M.A. Water absorption in chickpea (*C. arietinum*) and field pea (*P. sativum*) cultivars using the Peleg model. *J. Food Sci.* **1993**, *58*, 848–852. [CrossRef]
11. Sayar, S.; Turhan, M.; Gunasekaran, S. Analysis of chickpea soaking by simultaneous water transfer and water-starch reaction. *J. Food Eng.* **2001**, *50*, 91–98. [CrossRef]
12. Maharaj, V.; Sankat, C.K. Rehydration characteristics and quality of dehydrated dasheen leaves. *Can. Agric. Eng.* **2000**, *42*, 81–85.
13. Sopade, P.A.; Obekpa, J.A. Modeling water absorption in soybean, cowpea and peanuts at three temperatures using Peleg's equation. *J. Food Sci.* **1990**, *55*, 1084–1087. [CrossRef]
14. Sopade, P.A.; Ajisegiri, E.S.; Badau, M.H. The use of Peleg's equation to model water absorption in some cereal grains during soaking. *J. Food Eng.* **1992**, *15*, 269–283. [CrossRef]
15. Cenkowski, S.; Sosulski, F.W. Cooking characteristics of split peas treated with infrared heat. *Trans. ASAE* **1998**, *41*, 715–720. [CrossRef]
16. Khattab, R.Y.; Arntfield, S.D. Nutritional quality of legume seeds as affected by some physical treatments 2. Antinutritional factors. *Lebensm. Wiss. Technol.* **2009**, *42*, 1113–1118. [CrossRef]
17. Krajewska, M.; Ślaska-Grzywna, B.; Andrejko, D. Effect of Infrared Thermal Pre-Treatment of Sesame Seeds (*Sesamum Indicum* L.) on Oil Yield and Quality. *Ital. J. Food Sci.* **2018**, *30*, 487–496. [CrossRef]
18. Sobczak, P.; Zawiślak, K.; Panasiewicz, M.; Mazur, J.; Kocira, S.; Żukiewicz-Sobczak, W. Impact of Heat Treatment on the Hardness and Content of Anti-Nutritious Substances in Soybean Seeds. *Carpath. J. Food Sci. Technol.* **2018**, *10*, 46–52.

19. Hefnawy, T.H. Effect of processing methods on nutritional composition and anti-nutritional factors in lentils (*Lens culinaris*). *Ann. Agric. Sci.* **2011**, *56*, 57–61. [CrossRef]
20. Boye, J.I.; Aksay, S.; Roufik, S.; Ribéreau, S.; Mondor, M.; Farnworth, E.; Rajamohamed, S.H. Comparison of the functional properties of pea, chickpea and lentil protein concentrates processed using ultrafiltration and isoelectric precipitation techniques. *Food Res. Int.* **2010**, *43*, 537–546. [CrossRef]
21. Chopra, R.; Prasad, D.N. Standardization of soaking conditions for soybean seeds/cotyledons for improved quality of soymilk. *Indian J. Anim. Sci.* **1994**, *64*, 405–410.
22. Pan, Z.; Tangratanavalee, W. Characteristics of soybeans as affected by soaking conditions. *Lebensm. Wiss. Technol.* **2003**, *36*, 143–151. [CrossRef]

10

Assessment of the Properties of Rapeseed Oil Enriched with Oils Characterized by High Content of α-linolenic Acid

Agnieszka Sagan [1], Agata Blicharz-Kania [1,*], Marek Szmigielski [1], Dariusz Andrejko [1], Paweł Sobczak [2], Kazimierz Zawiślak [2] and Agnieszka Starek [1]

[1] Department of Biological Bases of Food and Feed Technologies, University of Life Sciences in Lublin, 20-612 Lublin, Poland; agnieszka.sagan@up.lublin.pl (A.S.); marek.szmigielski@up.lublin.pl (M.S.); dariusz.andrejko@up.lublin.pl (D.A.); agnieszka.starek@up.lublin.pl (A.S.)

[2] Department of Food Engineering and Machines, University of Life Sciences in Lublin, 20-612 Lublin, Poland; pawel.sobczak@up.lublin.pl (P.S.); kazimierz.zawislak@up.lublin.pl (K.Z.)

* Correspondence: agata.kania@up.lublin.pl

Abstract: Functional foods include cold-pressed oils, which are a rich source of antioxidants and bioactive n-3 and n-6 polyunsaturated fatty acids. The aim of this study was to assess the quality of rapeseed oils supplemented with Spanish sage and cress oils. Seven oil mixtures consisting of 70% of rapeseed oil and 30% of sage and/or cress oil were prepared for the analyses. The oil mixtures were analyzed to determine their acid value, peroxide value, oxidative stability, and fatty acid composition. In terms of the acid value and the peroxide value, all mixtures met the requirements for cold-pressed vegetable oils. The enrichment of the rapeseed oil with α-linolenic acid-rich fats resulted in a substantially lower ratio of n-6 to n-3 acids in the mixtures than in the rapeseed oil. The mixture of the rapeseed oil with the sage and cress oils in a ratio of 70:10:20 exhibited higher oxidative stability than the raw materials used for enrichment and a nearly 20% α-linolenic acid content. The oils proposed in this study can improve the ratio of n-6:n-3 acids in modern diets. Additionally, mixing the cress seed oils with rapeseed oil and chia oil resulted in a reduction in the content of erucic acid in the finished product. This finding indicates that cress seeds, despite their high content of erucic acid, can be used as food components. The production of products with a positive effect on human health is one of the most important factors in the sustainable development of agriculture.

Keywords: cold-pressed oils; functional food; oxidative stability; rapeseed oil; Spanish sage seed oil; cress seed oil

---

1. Introduction

The results available from health and nutrition research increase public awareness of the relationship between health and diet. Therefore, functional food is becoming increasingly popular among consumers. The idea of functional food originates from Japan, where a national program of studies on the relationship between food science and medicine was introduced in 1984. Functional food is defined as follows: "a food can be regarded as 'functional' if it is satisfactorily demonstrated to affect beneficially one or more target functions in the body, beyond adequate nutritional effects, in a way that is relevant to either an improved state of health and well-being and/or reduction of risk of disease. Functional foods must remain foods and they must demonstrate their effects in amounts that can normally be expected to be consumed in the diet: these are not pills or capsules, but part of a normal food pattern" [1–3]. The production of such food may be one of the most important aspects of strategies for sustainable development. The essence of the sustainable development of food production is the production of products with a positive effect on human health [4].

Functional foods include cold-pressed oils, as they are a rich source of antioxidants such as tocopherols and polyphenols, and hence exhibit high antioxidant activity. These also contain polyunsaturated fatty acids from the n-3 and n-6 groups as well as sterols, which exert a bioactive effect [5,6]. Seeds or fruits of plants that are traditionally regarded as oil-bearing species, e.g., rape, soybean, sunflower, olives, etc., are used for pressing oils. Also, oils pressed from unusual plant materials such as nuts and sage or garden cress seeds are gaining popularity.

Rapeseed oil is one of the most frequently consumed vegetable oils, and it is one of the most valuable edible fats, mainly due to the high content (approx. 90%) of 18-carbon unsaturated acids. It is produced from rapeseed varieties with a low level of erucic acid and is characterized by the presence of fatty acids that are desirable in human and animal diets and are present in ratios that are similar to those in the most valuable oils, e.g., olive oil. Furthermore, it is rich in many bioactive compounds, whose presence in food is the current focus of attention. These are primarily antioxidant compounds. The oil is also a source of essential unsaturated fatty acids from the n-6 and n-3 groups. The content of linoleic and α-linolenic acids in rapeseed oil is usually approx. 20% and 10%, respectively [7,8]. It also exhibits good oxidative stability, which is better than that of soybean and sunflower oils [9]. The oxidative stability of rapeseed oil can be improved by supplementation with natural antioxidants contained in spices (dried rosemary, oregano) and in resin from trees of the *Burseraceae* family [10,11]. Research was also conducted on the stability of mixtures of rapeseed oil with linseed oil. Mixtures of rapeseed oil with 25% and 50% linseed oil were found to have the best properties [12].

It is important to maintain an appropriate n-6:n-3 ratio of polyunsaturated fatty acids. Most reports in the literature indicate that the n-6:n-3 ratio should range from 1:1 to 4:1. However, modern diets usually contain up to 20-fold higher levels of n-6 than n-3 fatty acids [13]. Therefore, in terms of sustainable development strategies, it is vital to enrich food with n-3 acids, e.g., α-linolenic acid, which originates from vegetable oils such as Spanish sage or garden cress oils [14,15]. The production of products that have a positive effect on human health is one of the most important element in the sustainable development of agriculture.

Seeds of Spanish sage (*Salvia hispanica* L.) contain large amounts of fat (30%–40%), which is rich in essential fatty acids, in particular α-linolenic acid. This may account for 68% of all fatty acids [14,16]. Sage seeds are also characterized by a high level of antioxidants, which may be helpful in preventing diseases related to oxidative DNA damage [17,18].

Garden cress (*Lepidium sativum* L.) is a good source of n-3 fatty acids. It is an annual plant from the family Brassicaceae (*Cruciferae*). Its seeds contain large amounts of fat (28%). Its main fatty acids are oleic acid (30%) and α-linolenic acid (32%) representing the n-3 group. Cress seeds also contain a high level of polyphenols and are a source of bioactive glycosides [15,19]. It has been found that the isoflavone glycoside isolated from *L. sativum* seeds can improve liver function and the serum lipid profile. It can also reduce the generation of free radicals through induction of an antioxidant defense mechanism and acts as a potential antioxidant against paracetamol poisoning [20].

Polyunsaturated fatty acids from the n-3 group enhance the nutritional value of food products but also increase the susceptibility to fat oxidation. Spanish sage seeds are rich in essential fatty acids; yet, the low oxidative stability of the oil pressed from this raw material is a shortcoming. Additionally, the nutritional value of cress seed oil may be reduced by the elevated content of erucic acid.

The aim of this study was to verify the quality of a composition of rapeseed oil supplemented with oils from Spanish sage and cress seeds in various proportions.

2. Materials and Methods

2.1. Research Material

The research material included seven different oil mixtures based on rapeseed oil. Seeds of winter oilseed rape (*Brassica napus* L.), Spanish sage (*Salvia hispanica* L.), and garden cress (*Lepidium sativum* L.) were used to obtain the experimental oils. The seeds were analyzed for moisture [21] and fat content [22]. The moisture contents were approximately 7% and the fat contents were typical for the seeds of the plants (Table 1).

Table 1. Humidity and fat content in the seeds used for oil pressing and pressing yields.

Seeds	Humidity (%)	Fat Content (%)	Pressing Yield (%)
Rape	6.97 ^{a} ± 0.01	38.40 ^{b} ± 0.06	79.0
Spanish sage	7.11 ^{b} ± 0.01	36.59 ^{a} ± 0.31	60.7
Garden cress	7.61 ^{c} ± 0.03	20.53 ^{c} ± 0.01	41.9

^{abc}—Statistically significant differences in columns ($p \leq 0.05$). The results are expressed as mean ± SD (n = 3).

Individual batches of seeds were cold pressed in a Farmet DUO screw press (Czech Republic) with a capacity of 18–25 kg·h^{-1}, engine power 2.2 kW, and screw speed 1500 rpm equipped with a 10-mm diameter nozzle. Before starting, the press was heated to 50 °C. After pressing, the oil was left for natural sedimentation for 5 days in refrigerated conditions (10 ± 1 °C) and then decanted. The pressing efficiency was calculated based on the weight of the seeds used for the process, the fat content of the seeds, and the weight of the pressed oil. The following formula was used to calculate the pressing efficiency (W):

$$W = \frac{m_o \cdot 10^4}{f \cdot m_s} \ (\%) \quad (1)$$

where:

m_o—weight of oil (kg),
m_s—weight of seeds (kg),
f—fat content in seeds (%)

The highest yield of 79% was obtained from the rapeseeds, whereas the lowest value, i.e., approximately 42%, was noted for the cress seeds (Table 1).

The acid value (AV), peroxide value (PV), oxidative stability, and fatty acid composition in the pressed oil were also determined.

2.2. Research Methods

The oils were mixed in 500-ml amber glass bottles at appropriate weight ratios. Seven mixtures of oil marked from M1 to M7 were prepared for the analyses with 70% of rapeseed oil (R) and 30% of Spanish sage and/or cress oils. The proportions of the individual oils and the symbols denoting the individual mixtures are presented in Table 2. The oil mixtures were analyzed for their acid value (AV), peroxide value (PV), oxidative stability, and fatty acid composition.

Table 2. Experimental setup.

Oil	Proportions of the Individual Oils in the Mixtures (%)						
	M1	M2	M3	M4	M5	M6	M7
Rapeseed oil	70	70	70	70	70	70	70
Spanish sage seed oil	-	5	10	15	20	25	30
Cress seed oil	30	25	20	15	10	5	-

The acid value was determined with the titration method using a cold solvent and expressed in mg KOH per 1 g oil [23]. The peroxide value was estimated by titration with iodometric determination of the end point and expressed in mmol O_2 per 1 kg of oil [24]. Oxidative stability was assessed in the Rancimat accelerated oxidation test [25]. The test was carried out using a Metrohm 893 Professional Biodiesel Rancimat device. Oil samples (3.00 ± 0.01 g) were weighed, placed in a measuring vessel, and exposed to air at a flow rate of 20 l/h at 120 °C. The results were expressed as the induction time determined from the curve inflection point using the StabNet1.0 software provided by the manufacturer. The fatty acid composition was determined using gas chromatography (a Bruker 436GC chromatograph with an FID detector) following the relevant standards [26,27]. Fatty acid methyl esters were separated on a BPX 70 capillary column (60 m × 0.25 mm, 25 μm) with nitrogen as the carrier gas.

2.3. Statistical Analysis

All determinations were made in triplicate. The arithmetic mean of the replicates was assumed as the result. The results were analyzed statistically using the StatSoft Polska STATISTICA 10.0 program. The significance of differences between the mean values of the determined parameters was verified using Tukey's test. The calculations were made at the significance level $\alpha = 0.05$.

3. Results and Discussion

The determined acid and peroxide values, which determine the quality of pressed oils, are presented in Table 3.

Table 3. Chemical properties of the analyzed oils.

Oil	Acid Number ($mg\ KOH \cdot g^{-1}$)	Peroxide Number ($mmol\ O_2 \cdot kg^{-1}$)	Induction Time (h)
Rapeseed oil	$0.66^{a} \pm 0.02$	$1.85^{a} \pm 0.25$	$4.40^{a} \pm 0.09$
Spanish sage seed oil	$4.77^{b} \pm 0.02$	$1.00^{b} \pm 0.01$	$0.65^{c} \pm 0.07$
Cress seed oil	$0.64^{a} \pm 0.03$	$1.24^{ab} \pm 0.07$	$2.67^{b} \pm 0.04$

^{abc}—Statistically significant differences in columns ($p \leq 0.05$). The results are expressed as mean ± SD (n = 3).

The highest acid value (4.77 mg $KOH \cdot g^{-1}$ fat) was determined for the chia seed oil. This value was only slightly higher than the requirements specified for this parameter (LK ≤ 4 $mg \cdot g^{-1}$) in the Codex Alimentarius [28]. As demonstrated by Krygier et al. [29], the acid value of oil depends on the quality of seeds used for pressing and increases with the increasing amounts of damaged seeds, which is explained by the authors by the greater activity of lipolytic enzymes present in such material.

The peroxide value, which reflects the amount of primary oxidation products, ranged from 1.00 to 1.85 mmol $O_2 \cdot kg^{-1}$ in the analyzed oils and did not exceed the maximum value of 10 mmol $O_2 \cdot kg^{-1}$ specified in the Codex Alimentarius [28].

Sample gas chromatograms of each oil are shown in Figure 1. The analyzed rapeseed oil had a typical fatty acid composition [30]. The content of unsaturated fatty acids exceeded 91% (Table 4).

Oleic acid was the main fatty acid (55.22%). The rapeseed oil contained the highest amount of linoleic acid (24.24%) and the lowest level of α-linolenic acid (10.34%). The ratio of n-6 to n-3 acids of 2:3 was equally low.

The Spanish sage oil had a high level of unsaturated fatty acids (89.43%), which is consistent with results reported by other authors [31,32]. The main acids were represented by α-linolenic acid (63.40%) followed by linoleic acid (20.80%). Such a high amount of α-linolenic acid resulted in a very low ratio of n-6 to n-3 acids in chia seed oil, i.e., 0.3.

The cress seed oil exhibited the greatest variety of fatty acids. Palmitic acid (9.02%), stearic acid (3.06%), and arachidonic acid (3.64%) represented saturated fatty acids. Mono-unsaturated fatty acids were represented by oleic acid (20.68%) as well as eicosenoic acid (13.8%) and erucic acid (5.98%). The content of essential unsaturated fatty acids (linoleic and linolenic) accounted for 9.08% and 30.03% of the sum of fatty acids, respectively. The lower content of linoleic acid than that determined in the other oils resulted in a low ratio of n-6 to n-3 acids, as in the Spanish sage oil (0:3). Additionally, there were small amounts of eicosadienoic, eicosatrienoic, lignoceric, and nervonic acids, which were not present in the other oils. The results obtained in the present study are similar to those reported by other authors investigating cress seeds [33,34], although the analyzed oil had almost two-fold higher content of erucic acid than the value determined by Moser et al. [34]. Due to its properties, erucic acid is classified as an anti-nutritional compound. The presence of erucic acid in cress oil is a drawback and limits the application of cress seeds as a raw material for oil pressing.

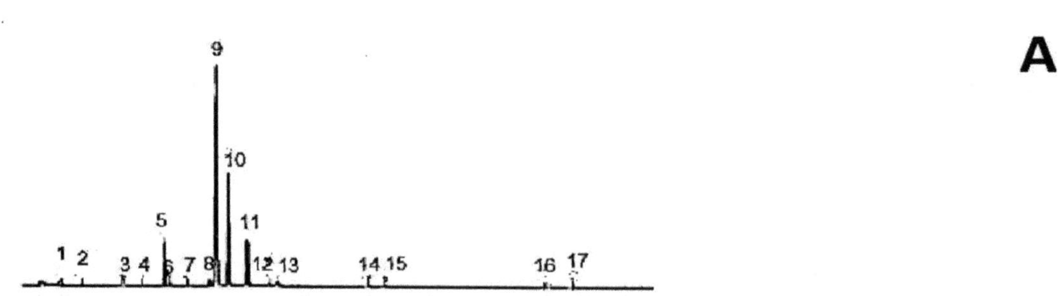

rapeseed oil metyl esters: 1 - capric (10:0), 2 - lauric (12:0), 3 - myristic(14:0), 4 - pentadecanic (15:0), 5 - palmitic (16:0), 6 - palmitoleic (16:1), 7 - heptadecanic (17:0), 8 - stearic (18:0), 9 - oleic (18:1), 10 - linoleic (18:2), 11 - linolenic (18:3), 12 - arachidic (20:0), 13 - eicosenoic (20:1), 14 - behenic (22:0), 15 - eruic (22:1), 16 - lignoceric (24:0), 17 - nervonic (24:1)

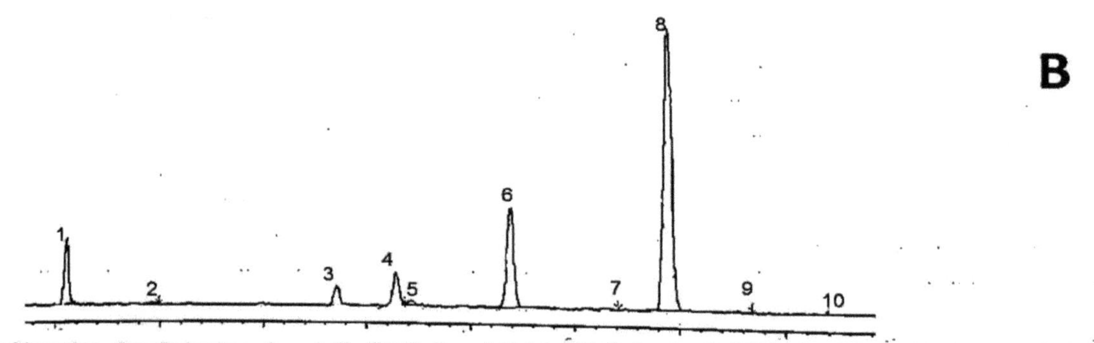

salvia hispanica oil methyl esters: 1 - palmitic (16:0), 2 - palmitoleic (16:1), 3 - stearic (18:0), 4 - oleic (18:1), 5 - wakcenic (18:1), 6 - linoleic (18:2), 7 - γ - linolenic (γ 18:3), 8 - α - linolenic (α 18:3), 9 - arachidic (20:0), 10 - eicosenic (20:1)

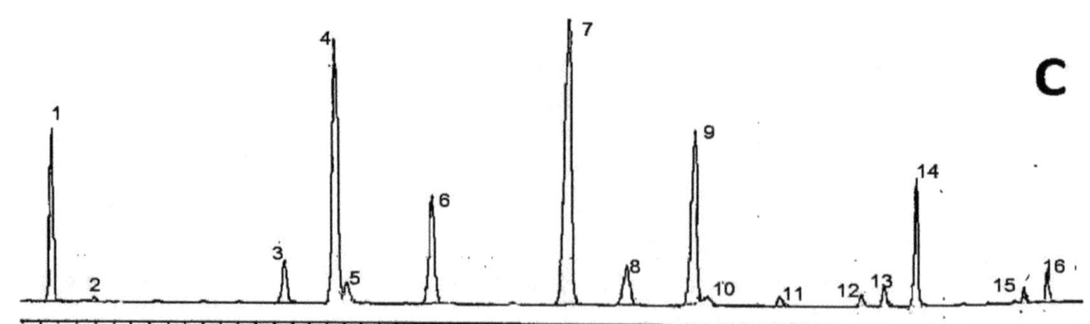

garden cress seeds oil methyl esters: 1 - palmitic (16:0), 2 - palmitoleic (16:1), 3 - stearic (18:0), 4 - oleic (18:1), 5 - wakcenic (18:1), 6 - linoleic (18:2), 7 - α - linolenic (α - 18:3), 8 - arachidic (20:0), 9 - eicosenic (c 20:1), 10 - eicosenic (t 20:1), 11 eicosadienic (20:2) 12 - eicosatrienic (20:3), 13 - behenic (22:0), 14 - eruic (22:1), 15 - lignoceric (24:0), 16 - nervonic (24:1)

Figure 1. Typical gas chromatogram of: A—rapeseed oil, B—Spanish sage seed oil, C—cress seed oil.

Table 4. Fatty acid composition of the oils (% of FA sum).

Fatty Acids	Rapeseed Oil	Spanish Sage Seed Oil	Cress Seed Oil
Myristic acid 14:0	0.25 ± 0.01	-	-
Palmitic acid 16:0	6.06 ± 0.18	7.46 ± 0.02	9.02 ± 0.17
Palmitoleic acid16:1	0.28 ± 0.01	0.15 ± 0.01	0.18 ± 0.01
Stearic acid 18:0	2.08 ± 0.09	2.79 ± 0.01	3.06 ± 0.05
Oleic acid 18:1	55.22 ± 0.85	4.71 ± 0.02	20.68 ± 0.02
Linoleic acid18:2	24.24 ± 1.13	20.80 ± 0.03	9.08 ± 0.01
α- linolenic acid 18:3 (n-3)	10.34 ± 0.91	63.40 ± 0.02	30.03 ± 0.25
γ- linolenic acid18-3 (n-6)	-	0.19 ± 0.01	-
Arachidic acid 20:0	0.27 ± 0.01	0.32 ± 0.01	3.64 ± 0.01
Eicosaenoic acid 20:1	1.00 ± 0.03	0.18 ± 0.01	13.83± 0.11
Eicosadienoic acid 20:2	-	-	0.54 ± 0.01
Eicosatrienoic acid 20:3	-	-	0.75 ± 0.01
Behenic acid 22:0	0.23 ± 0.01	-	1.11 ± 0.03
Erucic acid 22:1	0.03 ± 0.01	-	5.98 ± 0.12
Lignoceric acid 24:0	-	-	0.58 ± 0.01
Nervonic acid 24:1	-	-	1.52 ± 0.08
ΣSFA [1]	8.89	10.57	17.41
ΣMUFA [2]	56.53	5.04	42.19
ΣPUFA [3]	34.58	84.39	40.40
n-6/n-3	2.3	0.3	0.3

[1] SFA—saturated fatty acids; [2] MUFA—monounsaturated fatty acids; [3] PUFA—polyunsaturated fatty acids.

The oils used in the analyses had different values of oxidative stability (Table 3). The longest induction time was recorded for the rapeseed oil (4.40 h). It was similar to values reported by other authors [30,35]. The oxidative stability of vegetable oils is determined by the fatty acid composition: the greater the number of unsaturated bonds, the greater the susceptibility to oxidation. The chia seed oil exhibited the highest content of polyunsaturated fatty acids (over 84%) and hence the lowest oxidative stability (0.65 h).

The chemical properties of the oil mixtures are shown in Figures 2–4 and Table 5. The oils differed in their acid value and oxidative stability. The acid value ranged from 0.79 to 2.00 $mg \cdot g^{-1}$ and increased with higher amounts of the Spanish sage oil.

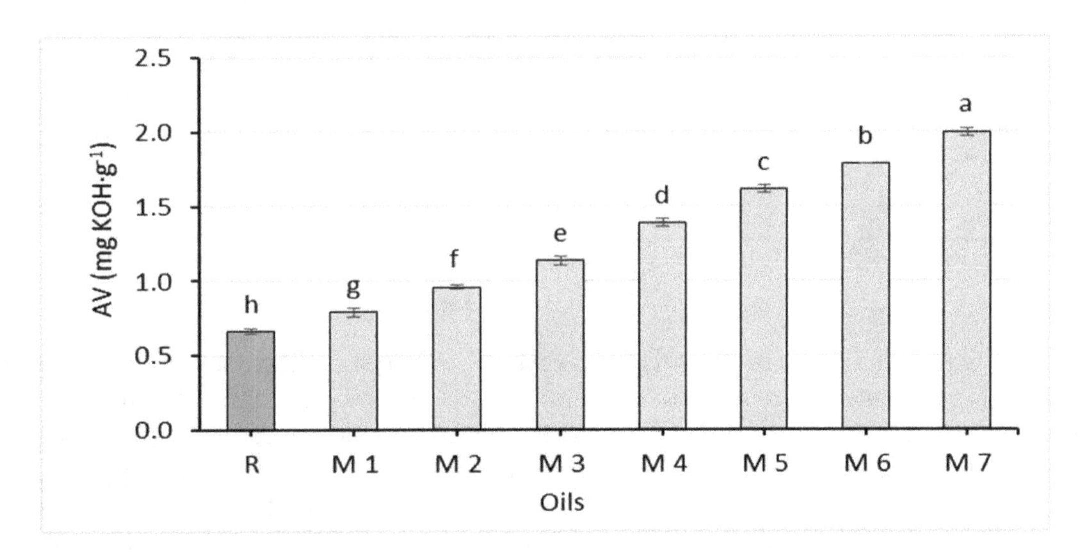

Figure 2. Comparison of the acid value of rapeseed oil (R) and oil mixtures (abc—statistically significant differences, $p \leq 0.05$).

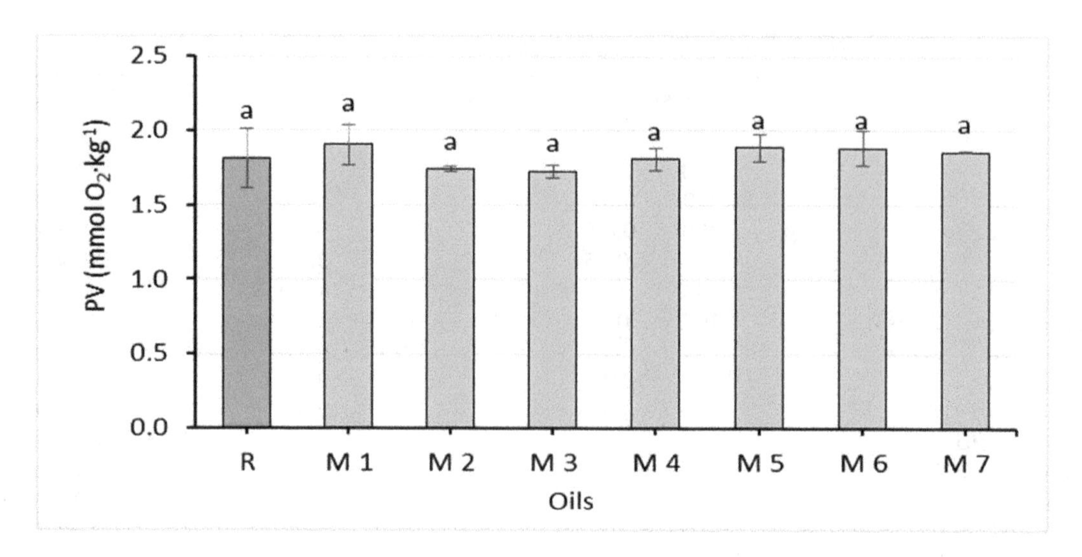

Figure 3. Comparison of the peroxide value of rapeseed oil (R) and oil mixtures (abc—statistically significant differences, $p \leq 0.05$).

Table 5. Fatty acid composition of the oil mixtures (% of FA sum).

Fatty Acids	Oil Mixtures						
	M1	M2	M3	M 4	M5	M6	M7
Myristic acid 14:0	0.13	0.18	0.18	0.18	0.18	0.16	0.18
Palmitic acid 16:0	7.01	6.73	6.62	6.85	6.68	6.60	6.45
Palmitoleic acid16:1	0.28	0.28	0.27	0.26	0.26	0.26	0.27
Stearic acid 18:0	2.27	2.18	2.23	2.35	2.34	2.12	2.33
Oleic acid 18:1	44.94	44.05	43.37	42.36	41.59	40.93	40.15
Linoleic acid 18:2	19.83	20.31	20.74	21.40	22.07	22.65	23.02
α- linolenic acid 18:3 (n-3)	16.18	17.82	19.72	21.26	22.92	24.61	26.30
γ- linolenic acid 18-3 (n-6)	-	0.01	0.02	0.03	0.04	0.05	0.06
Arachidic acid 20:0	1.29	1.11	1.02	0.75	0.61	0.45	0.29
Eicosaenoic acid 20:1	4.62	4.30	3.52	2.80	2.09	1.44	0.77
Eicosadienoic acid 20:2	0.16	0.14	0.11	0.08	0.05	0.03	-
Eicosatrienoic acid 20:3	0.23	0.19	0.15	0.12	0.09	0.04	-
Behenic acid 22:0	0.52	0.46	0.36	0.32	0.25	0.22	0.16
Erucic acid 22:1	1.81	1.69	1.25	0.92	0.62	0.33	0.02
Lignoceric acid 24:0	0.18	0.15	0.12	0.09	0.07	0.03	-
Nervonic acid 24:1	0.55	0.40	0.32	0.23	0.14	0.08	-
ΣSFA [1]	11.40	10.81	10.53	10.54	10.13	9.58	9.41
ΣMUFA [2]	52.20	50.72	48.73	46.57	44.70	43.04	41.21
ΣPUFA [3]	36.40	38.47	40.74	42.89	45.17	47.38	49.38
n-6/n-3	1.2	1.1	1.1	1.0	1.0	0.9	0.9

[1] SFA—saturated fatty acids; [2] MUFA—mono-unsaturated fatty acids; [3] PUFA—polyunsaturated fatty acids.

Since the oils used to prepare the mixtures did not differ significantly in their peroxide value, the values of this parameter were similar in the oil mixtures as well (1.73–1.91 mmol $O_2 \cdot kg^{-1}$). All mixtures met the requirements of acid and peroxide values for cold-pressed vegetable oils [28].

The fatty acid profile of the prepared oil mixtures reflected their raw material composition (Table 5). The highest content of saturated and mono-unsaturated fatty acids (11.40% and 52.20%, respectively) was determined for the M1 mixture consisting of 70% of rapeseed oil and 30% of cress seed oil, which was reflected in it showing the highest oxidative stability of all the oil mixtures. The induction time was only 0.73 h shorter than that of the rapeseed oil. The content of α-linolenic acid in the oil mixtures ranged from 16.18% to 23.02%.

The highest value for α-linolenic acid was determined for the M7 mixture containing 30% Spanish sage oil. Enrichment with this acid resulted in a substantially smaller ratio of n-6 to n-3 acids, i.e., in the range of 1.2 to 0.9, in the oil mixtures than in the rapeseed oil. As suggested by nutritional recommendations, there is a need to increase the intake of long-chain fatty acids from the n-3 group in the diet [36,37]. The oil mixtures proposed in this study can be used to enrich food with this component. According to the EU Commission Regulation of 2014 [38], the maximum allowable level of erucic acid in vegetable oils and fats is 50 $g \cdot kg^{-1}$, i.e., 5%. By mixing the cress seed oil (containing 5.98% of this acid) with the rapeseed and Spanish sage oils, the erucic acid content was reduced below 2% in all of the analyzed mixtures.

The oxidative stability of the oils, which was expressed as the induction time ranged from 3.67 to 1.92 h and decreased with the increasing content of the Spanish sage oil. However, there were no statistically significant differences in the values for this parameter between mixtures with 5 and 10%, 15 and 20%, as well as 20 and 25% of the chia seed oil (Figure 4).

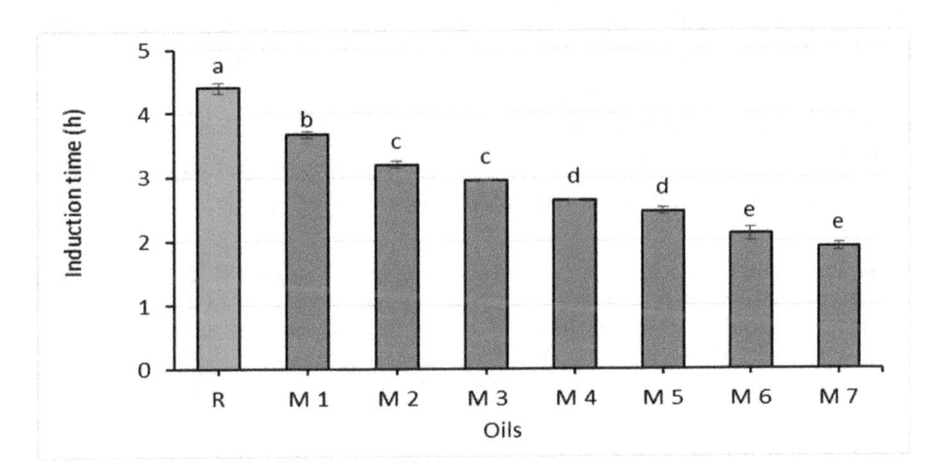

Figure 4. Comparison of the oxidative stability of rapeseed oil (R) and oil mixtures (abc—statistically significant differences, $p \leq 0.05$).

Fat oxidation susceptibility increases proportionally to the number of unsaturated bonds in individual fatty acids [30]. The dependence of oxidative stability of the experimental oil mixtures on the content of polyunsaturated fatty acids is presented graphically (Figure 5) as a linear function.

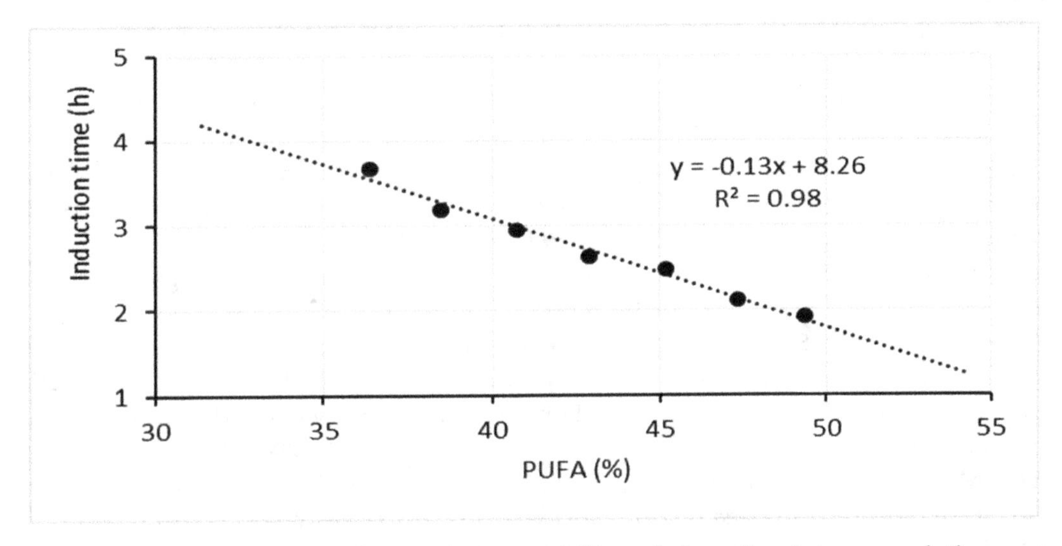

Figure 5. Correlation between the oxidative stability of the oil mixtures and the content of polyunsaturated fatty acids.

As shown by the equation, a 5% increase in the PUFA content in the oil mixtures reduces oxidative stability by shortening the induction time by 0.65 h.

4. Conclusions

The oil mixtures proposed in this study are characterized by higher contents of n-3 fatty acids compared to rapeseed oil, and can therefore improve the n-6 to n-3 ratio in the modern diet. This is beneficial in terms of consumers' health and complies with the principles of sustainable development of food production. The regression equation, developed using the study results, facilitates the precise determination of the properties of rapeseed, cress, and chia oil mixtures. The results indicated that a mixture of rapeseed, chia, and cress oils at a ratio of 70:10:20 exhibits higher oxidative stability than the individual enrichment raw materials, and a nearly 20% level of α-linolenic acid.

Furthermore, the mixture of cress seed oil with rapeseed and Spanish sage oils ensured reduced erucic acid content in the finished product. Hence, cress seeds, which are characterized by a high content of erucic acid can be used in the food industry, which is important for the sustainable development of agriculture and extensive use of its resources. This issue is also related to food security, which is one of the specific objectives of the strategy for sustainable development of rural areas, agriculture and fisheries for the years 2012–2020 in Poland. Within this objective, one of the priorities is the production of high-quality, agri-food products, which are safe for consumers [39].

Author Contributions: Conceptualization, A.S. (Agnieszka Sagan), D.A. and K.Z.; methodology, A.S. (Agnieszka Sagan), M.S. and D.A.; investigation, M.S., P.S. and A.S. (Agnieszka Starek); data curation, A.S. (Agnieszka Sagan), A.B.-K. and M.S.; writing—original draft preparation, A.S. (Agnieszka Sagan), A.B.-K. and M.S.; writing—review and editing, D.A., P.S. and K.Z.; visualization, A.S. (Agnieszka Sagan), A.B.-K. and A.S. (Agnieszka Starek); supervision, D.A.

References

1. Cencic, A.; Chingwaru, W. The role of functional foods, nutraceuticals, and food supplements in intestinal health. *Nutrients* **2010**, *2*, 611–625. [CrossRef] [PubMed]
2. Bigliardi, B.; Galati, F. Innovation trends in the food industry: The case of functional foods. *Trends Food Sci. Tech.* **2013**, *31*, 118–129. [CrossRef]
3. Ozen, A.E.; Pons, A.; Tur, J.A. Worldwide consumption of functional foods: A systematic review. *Nutr. Rev.* **2012**, *70*, 472–481. [CrossRef] [PubMed]
4. Tiwari, B.K.; Norton, T.; Holden, N.M. Introduction. In *Sustainable Food Processing*; Tiwari, B.K., Norton, T., Holden, N.M., Eds.; John Wiley & Sons: Hoboken, NJ, USA, 2014; pp. 1–7.
5. Obiedzińska, A.; Waszkiewicz-Robak, B. Oleje tłoczone na zimno jako żywność funkcjonalna. *Zywn Nauk. Technol. Ja.* **2012**, *1*, 27–44.
6. Orsavova, J.; Misurcova, L.; Ambrozova, J.V.; Vicha, R.; Mlcek, J. Fatty acids composition of vegetable oils and its contribution to dietary energy intake and dependence of cardiovascular mortality on dietary intake of fatty acids. *Int. J. Mol. Sci.* **2015**, *16*, 12871–12890. [CrossRef]
7. Gugała, M.; Zarzecka, K.; Sikorska, A. Prozdrowotne właściwości oleju rzepakowego. *Postępy Fitoter.* **2014**, *2*, 100–103.
8. Wang, Y.; Meng, G.; Chen, S.; Chen, Y.; Jiang, J.; Wang, Y.-P. Correlation analysis of phenolic contents and antioxidation in yellow- and black-seeded *Brassica Napus*. *Molecules* **2018**, *23*, 1815. [CrossRef]
9. Wroniak, M.; Łukasik, D.; Maszewska, M. Porównanie stabilności oksydatywnej wybranych olejów tłoczonych na zimno z olejami rafinowanymi. *Zywn Nauk. Technol. Ja.* **2006**, *1*, 214–221.
10. Krajewska, M.; Ślaska-Grzywna, B.; Szmigielski, M. The effect of the oregano addition on the chemical properties of cold-pressed rapeseed oil. *Przem. Chem.* **2018**, *97*, 1953–1956.
11. Starek, A.; Sagan, A.; Kiczorowska, B.; Szmigielski, M.; Ślaska-Grzywna, B.; Andrejko, D.; Kozłowicz, K.; Blicharz-Kania, B.; Krajewska, M. Effects of oleoresins on the chemical properties of cold-pressed rapeseed oil. *Przem. Chem.* **2018**, *97*, 771–773.

12. Marciniak-Łukasiak, K.; Zbikowska, A.; Krygier, K. Wpływ stosowania azotu na stabilność oksydacyjna mieszanin oleju rzepakowego z olejem lnianym. *Żywn Nauk Technol. Ja.* **2006**, *13*, 206–215.
13. Asif, M. Health effects of omega-3,6,9 fatty acids: *Perilla frutescens* is a good example of plant oils. *Orient Pharm. Exp. Med.* **2011**, *11*, 51–59. [CrossRef] [PubMed]
14. Kulczyński, B.; Kobus-Cisowska, J.; Taczanowski, M.; Kmiecik, D.; Gramza-Michałowska, A. The chemical composition and nutritional value of chia seeds—Current state of knowledge. *Nutrients* **2019**, *11*, 1242. [CrossRef] [PubMed]
15. Zia-Ul-Haq, M.; Ahmad, S.; Calani, L.; Mazzeo, T.; del Rio, D.; Pellegrini, N.; de Feo, V. Compositional study and antioxidant potential of *ipomoea hederacea* jacq. and *lepidium sativum* l. seeds. *Molecules* **2012**, *17*, 10306–10321. [CrossRef] [PubMed]
16. Grimes, S.J.; Phillips, T.D.; Hahn, V.; Capezzone, F.; Graeff-Hönninger, S. Growth, yield performance and quality parameters of three early flowering chia (*Salvia hispanica* l.) genotypes cultivated in southwestern Germany. *Agriculture* **2018**, *8*, 154. [CrossRef]
17. Guevara-Cruz, M.; Tovar, A.R.; Aguilar-Salinas, C.A.; Medina-Vera, I.; Gil-Zenteno, L.; Hernandez-Viveros, I.; Lopez-Romero, P.; Ordaz-Nava, G.; Canizales-Quinteros, S.; Guillen Pineda, L.E.; et al. A dietary pattern including nopal, chia seed, soy protein, and oat reduces serum triglycerides and glucose intolerance in patients with metabolilic syndrome. *J. Nutr.* **2012**, *142*, 64–69. [CrossRef]
18. Nowak, K.; Majsterek, S.Z.; Ciesielska, N.; Sokołowski, R.; Klimkiewicz, K.; Zukow, W. The role of chia seeds in nutrition in geriatric patients. *J. Educ. Health Sport* **2016**, *6*, 35–40.
19. Fan, Q.L.; Zhu, Y.D.; Huang, W.H.; Qi, Y.; Guo, B.L. Two new acylated flavonol glycosides from the seeds of *Lepidium sativum*. *Molecules* **2014**, *19*, 11341–11349. [CrossRef]
20. Sakran, M.; Selim, Y.; Zidan, N. A new isoflavonoid from seeds of *Lepidium sativum* L. and its protective effect on hepatotoxicity induced by paracetamol in male rats. *Molecules* **2014**, *19*, 15440–15451. [CrossRef]
21. PN-EN ISO 665:2004. *Nasiona Oleiste. Oznaczanie Wilgotności i Zawartości Substancji Lotnych*; Polski Komitet Normalizacyjny: Warsaw, Poland, 2004.
22. PN-EN ISO 659:2010. *Nasiona Oleiste. Oznaczanie Zawartości Oleju (Metoda Odwoławcza)*; Polski Komitet Normalizacyjny: Warsaw, Poland, 2010.
23. PN-EN ISO 660:2010. *Oleje i Tłuszcze Roślinne oraz Zwierzęce—Oznaczanie Liczby Kwasowej i Kwasowości*; Polski Komitet Normalizacyjny: Warsaw, Poland, 2010.
24. PN-EN ISO 3960:2012. *Oleje i Tłuszcze Roślinne oraz Zwierzęce—Oznaczanie Liczby Nadtlenkowej—Jodometryczne (Wizualne) Oznaczanie Punktu Końcowego*; Polski Komitet Normalizacyjny: Warsaw, Poland, 2012.
25. Mathäus, B. Determination of the oxidative stability of vegetable oils by rancimat and conductivity and chemiluminescence measurements. *J. Am. Oil Chem. Soc.* **1996**, *73*, 1039–1043.
26. PN-EN ISO 5508:1996. *Oleje i Tłuszcze Roślinne oraz Zwierzęce—Analiza Estrów Metylowych Kwasów Tłuszczowych Metodą Chromatografii Gazowej*; Polski Komitet Normalizacyjny: Warsaw, Poland, 1996.
27. PN-EN ISO 5509:2001. *Oleje i Tłuszcze Roślinne oraz Zwierzęce—Przygotowanie Estrów Metylowych Kwasów tłuszczowych*; Polski Komitet Normalizacyjny: Warsaw, Poland, 2001.
28. FAO/WHO. Codex Standard for Named Vegetable Oils. In *Codex Alimentarius*; ALINORM: Santa Croce, Italy, 2009.
29. Krygier, K.; Wroniak, M.; Grześkiewicz, S.; Obiedziński, M. Badanie wpływu zawartości nasion uszkodzonych na jakość oleju rzepakowego tłoczonego na zimno. *Oilseed Crops* **2000**, *41*, 587–596.
30. Cichosz, G.; Czeczot, H. Stabilność oksydacyjna tłuszczów jadalnych—Konsekwencje zdrowotne. *Bromatol. Chem. Toksyk.* **2011**, *44*, 50–60.
31. Krajewska, M.; Zdybel, B.; Andrejko, D.; Ślaska-Grzywna, B.; Tańska, M. Właściwości chemiczne wybranych olejów tłoczonych na zimno. *Acta Agroph.* **2017**, *24*, 579–590.
32. Segura-Campos, M.R.; Ciau-Solis, N.; Rosado-Rubio, G.; Chel-Guerrero, L.; Betancur-Ancona, D. Physicochemical characterization of chia (*Salvia hispanica*) seed oil from Yucatán, México. *Agric. Sci.* **2014**, *5*, 220–226.
33. Gokavi, S.S.; Malleshi, N.G.; Guo, M. Chemical composition of garden cress (*Lepidium sativum*) seeds and its fractions and use of bran as a functional ingredient. *Plant Food Hum. Nutr.* **2004**, *59*, 105–111. [CrossRef]
34. Moser, B.R.; Shah, S.N.; Winkler-Moser, J.K.; Vaughn, S.F.; Evangelista, R.L. Composition and physical properties of cress (*Lepidium sativum* L.) and field pennycress (*Thlaspi arvense* L.) oils. *Ind. Crop. Prod.* **2009**, *30*, 199–205. [CrossRef]

35. Kruszewski, B.; Fąfara, P.; Ratusz, K.; Obiedziński, M. Ocena pojemności przeciwutleniającej i stabilności oksydacyjnej wybranych olejów roślinnych. *Zesz. Probl. Post. Nauk Roln.* **2013**, *572*, 43–52.
36. Marciniak-Łukasik, K. Rola i znaczenie kwasów tłuszczowych omega-3. *Zywn Nauk. Technol. Ja.* **2011**, *6*, 24–35.
37. Sawada, N.; Inoue, M.; Iwasaki, M.; Sasazuki, S.; Shimazu, T.; Yamaji, T.; Takachi, R.; Tanaka, Y.; Mizokami, M.; Tsugane, S. Consumption of n-3 fatty acids and fish reduces risk of hepatocellular carcinoma. *Gastroenterology* **2012**, *142*, 1468–1475. [CrossRef]
38. Commission Regulation (EU) No 696/2014 of 24 June 2014 amending Regulation (EC) No 1881/2006 as regards maximum levels of erucic acid in vegetable oils and fats and foods containing vegetable oils and fats. *Off. J. EU* **2014**, *L184*, 1–2.
39. Żmija, D. Zrównoważony rozwój rolnictwa i obszarów wiejskich w Polsce. *Studia Ekonomiczne* **2014**, *166*, 149–158.

11

Effect of Press Construction on Yield and Quality of Apple Juice

Kamil Wilczyński [1], Zbigniew Kobus [2,*] and Dariusz Dziki [3]

1 Department of Food Engineering and Machines, University of Life Sciences postcode, 20-612 Lublin, Poland
2 Department of Technology Fundamentals, University of Life Sciences, 20-612 Lublin, Poland
3 Department of Thermal Technology and Food Process Engineering, University of Life Sciences, 20-612 Lublin, Poland
* Correspondence: zbigniew.kobus@up.lublin.pl

Abstract: The paper presents the possibility of applying different press constructions for juice extraction in small farms. The research was carried out with three different varieties of apples, namely, Rubin, Mutsu, and Jonaprince. Two types of presses were tested: a basket press and a screw press. Generally, application of the screw press makes it possible to obtain a higher yield of extraction compared to the basket press. In our study, the differences in the pressing yield among press machines also depended on the apple variety used. The juices obtained on the screw press were found to be of a higher quality characterized by a higher content of soluble solids, higher viscosity, higher total content of polyphenols, higher antioxidant activity, and lower acidity. Thus, the selection of an appropriate press is the key to producing high-quality apple juice with health-promoting properties for manufacturers of apple juice at the local marketplace.

Keywords: sustainable production; screw press; basket press; polyphenols; antioxidant activity; texture properties

---

1. Introduction

Apples (*Malus domestica*) are the most commonly used fruit for juice extraction in the European Union (EU). They are a rich source of nutrients and polyphenols and possess antioxidant properties that have beneficial effects on human health [1,2]. The yield of apples is estimated to be 12.59 million tons per year in Europe and 3.6 million tons in Poland [3].

Apples are mainly processed into concentrates, which contributes to the reduction of volume and facilitates storage. Poland is the largest producer and exporter of concentrated juices extracted from fruits grown in the temperate zone of the EU [4]. The European share of the juice yield was estimated to account for 66% in 2016.

Currently, there is a growing trend towards healthy eating and an increased interest in ecological produced food. This in turn contributes to the increasing interest of scientists in functional foods and new methods of production to preserve their quality and high level of bioactive compounds [5]. Some examples of these kinds of foods include juices that are not obtained from concentrate (NFC) and freshly squeezed non-pasteurized juices (FS). These juices are obtained from the fruit tissue by pressing and centrifugation of the pulp. Cloudy juices are classified as products obtained with a low degree of processing. They contain higher amounts of bioactive compounds, such as polyphenols or flavonoids, than clarified ones due to the omission of enzymatic and clarifying treatments. They are also richer in dietary fibre, which is necessary for the proper functioning of the digestive system, and several mineral compounds [6]. In their studies, Paepe et al. [7] and Markowski et al. [8] showed that cloudy juices contain substantially higher amounts of beneficial compounds such as polyphenols and also exhibit considerably higher antioxidant activity than clarified juices. Additionally, apple pomace obtained after low-degree processing can be used as animal feed after drying or pickling. The natural nutrients contained in the pomace can serve as a valuable source of nutrition for animals in organic farms [9].

Fruit juices are extracted on an industrial scale using different devices, depending on the type of operation (periodic or continuous) and raw material [10]. Many different press construction solutions are used in the industry for obtaining apple juice. Among them, belt press, water press [11], decanter, rack-and-frame press, hydraulic press [12], basket press [13], and screw press [14] are the most distinguished.

A basket press ensures high pressing yields of up to 60%, but it is less often used on an industrial scale. In one study, Nadulski [15] confirmed the usefulness of the basket press in juice extraction from Jonagold apples. In another study, the same author and his group [10] showed that not all apple cultivars are suitable for juice pressing under farm conditions. This suggests that the use of a suitable fruit variety for juice pressing may reduce the cost of pressing and increase the quality parameters of the juice produced.

A screw press is mainly used for oil extraction from oilseeds such as rape [16,17], flax [18], sunflower [19], cumin [20], tobacco seeds [21], and pistachio nuts [22]. Currently, screw presses are becoming increasingly popular, especially on small farms, among local entrepreneurs and for domestic applications. Based on the construction solution, screw presses are classified as single- and double-screw type. These presses are characterized by several extracting processes, where the pulp is subjected to energetic grinding and mixing. The main advantage of a screw press is that the juice obtained has a much higher amount of soluble solids and bioactive compounds [23,24].

The programmes implemented by the EU such as 'Agricultural and rural development 2014–2020' and 'Promotion of farm products' help producers develop new products based on the policy of sustainable agriculture [25,26]. These programmes allow appropriate usage of resources (raw material, energy) and increase the effectiveness of production without affecting the environment [27]. The protection and promotion of regional and traditional products is one of the most important factors supporting the sustainable development of rural areas (as it increases the income of agricultural producers, prevents depopulation and enhances the attractiveness of rural areas). In addition, this may be considered as potentially influencing the development of agricultural products. In the context of sustainable agriculture, the use of a new press construction may allow the farmers to obtain juices with higher quality and a large amount of bioactive compounds. Moreover, fresh pressing of apple juice may help open new markets.

Considering the above, it appears viable that local farms can use small presses for juice production. In this view, the aim of the present study is to compare the efficiency of screw and basket presses in the extraction of apple juice. This includes the determination of parameters affecting the juice quality such as the content of soluble solids, pH, viscosity, total phenolic content (TPC) and antioxidant activity.

2. Materials and Methods

The research material included three varieties of apples, namely Rubin, Mutsu, and Jonaprince, all obtained from the 2017 harvest. The fruits were delivered by Groups of Fruit Producers, which is related to the company Rylex Sp. z o.o with office registered in the village of Błędów (near Grójec, Poland; GPS coordinates: 51°47′N, 20°42′E), to the local Auchan store in Lublin, Poland.

The experimental flowchart is presented in Figure 1.

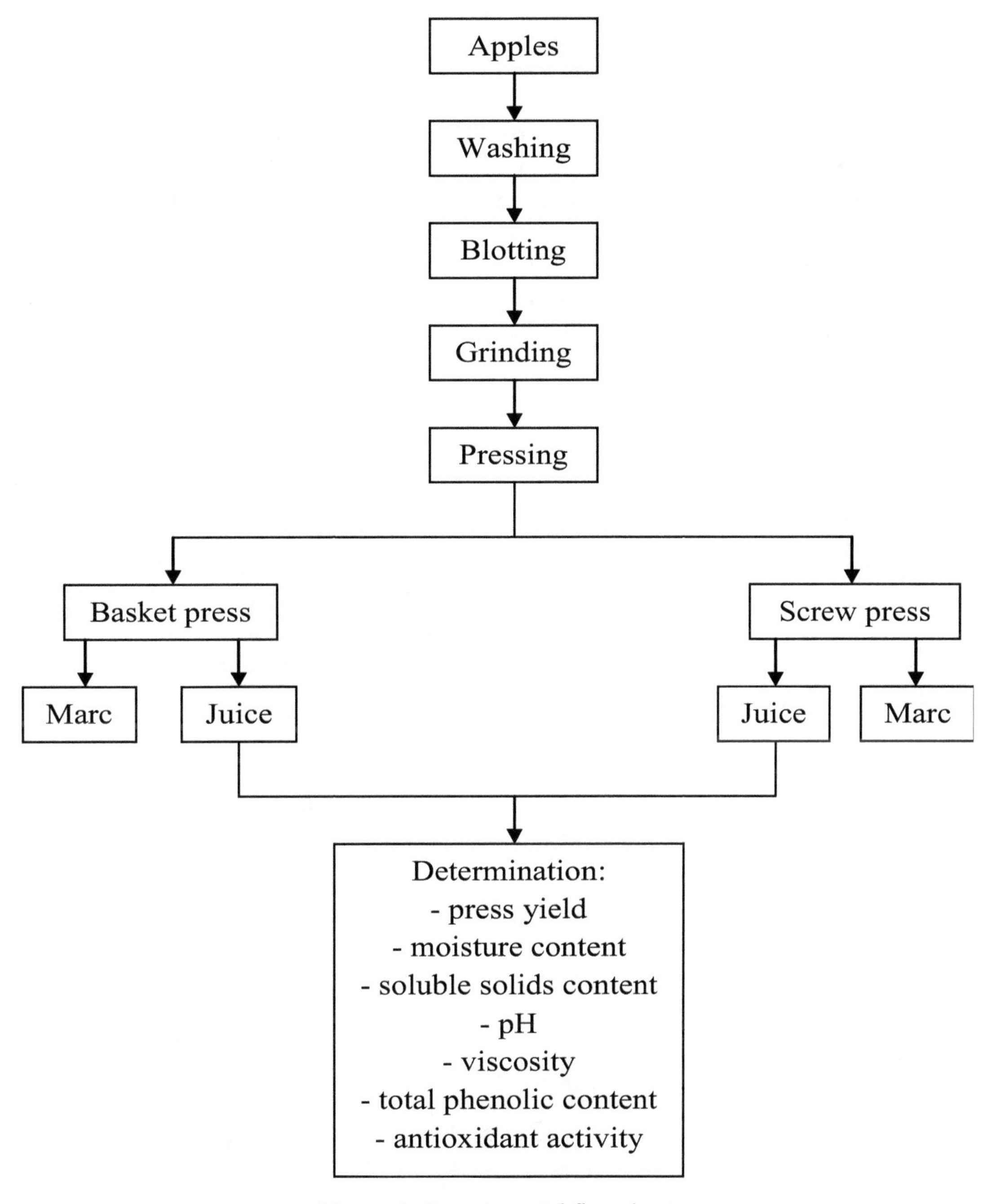

Figure 1. Experimental flowchart.

2.1. *Washing and Blotting*

The apples were washed in tap water and blotted dry using laboratory tissue paper.

2.2. *Grinding*

The apples were crushed using a shredding machine (MKJ250; Spomasz Nakło, Poland) equipped with a standard grating disc with 8 mm holes. The rotational speed of the disc was set at 170 rpm. The obtained mash was divided into portions weighing 300 g and placed in plastic containers.

2.3. *Determination of Moisture Content*

The moisture content of the apple mash was measured before pressing by drying 3 g of mash at a temperature of 105 °C for 3 hours [28]. The measurements were carried out in five replicates.

2.4. *Analysis of Texture Properties*

The texture of the apples was examined using a texture analyser (model TA.XT Plus Texture Analyzer) equipped with a measuring head that has a working range of up to 0.5 kN. A double-compression test and a cutting test were performed. The speed of the measuring head was set at 0.83 mm·s^{-1} for the double-compression test (TPA) and 1 mm·s^{-1} for the cutting test. For TPA, the apple samples with skin were cut into cylinders of 15 mm diameter and 10 mm height and were then placed with their diameter dimension parallel to the device base. Hardness was defined as the maximum force recorded during the first compression cycle (Fh). For the cutting test, a Warner–Bratzler blade was used. The samples with skin were cut into cylinders of 20 mm diameter and 20 mm height. The maximum value of the cutting force (Fc) was determined. The texture analysis was repeated ten times.

2.5. *Pressing*

Two kinds of presses were used to extract apple juices: basket press and screw press.

2.5.1. Types of Presses

The basket press (Figure 2) consisted of a perforated cylinder with holes of 3 mm diameter, piston, construction frame, and hydraulic system (UHJG 20/C/2; Hydrotech, Lublin, Poland), which allowed for maintaining a pressure of 4.5 MPa. The press was equipped with a tensometric sensor system for measuring pressure (EMS50; WObit, Poznan, Poland) combined with a digital recorder (MG-TAE1; WObit, Poznan, Poland).

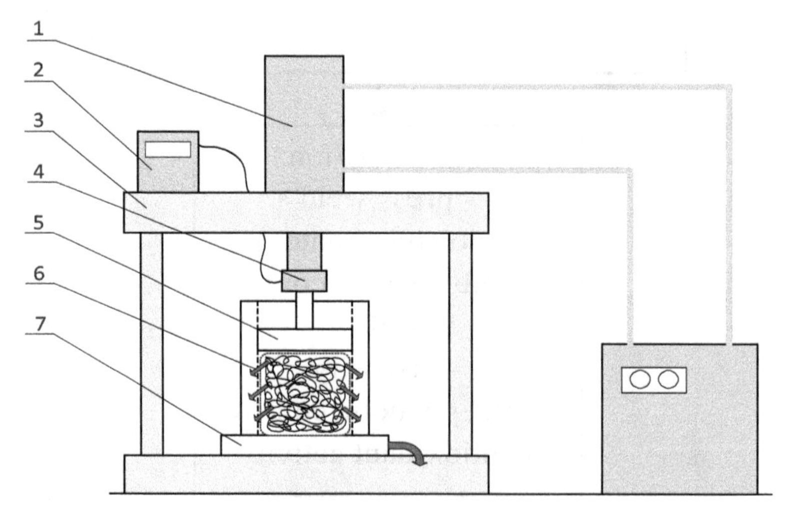

Figure 2. Construction and principle underlying the operation of the hydraulic press: 1 – hydraulic ram, 2 – measurement system, 3 – frame, 4 – tensometer, 5 – piston, 6 – cylinder and 7 – base.

The screw press used was a twin-screw type (Green Star Elite 5000; Tribest), which had a rated power of 260 W and was equipped with a sieve with holes of 0.4/0.5 mm. The press has two rotating gears. The dual stainless steel gears contain magnets and utilize bioceramic technologies that pull more nutrients into the juice. The press gears at a low 110 rpm and generates minimal heat while juicing.

2.5.2. Extraction Procedure

The sample materials weighed 300 g. They were pressed in ten replicates in each type of press.

In the case of the basket press, the material was put in a special bag made of pressing cloth, placed in the cylinder of the press, and subjected to the force applied by the piston. Once a pressure of 4.0 ± 0.1 MPa was reached, the extraction process was stopped.

In the case of the screw press, the material was directly put into the press chamber.

The extracted juice was collected in plastic containers, filtered using a Whatman No. 1 filter and stored in a refrigerator at a temperature of 4 °C.

2.5.3. Pressing Yield

The efficiency of pressing was calculated using the formula:

$$W_j(\%) = \frac{M_j}{M_i} \cdot 100 \quad (1)$$

where

W_j is the efficiency of pressing (%),
M_j is the mass of juice after pressing (kg), and
M_i is the mass of input material (kg).
The calculation was performed for all ten replicated samples.

2.6. Determination of Soluble Solids Content (Brix) and pH

After each extraction, the weight of the obtained juices was measured. The pH [29] and content of soluble solids [30] were also calculated. To determine the pH of the juices, a CP-411 pH-meter (Elmetron, Zabrze, Poland) was used. The content of soluble substances in fruit juices was measured using a PAL-3 refractometer (Atago, Tokyo, Japan). The parameters were determined in three replicates for each sample of extracted juice.

2.7. Determination of Viscosity

Viscosity was measured using a Brookfield viscometer (model LVDV-II + PRO; Brookfield Engineering Laboratories) with Rheolac 3.1 software. A 16 mL sample of juice was taken in ULA – 10EY-baker for all experiments. The spindle speed was set at 10–80 rpm, which corresponded to a shear rate of 12.24–97.84$\cdot s^{-1}$. The temperature was kept constant at 20 °C using a water bath (Brookfield TC-502P). The measurement was performed in three replicates for each sample.

2.8. Determination of TPC

The total phenolic content (TPC) of apple juices was determined according to the FC method [31] with slight modification. Gallic acid was used as the standard and was diluted with methanol (1 mg/1 mL) to give the appropriate concentrations required for plotting a standard curve. First, the sample extract (0.2 mL) was mixed with 2 mL of methanol in a 25 mL volumetric flask. Then, Folin–Ciocalteu reagent (2 mL, diluted 1:10) was added and allowed to react for 3 minutes. Next, 2 mL of Na_2CO_3 solution were added, and the mixture was made up to 25 mL with distilled water. After leaving the mixture for 30 minutes at room temperature in the dark, the absorbance at 760 nm was measured using a spectrophotometer (UV-1800; Shimadzu, Japan). The results were expressed as mg gallic acid equivalent per 100 mL of fresh juice (mg GAE 100 mL^{-1}). The measurement was performed in three replicates for each sample.

2.9. Determination of Antioxidant Activity

The antioxidant activity of apple juices was evaluated using DPPH (2,2-diphenyl-1-picrylhydrazyl) assay. For this analysis, 0.2 mL of apple juice was mixed with an aliquot of 5.8 mL of freshly prepared $6 \cdot 10^{-5}$ M DPPH radical in methanol. After allowing to stand for 30 minutes at room temperature, the spectrophotometric absorbance of the juice at 516 nm was measured using methanol as a blank. The measurement was performed in three replicates for each sample. Antioxidant activity was expressed as percentage inhibition of the DPPH radical calculated using the following equation [32]:

$$AA(\%) = \frac{\text{Absorbanceofcontrol} - \text{Absorbanceofsample}}{\text{Absorbanceofcontrol}} \cdot 100 \quad (2)$$

2.10. Statistical Analysis

Statistical analysis of the data was performed with Statistica software (Statistica 12; StatSoft Inc., Tulsa, OK, USA) using analysis of variance for factorial designs. The significance of differences was tested using Tukey's least significant difference (LSD) test ($p \leq 0.05$) and using T-test ($p \leq 0.05$) for data in Tables 1 and 2.

Table 1. Total soluble solids (Brix) of the apple juice depending on the type of press machine.

Variety	Solid Soluble Content (in Brix)	
	Basket Press	Screw Press
Rubin	11.9 ± 0.00a	13.1 ± 0.00b
Mutsu	12.9 ± 0.00a	12.9 ± 0.00a
Jonaprince	11.2 ± 0.00a	11.9 ± 0.00b

a, b and c – average values in the row marked with the same letter are not statistically significantly different (T-test, $p \leq 0.05$).

Table 2. Acidity (pH) of the apple juice depending on the type of press machine.

Variety	Acidity (pH)	
	Basket Press	Screw Press
Rubin	3.61 ± 0.012a	3.45 ± 0.125b
Mutsu	3.72 ± 0.008a	3.62 ± 0.08b
Jonaprince	3.81 ± 0.008a	3.73 ± 0.012a

a, b and c – average values in the row marked with the same letter are not statistically significantly different (T-test, $p \leq 0.05$).

3. Results and Discussion

3.1. Moisture Content of Fresh Apples

The moisture content of the apples ranged from 84.0% to 85.27%, and no significant differences were observed between the tested varieties. Moisture content is an important parameter to be determined to carry out further investigations because it influences the yield of juice processing.

3.2. Effect of the Press Construction on the Yield of Pressing

The effect of the type of press on pressing yield is shown in Figure 3.

The yield of pressing ranged from 61.9% to 71.6%. Generally, higher pressing yields were obtained with screw press. In addition, statistical analysis showed significant differences in the efficiency of juice extraction from the pulp of the different varieties of apples. A statistically significant effect of the press type on the pressing yield was observed in the case of the Mutsu and Jonaprince varieties, whereas no significant differences were noted in the case of Rubin.

In the case of the screw press, pieces of apple are fed into the cylinder and thrown to the perforated wall by the centrifugal action of the gears. These gears crush, cut, and squeeze the pieces at the same time. This maximizes the yield as well as the quality of juice. The yield of the recovered juice depends on the diameter of the perforations, the speed of rotation of the gears and the gap between the knob and the pulp discharge casing.

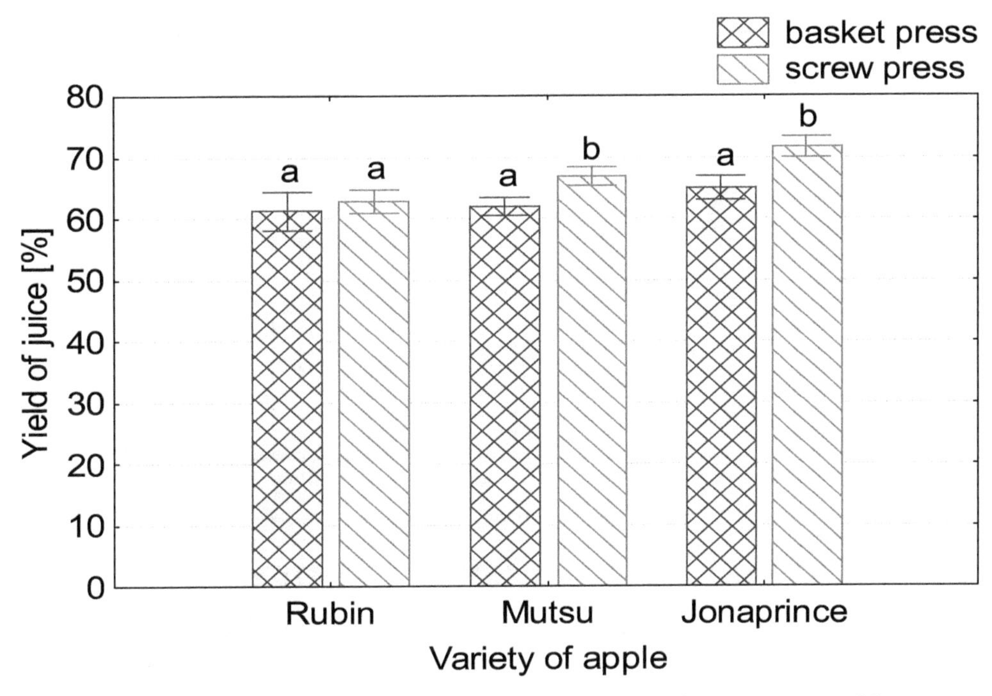

Figure 3. Effect of the type of press machine on the pressing yield.

In the case of the basket press, the crushed material is wrapped in the pressing cloth and gradually compressed by the hydraulic piston. The pressing cloth serves both as a package for the mash and as a filter for the juice.

In the case of the screw press, the correct course of the process is influenced by the texture of apples. Apples with a greater hardness are better for juice extraction on screw press, whereas those with a lower hardness are shredded to very small-sized particles, creating a layer of mousse inside the chamber, which clogs the sieve openings. This leads to a drop in the pressing yield or the passage of the mousse to the liquid phase.

To support the hypothesis about the relationship between hardness and yield of pressing, the texture properties of the apples were analysed.

Figures 4 and 5 show the hardness and cutting force of the tested apple varieties, respectively.

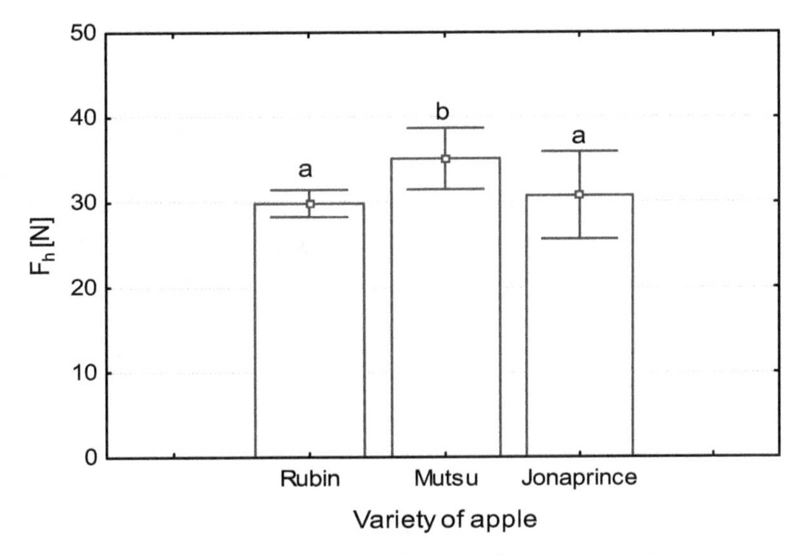

Figure 4. Hardness of the tested apple varieties.

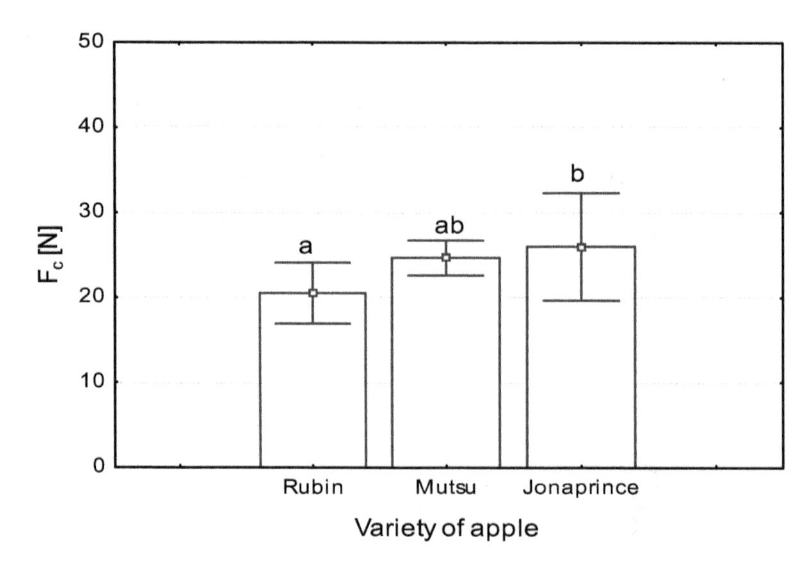

Figure 5. Cutting force of the tested apple varieties.

The texture analysis showed that, among the studied varieties, Rubin had the lowest hardness and cutting force. Probably due to these features, no statistically significant differences in the pressing yield were observed between the basket press and the screw press for this cultivar.

High efficiency of pressing was also reported by Takenaka et al. [33] when screw press was used for the extraction of citrus juice. In their experiment, the authors demonstrated that, of the three tested presses (belt, centrifugal and screw press), screw press provided the highest yield of pressing. The extraction yield obtained with the screw press was approximately 54% higher as compared to the belt press and approximately 18% higher as compared to the centrifugal extractor.

3.3. Effect of the Press Construction on the Soluble Solids Content

The content of soluble solids mainly reflects the amount of sugars and organic acids and is a very important parameter of juice quality. A study conducted by Eisele and Drake [34] showed that the content of soluble solids in juices obtained from various apple cultivars varied from 10.26 to 21.62 Brix.

In the present study, the content of soluble solids in the juices ranged from 11.2 to 13.1 °Brix (Table 1). This is similar to the content of soluble solids in the pressure-extracted apple juice produced in the Polish fruit and vegetable industry (i.e., 11.0–12.4 Brix) [35].

Generally, the juices obtained on the screw press had statistically significantly higher content of soluble solids than the ones obtained on the basket press. This was due to the wider opening of cell membranes and release of greater amounts of the deep-seated nutrients.

It was noted that the press type had no statistically significant effect on the content of soluble solids in the case of the Mutsu variety, probably due to the structure of the apple tissue. In fact, the flesh of the Mutsu apples is characterized by a coarse structure, while the Rubin and Jonaprince apples have a fine-grained structure. It was likely that the basket press enabled the breakdown of coarse-grained apples to the same extent as the screw press. Therefore, no effect of the type of press on the extract content was found in the case of the Mutsu variety. On the other hand, due to the wider opening of the cell membranes, the screw press enables a better breakdown of fine-grained structure. Therefore, there were differences in the content of soluble solids between presses in the case of the Rubin and Jonaprince varieties.

3.4. Effect of the Press Construction on Acidity

The apple juices were characterized by different levels of acidity (Table 2).

The pH of the juices obtained in this study ranged from 3.45 to 3.81. According to the literature, acidity (pH) of the apple juice varies from 3.37 to 4.24 [34]. Kobus et al. [14] found that the pH of the apple juice obtained on the screw press was in the range of 3.66 ± 0.09.

The press construction had a significant effect on the pH of the tested apple juice. It was observed that the juice extracted on the screw press had a slightly higher acidity (lower pH) compared to the juice from the basket press. The acidity was higher by 2.2%, 2.8%, and 4.6% for Jonaprince, Mutsu, and Rubin, respectively. A probable reason for the higher acidity of the juices from the screw press was increased migration of microelements, organic acids, and secondary plant metabolites such as polyphenols caused by the greater disintegration of membranes and cell walls.

Acidity is one of the most important traits of freshly squeezed apple juice. It influences the flavour, clarity, colour, aroma, and overall sensory satisfaction [36]. In addition, high acidity acts as a natural barrier against contamination by most microorganisms [37]. Consequently, it is reasonable to employ the screw press to produce juice with relatively high acidity.

3.5. Effect of the Press Construction on Viscosity

The viscosity of the juices obtained in this study ranged from 3.15 to 4.35 mPa·s (Figure 6). The viscosity of cloudy apple juice showed a wide range from 1.7 to 9.6 mPa·s, depending on the apple variety, method of processing, and concentration of soluble solids. Genovese and Lozano [38] found that the viscosity of apple juice from Granny Smith cv. ranged between 1.71 mPa·s at 10 °Brix and 3.65 mPa·s at 20 °Brix. Will et al. [39] reported that the values of viscosity ranged between 1.74 (Topaz cv.) and 2.15 mPa·s (Boskoop cv.), and the values determined by Teleszko et al. [40] ranged between 2.40 and 9.60 mPa·s.

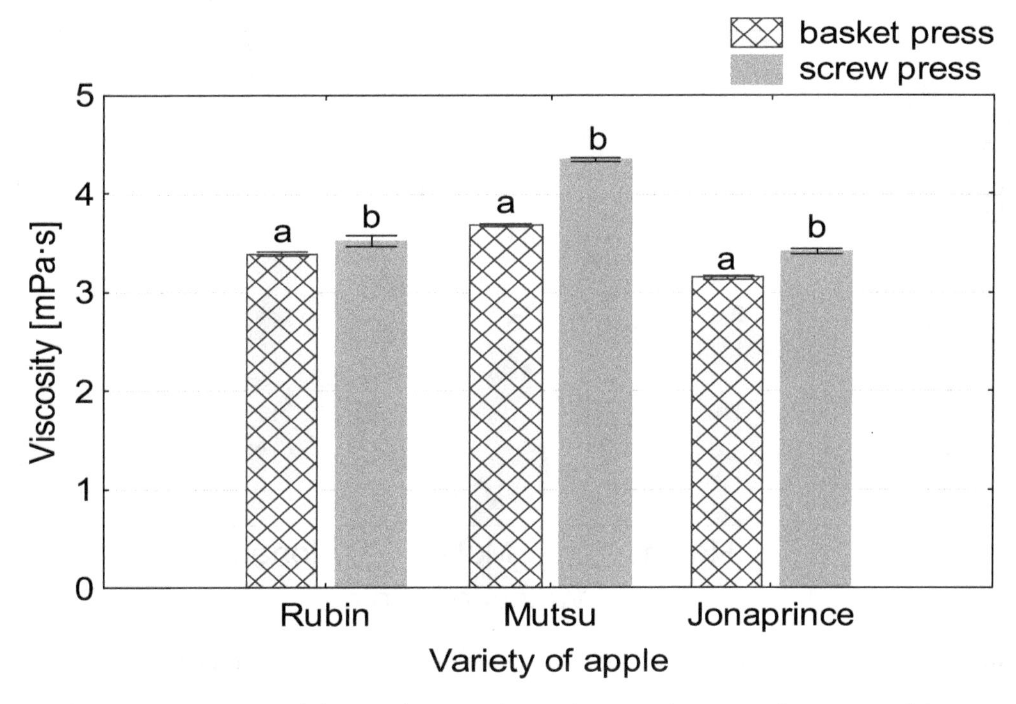

Figure 6. Viscosity of the apple juice depending on the type of press machine.

In the present study, the press construction was found to have a significant effect on the viscosity of the apple juice. In the case of Rubin, the viscosity of apple juice obtained on the screw press was 3.8% higher compared to the juice from the basket press. In the case of the Jonagold, the difference was greater and amounted to 8.3%, while the highest difference amounting to 18.2% was observed in the case of Mutsu.

The rheological behaviour of a cloudy juice is governed by both liquid viscosity and size characteristics of the solids [41].

In the case of clear juice, the viscosity depends on the content of soluble solids, with sugars playing the main role [42]. The juice from the Jonaprince variety obtained on the basket press was characterized by the lowest content of soluble solids and lowest viscosity. By contrast, the juice obtained on the screw press had a higher content of soluble solids, which probably resulted in higher juice viscosity. Thus, in the case of Jonaprince, the higher viscosity of the juice obtained from the screw press could have resulted from the greater disintegration and release of soluble phytochemical ingredients from apples.

There were no statistically significant differences found in the content of soluble solids in the juices obtained from the Mutsu cultivar between the tested presses. The higher viscosity observed in the case of this variety may be caused by the higher amount of total suspended solids in the juice obtained on the screw press, and due to the use of the pressing cloth, which serves both as a package for the mash and as a filter for the juice, for extraction on the basket press.

In the case of Rubin, the higher viscosity of the juice obtained on the screw press was mainly due to the higher content of soluble solids.

3.6. Effect of the Press Construction on TPC

Polyphenols play an important role in fruit juices because they influence the colour and flavour [11]. The content of polyphenols in juices varies depending on both the fruit variety and production technique used.

In the present study, the content of polyphenols in the juices ranged from 29.89 to 60.96 mg GAE 100 mL^{-1} (Figure 7).

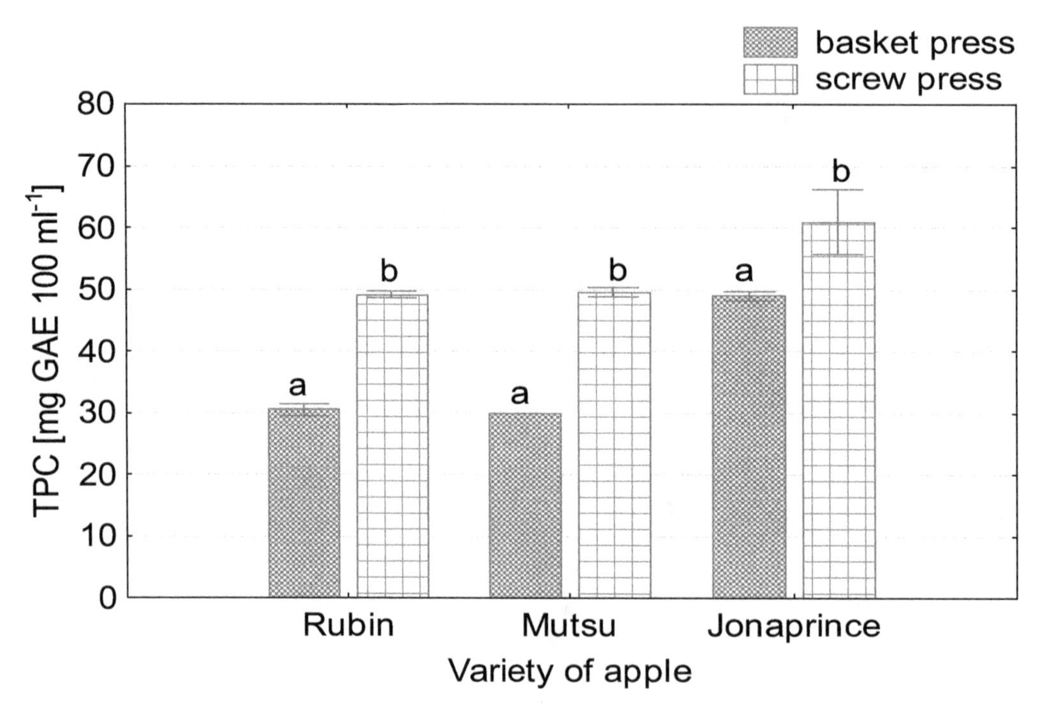

Figure 7. Total phenolic content of the apple juice depending on the type of press machine.

The TPC in juices produced in Europe varies within a broad range from 10 to 300 mg GAE 100 mL^{-1} [43]. Kobus et al. [14] showed that the content of polyphenols in juices obtained on the screw press ranged from 44.2 to 58.3 mg GAE 100 mL^{-1}.

In this study, the statistical analysis revealed a significant effect of the press type on the content of polyphenols in the obtained juices. The juices from the screw press were statistically significantly richer in polyphenols than the juices from the basket press. The higher content of polyphenols in the juice from the screw press was related to the higher percentage of solid particles and probably better grinding of apple tissue, especially skin.

It is generally known that apple skin is characterized by a higher content of polyphenols compared to the flesh. There is up to a three- to fourfold difference between these two tissue types [44]. Thus, with the complex action of the gears, the screw press ensures better disintegration of apple skin and greater release of the cell contents to the extracted juice.

It is also known that the content of phenolic compounds decreases gradually during grinding and pressing due to the oxidation of pulp and freshly extracted juices [11,45]. So, our results can also be attributed to the rate of mash oxidation. Since the apples were disintegrated directly in the screw press, the entire process of juice extraction was faster and the rate of mash oxidation was lower than in the case of the basket press.

Noteworthy, for varieties with similar contents of polyphenols in the juice extracted on the basket press (Rubin – 30.47 mg GAE 100 mL^{-1} and Mutsu – 29.89 mg GAE 100 mL^{-1}), a similar percentage of increase in polyphenols was also observed in the juice obtained on the screw press (62% for Rubin and 65% for Mutsu).

Jaeger et al. [12] found that the release of polyphenols from coarse mash was lower than that from fine mash. Since the screw press crushes apples more finely than the basket press, more polyphenols are extracted to the juice. A similar observation was reported by Heinmaa et al. [11], who noticed a higher amount of individual polyphenols in the juice extracted on a belt press (which also crushes the apples during pressing) as compared to the rack-and-frame press.

The efficiency of polyphenol extraction on individual presses may also be influenced by the textural characteristics of the apples tested. It is worth mentioning that the greatest differences were observed in the Mutsu and Rubin varieties, which are characterized by lower cutting forces.

3.7. Effect of the Press Construction on Antioxidant Activity

The antioxidant capacity of a substance is defined as its ability to scavenge reactive oxygen species and electrophiles [46]. Antioxidant activity is a very important parameter of juice quality.

The effect of press construction on the antioxidant capacity of juices is presented in Figure 8.

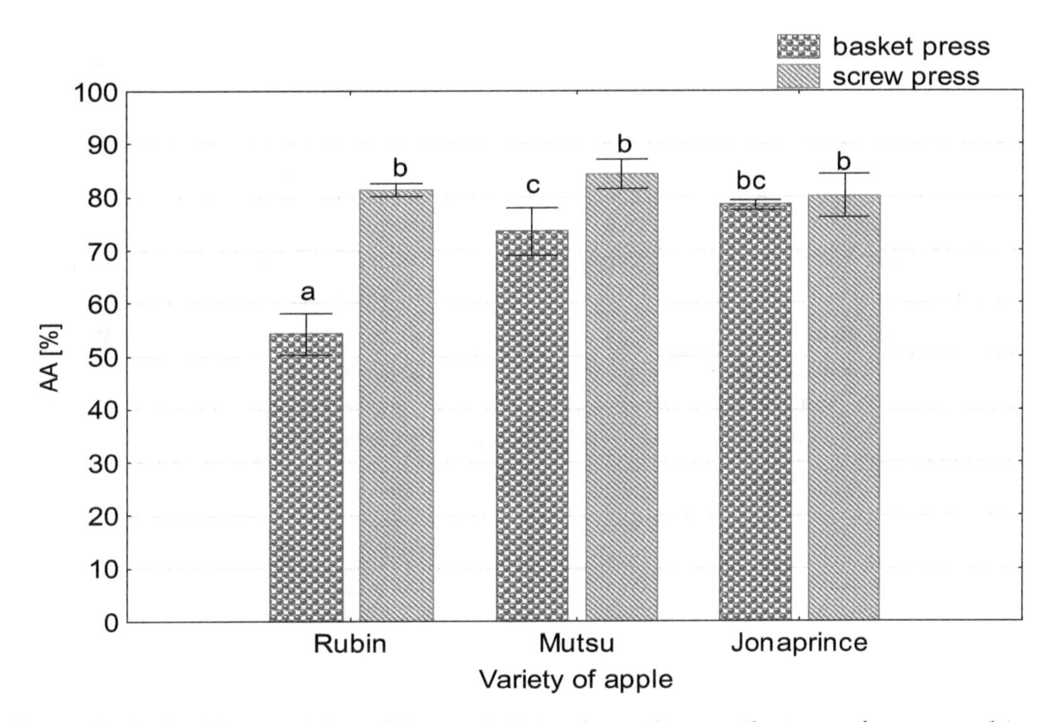

Figure 8. Antioxidant activity of the apple juice depending on the type of press machine.

The study showed that press construction had a significant effect on the antioxidant activity of the juice depending on the apple variety used. In the case of Rubin and Mutsu, the differences in antioxidant activity were statistically significant, whereas no statistically significant differences were found for Jonagold. Among the cultivars, the juice obtained from Mutsu showed the highest antioxidant activity.

The ability to scavenge free radicals was very closely related to the content of phenolic compounds in the apple juice. For the Jonagold, only small differences in the content of polyphenols in the juices were observed between the presses, which probably resulted in the absence of statistically significant differences in the antioxidant activity. Numerous studies have demonstrated a strong relationship between the polyphenols concentration and the antioxidant capacity of foods [44,47–50].

On the other hand, Wolfe et al. [51] did not find any relation between the antioxidant capacity and the TPC in apples. The lack of a correlation between these two properties may be related to the extraction process, apple varieties, percentage share of individual polyphenols, and the content of other compounds such as vitamin C, which may influence the antioxidant capacity.

Another determinant of the antioxidant activity is the degree of disintegration of the raw material. After cutting, defence metabolism is activated and synthesis and oxidation of phenolics occur simultaneously, modifying the initial phenolic composition of the fruit [46].

An effect of the processing technology on antioxidant capacity was also observed in other studies. For instance, Heinmaa et al. [11] reported that antioxidant capacity was highest in juices extracted on water press, followed by those extracted on belt press and rack-and-frame press. Their results also showed that the degrees of correlation between individual polyphenols and antioxidant activity were different.

Both antioxidant activity and TPC indicate the quality of a product with respect to its biological properties, and hence can be used in the process of quality control in the production of apple juices [52]. As indicated by our results, small-scale manufacturers should be advised to make use of a screw press to produce healthy and high-quality apple juice.

4. Conclusions

The present research demonstrated the varied efficiency of extraction of apple juice depending on the design of the press and apple variety. Generally, application of the screw press ensured higher yields of pressing compared to the basket press. The differences in the yield of extracted juice observed between the presses were related to the wider opening of cell membranes and release of greater amounts of nutrients facilitated by the screw press.

This study also showed the effect of press construction on the quality of apple juice. The juices extracted on screw press had higher TPC and antioxidant activity. This was probably due to the more intensive grinding and mixing of the raw material and the greater release of valuable bioactive components into the extracted juice. Additionally, the apple juices extracted on the screw press were characterized by a higher content of soluble solids, higher viscosity, and lower acidity. The obtained results indicate the necessity of further research on the use of screw presses for the production of juices from various varieties of apples in the farm conditions.

Author Contributions: Z.K. conceptualization; K.W. methodology; Z.K. formal analysis; K.W. investigation; K.W. data curation; Z.K. and K.W. writing—original draft preparation; D.D. writing—review and editing; Z.K. and D.D. supervision.

References

1. Li, Z.; Teng, J.; Lyu, Y.; Hu, X.; Zhao, Y.; Wang, M. Enhanced Antioxidant Activity for Apple Juice Fermented with Lactobacillus plantarum ATCC14917. *Molecules* **2018**, *24*, 51. [CrossRef] [PubMed]
2. Kschonsek, J.; Wolfram, T.; Stöckl, A.; Böhm, V. Polyphenolic Compounds Analysis of Old and New Apple Cultivars and Contribution of Polyphenolic Profile to the In Vitro Antioxidant Capacity. *Antioxidants (Basel)* **2018**, *7*, 20. [CrossRef] [PubMed]
3. FAOSTAT. Food and agriculture organization: FAO-STAT 2016. Available online: http://www.fao.org/faostat/en/#data/QC (accessed on 23 January 2019).

4. Bugala, A. Zmiany w polskim handlu zagranicznym sokami zagęszczonymi w 2016 r. *Przemysł Fermentacyjny i Owocowo-Warzywny* **2017**, *61*, 19–21. [CrossRef]
5. Francini, A.; Sebastiani, L. Phenolic Compounds in Apple (Malus x domestica Borkh.): Compounds Characterization and Stability during Postharvest and after Processing. *Antioxidants (Basel)* **2013**, *2*, 181–193. [CrossRef] [PubMed]
6. Plocharski, W.; Markowski, J.; Groele, B.; Stos, K.; Koziol-Kozakowska, A. Soki, nektary, napoje - aspekty rynkowe i zdrowotne. *Przemysł Fermentacyjny i Owocowo-Warzywny* **2017**, *61*, 6–10. [CrossRef]
7. De Paepe, D.; Coudijzer, K.; Noten, B.; Valkenborg, D.; Servaes, K.; De Loose, M.; Diels, L.; Voorspoels, S.; Van Droogenbroeck, B. A comparative study between spiral-filter press and belt press implemented in a cloudy apple juice production process. *Food Chem.* **2015**, *173*, 986–996. [CrossRef] [PubMed]
8. Markowski, J.; Baron, A.; Le Quéré, J.-M.; Płocharski, W. Composition of clear and cloudy juices from French and Polish apples in relation to processing technology. *LWT - Food Sci. Technol.* **2015**, *62*, 813–820. [CrossRef]
9. Kruczek, M.; Drygaś, B.; Habryka, C. Pomace in fruit industry and their contemporary potential application. *World Sci. News* **2016**, *48*, 259–265.
10. Nadulski, R.; Kobus, Z.; Wilczyński, K.; Guz, T.; Ahmed, Z.A. Characterisation of selected apple cutivars in the aspect of juice production in the condition of farm. In *Farm Machinery and Processes Management in Sustainable Agriculture. Proceedings IX International Scientific Symposium Farm Machinery and Processes Management in Sustainable Agriculture, Lublin, Poland, 22–24 November 2017*; pp. 255–259. Available online: https://depot.ceon.pl/handle/123456789/14803 (accessed on 02 July 2019).
11. Heinmaa, L.; Moor, U.; Põldma, P.; Raudsepp, P.; Kidmose, U.; Lo Scalzo, R. Content of health-beneficial compounds and sensory properties of organic apple juice as affected by processing technology. *LWT - Food Sci. Technol.* **2017**, *85*, 372–379. [CrossRef]
12. Jaeger, H.; Schulz, M.; Lu, P.; Knorr, D. Adjustment of milling, mash electroporation and pressing for the development of a PEF assisted juice production in industrial scale. *Innov. Food Sci. Emerg. Technol.* **2012**, *14*, 46–60. [CrossRef]
13. Nadulski, R.; Kobus, Z.; Wilczyński, K.; Zawiślak, K.; Grochowicz, J.; Guz, T. Application of Freezing and Thawing in Apple (Malus domestica) Juice Extraction. *J. Food Sci.* **2016**, *81*, E2718–E2725. [CrossRef] [PubMed]
14. Kobus, Z.; Nadulski, R.; Anifantis, A.S.; Santoro, F. Effect of press construction on yield of pressing and selected quality characteristics of apple juice. Engineering for rural development. Jelgava, 23.-25.05.2018. Available online: https://www.researchgate.net/publication/325386311_Effect_of_press_construction_on_yield_of_pressing_and_selected_quality_characteristics_of_apple_juice (accessed on 24 Janauary 2019).
15. Nadulski, R. Ocena przydatności laboratoryjnej prasy koszowej do badań procesu tłoczenia soku z surowców roślinnych. *Inżynieria Rolnicza* **2006**, *6*, 73–80.
16. Łaska, B.; Myczko, A.; Golimowski, W. Badanie wydajności prasy ślimakowej i sprawności tłoczenia oleju w warunkach zimowych i letnich. *Problemy Inżynierii Rolniczej* **2012**, *4*, 163–170.
17. Wroniak, M.; Ptaszek, A.; Ratusz, K. Ocena wpływu warunków tłoczenia w prasie ślimakowej na jakość i skład chemiczny olejów rzepakowych. *Żywność: Nauka - technologia - jakość* **2013**, *20*, 92–104.
18. Bogaert, L.; Mathieu, H.; Mhemdi, H.; Vorobiev, E. Characterization of oilseeds mechanical expression in an instrumented pilot screw press. *Ind. Crop. Prod.* **2018**, *121*, 106–113. [CrossRef]
19. Isobe, S.; Zuber, F.; Uemura, K.; Noguchi, A. A new twin-screw press design for oil extraction of dehulled sunflower seeds. *J. Am. Oil Chem. Soc.* **1992**, *69*, 884. [CrossRef]
20. Bakhshabadi, H.; Mirzaei, H.; Ghodsvali, A.; Jafari, S.M.; Ziaiifar, A.M. The influence of pulsed electric fields and microwave pretreatments on some selected physicochemical properties of oil extracted from black cumin seed. *Food Sci. Nutr.* **2017**, *6*, 111–118. [CrossRef]
21. Sannino, M.; del Piano, L.; Abet, M.; Baiano, S.; Crimaldi, M.; Modestia, F.; Raimo, F.; Ricciardiello, G.; Faugno, S. Effect of mechanical extraction parameters on the yield and quality of tobacco (Nicotiana tabacum L.) seed oil. *J. Food Sci. Technol.* **2017**, *54*, 4009–4015. [CrossRef]
22. Rabadán, A.; Álvarez-Ortí, M.; Gómez, R.; Alvarruiz, A.; Pardo, J.E. Optimization of pistachio oil extraction regarding processing parameters of screw and hydraulic presses. *LWT - Food Sci. Technol.* **2017**, *83*, 79–85. [CrossRef]
23. Wilczyński, K.; Kobus, Z.; Guz, T. Analysis of efficiency and particles size distribution in apple juice obtained with different methods. *J. Res. Appl. Agric. Eng.* **2018**, *63*, 140–143.

24. Wilczyński, K. Charakterystyka oraz efektywność urządzeń wykorzystywanych w przemyśle do pozyskiwania soku z owoców i warzyw. In *Wybrane problemy z zakresu przemysłu spożywczego - teoria i praktyka*; Stoma, M., Dudziak, A., Kobus, Z., Eds.; Libropolis: Lublin, Poland, 2016; pp. 81–90.
25. Agriculture and rural development 2014-2020. Available online: https://ec.europa.eu/agriculture/rural-development-2014-2020_en (accessed on 28 May 2019).
26. Promotion of EU farm products. 2019. Available online: https://ec.europa.eu/info/food-farming-fisheries/key-policies/common-agricultural-policy/market-measures/promotion-eu-farm-products_en (accessed on 28 May 2019).
27. Juściński, S. The mobile service of agricultural machines as the element of the support for the sustainable agriculture. In *Farm Machinery and Processes Management in Sustainable Agriculture. Proceedings IX International Scientific Symposium Farm Machinery and Processes Management in Sustainable Agriculture, Lublin, Poland, 22–24 November 2017*; pp. 136–141. Available online: https://depot.ceon.pl/handle/123456789/14824 (accessed on 02 July 2019).
28. Horwitz, W. *Official Methods of Analysis of AOAC International*, 17th ed.; AOAC International: Gaithersburg, MD, USA, 2000.
29. PN-EN 1132:1999 - Polish version. 2019. Available online: http://sklep.pkn.pl/pn-en-1132-1999p.html (accessed on 24 January 2019).
30. PN-EN 12143:2000 - Polish version. 2019. Available online: http://sklep.pkn.pl/pn-en-12143-2000p.html (accessed on 24 January 2019).
31. Singleton, V.L.; Orthofer, R.; Lamuela-Raventós, R.M. [14] Analysis of total phenols and other oxidation substrates and antioxidants by means of folin-ciocalteu reagent. *Methods Enzymol.* **1999**, *299*, 152–178.
32. Sharma, S.; Kori, S.; Parmar, A. Surfactant mediated extraction of total phenolic contents (TPC) and antioxidants from fruits juices. *Food Chem.* **2015**, *185*, 284–288. [CrossRef] [PubMed]
33. Takenaka, M.; Nanayama, K.; Isobe, S.; Ozaki, K.; Miyagi, K.; Sumi, H.; Toume, Y.; Morine, S.; Ohta, H. Effect of extraction method on yield and quality of Citrus depressa juice. *Food Sci. Technol. Res.* **2007**, *13*, 281–285. [CrossRef]
34. Eisele, T.A.; Drake, S.R. The partial compositional characteristics of apple juice from 175 apple varieties. *J. Food Compos. Anal.* **2005**, *18*, 213–221. [CrossRef]
35. Kowalczyk, R. Wydajność tłoczenia i wskaźnik zużycia jabłek w procesie wytwarzania zagęszczonego soku jabłkowego. *Problemy Inżynierii Rolniczej* **2004**, *12*, 21–30.
36. Huang, Z.; Hu, H.; Shen, F.; Wu, B.; Wang, X.; Zhang, B.; Wang, W.; Liu, L.; Liu, J.; Chen, C.; et al. Relatively high acidity is an important breeding objective for fresh juice-specific apple cultivars. *Sci. Hortic.* **2018**, *233*, 29–37. [CrossRef]
37. Choi, L.H.; Nielsen, S.S. The Effects of Thermal and Nonthermal Processing Methods on Apple Cider Quality and Consumer Acceptability. *J. Food Qual.* **2005**, *28*, 13–29. [CrossRef]
38. Genovese, D.B.; Lozano, J.E. Contribution of colloidal forces to the viscosity and stability of cloudy apple juice. *Food Hydrocoll.* **2006**, *20*, 767–773. [CrossRef]
39. Will, F.; Roth, M.; Olk, M.; Ludwig, M.; Dietrich, H. Processing and analytical characterisation of pulp-enriched cloudy apple juices. *LWT - Food Sci. Technol.* **2008**, *41*, 2057–2063. [CrossRef]
40. Teleszko, M.; Nowicka, P.; Wojdyło, A. Chemical, enzymatic and physical characteristic of cloudy apple juices. *Agric. Food Sci.* **2016**, *25*, 34–43. [CrossRef]
41. Genovese, D.B.; Lozano, J.E. Effect of Cloud Particle Characteristics on the Viscosity of Cloudy Apple Juice. *J. Food Sci.* **2000**, *65*, 641–645. [CrossRef]
42. Juszczak, L.; Fortuna, T. Effect of temperature and soluble solids content on the viscosity of cherry juice concentrate. *Int. Agrophysics* **2004**, *18*, 17–21.
43. Gökmen, V.; Artık, N.; Acar, J.; Kahraman, N.; Poyrazoğlu, E. Effects of various clarification treatments on patulin, phenolic compound and organic acid compositions of apple juice. *Eur. Food Res. Technol.* **2001**, *213*, 194–199. [CrossRef]
44. Vieira, F.G.K.; Borges, G.S.; Copetti, C.; Gonzaga, L.V.; Nunes, E.C.; Fett, R. Activity and contents of polyphenolic antioxidants in the whole fruit, flesh and peel of three apple cultivars. *Arch. Latinoam. Nutr.* **2009**, *59*, 101–106. [PubMed]
45. Van Der Sluis, A.A.; Dekker, M.; Skrede, G.; Jongen, W.M.F. Activity and concentration of polyphenolic antioxidants in apple juice. 1. Effect of existing production methods. *J. Agric. Food Chem.* **2002**, *50*, 7211–7219. [CrossRef] [PubMed]

46. Queiroz, C.; Lopes, M.L.M.; Fialho, E.; Valente-Mesquita, V.L. Changes in bioactive compounds and antioxidant capacity of fresh-cut cashew apple. *Food Res. Int.* **2011**, *44*, 1459–1462. [CrossRef]
47. He, L.; Xu, H.; Liu, X.; He, W.; Yuan, F.; Hou, Z.; Gao, Y. Identification of phenolic compounds from pomegranate (Punica granatum L.) seed residues and investigation into their antioxidant capacities by HPLC–ABTS+ assay. *Food Res. Int.* **2011**, *44*, 1161–1167. [CrossRef]
48. Chinnici, F.; Bendini, A.; Gaiani, A.; Riponi, C. Radical scavenging activities of peels and pulps from cv. Golden Delicious apples as related to their phenolic composition. *J. Agric. Food Chem.* **2004**, *52*, 4684–4689. [CrossRef] [PubMed]
49. do Rufino, M.; Alves, R.E.; de Brito, E.S.; Pérez-Jiménez, J.; Saura-Calixto, F.; Mancini-Filho, J. Bioactive compounds and antioxidant capacities of 18 non-traditional tropical fruits from Brazil. *Food Chem.* **2010**, *121*, 996–1002. [CrossRef]
50. Drogoudi, P.D.; Michailidis, Z.; Pantelidis, G. Peel and flesh antioxidant content and harvest quality characteristics of seven apple cultivars. *Sci. Hortic.* **2008**, *115*, 149–153. [CrossRef]
51. Wolfe, K.; Wu, X.; Liu, R.H. Antioxidant activity of apple peels. *J. Agric. Food Chem.* **2003**, *51*, 609–614. [CrossRef] [PubMed]
52. Pavun, L.; Đurđević, P.; Jelikić-Stankov, M.; Đikanović, D.; Uskoković-Marković, S. Determination of flavonoids and total polyphenol contents in commercial apple juices. *Czech. J. Food Sci.* **2018**, *36*, 233–238.

12

The Effect of Selected Parameters on Spelt Dehulling in a Wire Mesh Cylinder

Andrzej Anders, Ewelina Kolankowska, Dariusz Jan Choszcz, Stanisław Konopka and Zdzisław Kaliniewicz *

Department of Heavy Duty Machines and Research Methodology, University of Warmia and Mazury in Olsztyn, 10-719 Olsztyn, Poland; andrzej.anders@uwm.edu.pl (A.A.); ewelina.kolankowska@uwm.edu.pl (E.K.); dariusz.choszcz@uwm.edu.pl (D.J.C.); stanislaw.konopka@uwm.edu.pl (S.K.)

* Correspondence: zdzislaw.kaliniewicz@uwm.edu.pl

Abstract: A spelt dehuller with an innovative structural design is described in the study. In the developed solution, spelt kernels are separated by the mechanical impact of friction in a wire mesh cylinder with 4 × 4 mm openings. The dehuller is powered by a motor with a rotating impeller with an adjustable blade angle. In the experimental part of the study, spelt kernels are dehulled at five rotational speeds of the shaft: 160 to 400 rpm at intervals of 60 rpm, and five rotor blade angles: 50° to 90° at intervals of 10°. The efficiency of spelt dehulling (removal of glumes and glumelles) is evaluated based on kernel separation efficiency, husk separation efficiency, and the proportion of damaged kernels.

Keywords: spelt; threshing; dehuller

---

1. Introduction

Sustainable agriculture is the production of healthy, high-quality foods in a way that protects the environment and provides economic benefits to farmers. Spelt as well as emmer and einkorn are ancient wheat species. Spelt is a relict species that has recently been "rediscovered" and is being widely used in the production of flour, groats, flakes, pasta, bread, vodka, and beer [1,2]. The growing popularity of spelt on the consumer market can be attributed to its unique flavor, health-promoting properties, environmental benefits, and high content of biologically active compounds including essential nutrients. Spelt grain contains high-quality protein; unsaturated fatty acids; B complex vitamins; PP vitamin (niacin, vitamin B3, and pellagra-preventing factor); and minerals such as zinc, potassium, calcium, and iron [3–6]. Spelt kernels are enveloped by tough husks that protect this grain against atmospheric pollution and radiation. Tough husks are difficult to separate, which poses a considerable problem during threshing and limits the processing suitability of spelt grain. Spelt is difficult to harvest because grain is not effectively separated from spikelets by combine harvesters. Spelt spikes are hard, awned or awnless, and nonfree-threshing. Loose spikes have a brittle rachis, which is broken during threshing into several fragments, and each fragment contains one spikelet. Spikelets typically have two flowers with two kernels per spikelet, and kernels are tightly enclosed by four glumelles and two glumes, which makes the species nonfree-threshing. Spelt spikelets rarely contain three kernels. Spelt kernels are vitreous, white to red in color, with distinctive brush hairs in the apical part [7,8].

The harvested grain cannot be directly used in the food processing industry due to considerable contamination with chaff. After harvest, spikelets require additional treatment in a process that is commonly referred to as threshing. The harvested material is a mixture of grain and spikelets. This mixture has to be separated before threshing, because repeated threshing only increases energy consumption, contributes to grain damage, and increases grain loss [9].

The popularity of spelt is on the rise in the agricultural sector, in particular in organic farms. According to Mieczysław Babalski, one of the leading spelt producers in Poland, and Józef Tyburski,

PhD, spelt can be effectively dehulled in a modified thresher, an abrasive discdehuller, or specialist equipment manufactured in Western Europe [7,8,10]. Shelling machines and tangential dehullers can also be used for this purpose [11]. Most commercially available dehulling machines feature a cylindrical sieve and a rotating impeller with beaters. However, their efficiency is not highly satisfactory because spelt grain is encased by tightly adhering husks. Dehulling machines are also very expensive, and many farmers make attempts to build their own dehulling equipment. For instance, Tudor (2012) designed and built a dehulling machine as part of the Farmer FNE11–731 project [12]. It should also be noted that the grain and spikelet mixture has to be cleaned several times during supplementary threshing.

Dehulling devices are not highly efficient in separating spelt grain from husks, and they are very expensive. In Polish farms, modified clover hullers are often used for the supplementary threshing of spelt. However, spelt grain is not effectively separated by modified clover hullers, and the processed material requires further cleaning and sorting.

To address these concerns, a spelt dehuller with an innovative structural design is proposed in this study [13]. In the developed solution, spelt kernels are separated by the mechanical impact of friction in a wire mesh cylinder with 4×4 mm openings. The dehuller is powered by a motor with a rotating impeller, and the blade angle can be adjusted from 50° to 90°, at intervals of 10° [14]. The aim of the study is to determine the effect of selected parameters (rotor blade angle and rotational speed of the shaft) on the efficiency of glume and glumelle removal from spelt kernels.

2. Materials and Methods

The experimental material comprised spikelets of spelt cv. *Schwabenkorn* purchased in an organic farm in Praslity (54°01′55″ N 20°21′29″ E), municipality of Dobre Miasto, Region of Warmia and Mazury in Poland. The experimental material (spelt) was purchased from an organic farm (not from a commercial seed center) immediately after harvest. In order to remove impurities, weeds, and kernels of other cereal species, the seeds were cleaned using screens/sieves and a pneumatic separator. The percentage share of seeds of the main species in an average sample of input material (purity of the examined material) was determined at 85%, and its relative moisture content was determined at 11.56 ± 2.00%.

The proposed dehuller (Figure 1) features a wire mesh cylinder with 4×4 mm openings (Figure 2) and an open area factor (as a ratio (percentage) of the area available for the material to pass (openings) to the total area of the screen) of 0.38. Cylindrical mesh sieves are generally applied in specialist machines for separating spelt grain from chaff and in modified clover hullers.

The experiment was conducted in two stages. The process of removing husks from spelt kernels was examined in the first stage, and the separation of grain from threshed spikelets was analyzed in the second stage.

The following parameters were adopted in the first stage of the experiment:

1. Fixed parameters:
   - Width of the grain inlet $s_z = 10$ mm;
   - Sample mass $m_p = 300$ g;
   - Relative moisture content of the sample—11.56 ± 2.00%;
   - Rubber impeller blades (properties T-REX40 red natural rubber (T-Rex Rubber International, Netherlands): hardness: 40 ± 5° ShA, min. 600% elongation at break, tensile strength—16 MPa, max. abrasion resistance—210 mm^3);

   (a) Distance between the cylinder and beaters $s_r = 2$ mm.
2. Independent variables:
   - Rotor blade angle αw—50° ÷ 90°, at intervals of 10°;
   - Rotational speed of the shaft nw—160 ÷ 400 rpm, at intervals of 60 rpm;
3. Dependent variables:

- Kernel separation efficiency η_z, %;
- Husk separation efficiency η_p, %;
- Proportion of damaged kernels U_z, %.

Figure 1. Prototype of a spelt dehuller: 1—dehulling chamber, 2—wire mesh cylinder, 3—rotating impeller, 4—receiving outlet, 5—gearmotor, 6—radial fan, and 7—impeller blades.

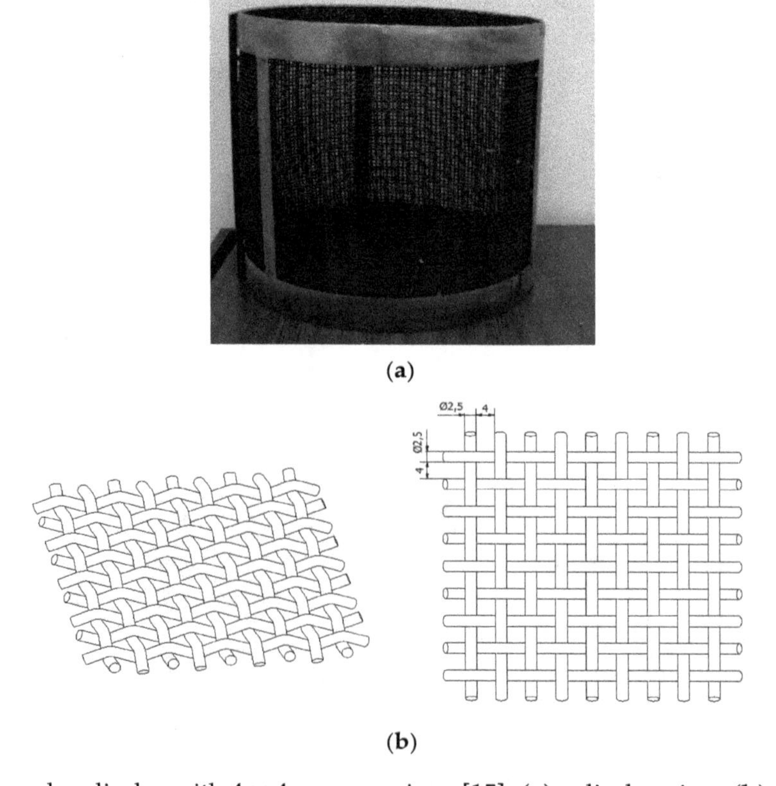

(a)

(b)

Figure 2. Wire mesh cylinder with 4 × 4 mm openings [15]: (**a**) cylinder view, (**b**) sieve structure.

Kernel and husk separation efficiency $\eta$and the proportion of damaged kernels U_z were determined based on the mass of the mixture components separated in each sample in the first stage of the experiment [16].

In each sample, kernel separation efficiency η_z was calculated with the use of the following formula:

$$\eta_z = \frac{M_z}{M_c} \cdot 100\% \quad (1)$$

where

M_z—mass of separated kernels (g),

M_c—mass share of kernels in a sample (g).

In each sample, husk separation efficiency η_p was calculated with the use of the following formula:

$$\eta_p = \frac{M_p}{M_m} \cdot 100\% \quad (2)$$

where

M_p—husk mass (g),

M_c—mass share of husks in a sample (g).

The proportion of damaged kernels U_z was calculated with the use of the following formula:

$$U_z = \frac{M_u}{M_c} \cdot 100\% \quad (3)$$

where

M_u—mass of damaged kernels (g).

In the first stage of the experiment, the dehuller's operating parameters were set before every measurement. Spelt spikelets were fed into the hopper, and the dehuller was turned on. Machine start-up time was around 30 s. The sliding gate in the hopper was opened, and husks were removed from spikelets by the mechanical impact of friction. The processed material was weighed on a laboratory scale to the nearest 0.01 g. A preweighed sample of spelt spikelets was always present in the feeder hopper to ensure the repeatability of measurements.

In the second stage of the experiment, the threshed material was separated into whole kernels, damaged kernels, spikelets, and husks in a pneumatic separator ([17], Figure 3) and a Fritsch Analysette 3 (Fritsch GmbH, Germany) vibratory sieve shaker. Based on initial measurements, the air flow rate in the pneumatic separator was set at 7.5 $m \cdot s^{-1}$. The vibratory sieve shaker consisted of 5 mesh screens with longitudinal openings (4.0 × 20 mm, 3.5 × 20 mm, 3.0 × 20 mm; 2.5 × 20 mm, and 2.0 × 20 mm). Processing time was 3 min, and the amplitude of vibration was 0.1 mm. Spelt kernels were considered damaged when they were broken, cracked, had a interior visible, or had damaged seed coats.

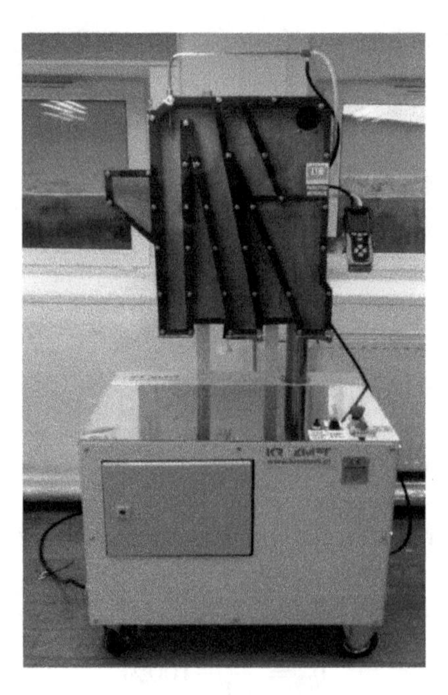

Figure 3. W.124984 pneumatic separator [15].

In all sub-stages of the experiment, the measurements for each combination of independent variables were performed in triplicate.

The results of all measurements were processed statistically by correlation analysis and stepwise multivariate polynomial regression analysis for a second-order polynomial model. Statistical analyses were conducted in Statistica PL v. 13.1 [18].

3. Results and Discussion

The results of the statistical analysis of the proportion of spelt kernels damaged at different rotor blade angles and rotational speeds are presented in Table 1.

The results of the multiple regression analysis indicate that the correlation coefficient for the proportion of spelt kernels damaged during threshing in a wire mesh cylinder with 4 × 4 mm openings at the tested rotor blade angles and rotational speeds ranged from 0.32 to 0.58. The multiple regression equations describing the proportion of damaged kernels generally fit the empirical data well, and the coefficient of determination for the wire mesh cylinder reached 0.70.

The equations describing the percentage of kernels damaged at different rotor blade angles and rotational speeds are presented graphically in Figure 4. In the wire mesh cylinder with 4 × 4 mm openings, the proportion of damaged kernels increased with a rise in rotational speed in the following range of rotor blade angles: $\alpha_w = 70° \div 80°$.

In the tested wire mesh cylinder, the efficiency of kernel separation was highly correlated with rotational speed, and the value of the correlation coefficient was determined at 0.84 (Table 2). The efficiency of kernel separation was less correlated with the rotor blade angle, and the value of the correlation coefficient was determined at 0.23. The multiple regression equation describing the efficiency of kernel separation was characterized by good and very good fit to empirical data. In the wire mesh cylinder with 4 × 4 mm openings, the coefficient of determination after stepwise elimination of non-significant variables was determined at 0.88.

The quadratic equation describing kernel separation efficiency at different rotor blade angles and rotational speeds is presented graphically in Figure 5. In the wire mesh cylinder with 4 × 4 mm openings, kernel separation efficiency increased with a rise in rotational speed. The rotor blade angle exerted a smaller and more ambiguous effect on kernel separation efficiency. Kernels were most efficiently separated at a rotational speed of 400 rpm and a rotor blade angle of 90°.

Table 1. A correlation analysis of the proportion of spelt kernels damaged during threshing in a wire mesh cylinder with 4 × 4 mm openings.

General Data: Correlation Coefficients are Significant at $\alpha < 0.05$ N = 25				
No.	**Variable**	**Mean**	**Standard Deviation**	**Coefficient of Variation (%)**
1.	Rotor blade angle α_w (°)	70.00	14.43	20.62
2.	Rotational speed n_w (rpm)	280.00	86.60	30.93
3.	Proportion of damaged kernels U_z (%)	4.74	3.96	83.60
Correlation Matrix				
		α_w	n_w	U_z
	α_w	1	0	0.584
	n_w		1	0.320
	U_z			1
Variable	**F-Statistic**	**Coefficient of Determination R^2**	**Standard Error of the Estimate**	***t*-Statistic**
Free term				−3.18
α_w				3.76
α_w^2				−3.84
n_w	8.81	0.70	2.44	−0.30
n_w^2				−0.04
$\alpha_w \cdot n_w$				1.14
Quadratic equation for two independent variables: $U_z = 1.601419 \cdot \alpha_w - 0.011224 \cdot \alpha_w^2 - 0.016039 \cdot n_w - 0.000003 \cdot n_w^2 + 0.000465 \cdot \alpha_w \cdot n_w - 54.454576$ Stepwise regression did not decrease the degree of the polynomial function for two independent variables				

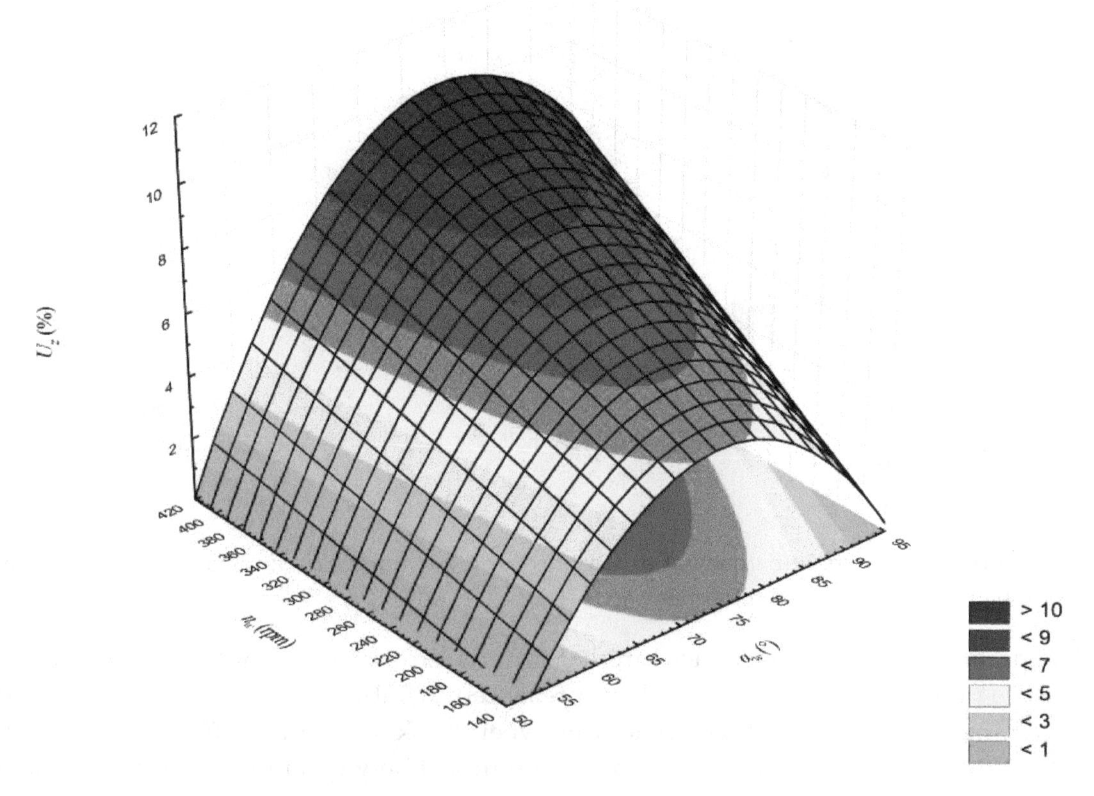

Figure 4. Proportion of kernels damaged at different rotor blade angles α_w and rotational speeds n_w in a wire mesh cylinder with 4 × 4 mm openings.

Table 2. A correlation analysis of kernel separation efficiency during threshing in a wire mesh cylinder with 4 × 4 mm openings.

General Data: Correlation Coefficients are Significant at $\alpha < 0.05$ N = 25				
No.	**Variable**	**Mean**	**Standard Deviation**	**Coefficient of Variation (%)**
1.	Rotor blade angle α_w (°)	70.00	14.43	20.62
2.	Rotational speed n_w (rpm)	280.00	86.60	30.93
3.	Kernel separation efficiency η_z (%)	52.14	8.58	16.46
Correlation Matrix				
		α_w	n_w	η_z
	α_w	1	0	0.226
	n_w		1	0.845
	η_z			1
Variable	**F-Statistic**	**Coefficient of Determination R^2**	**Standard Error of the Estimate**	***t*-Statistic**
Free term				5.65
α_w	53.77	0.88	3.11	−4.41
α_w^2				4.68
n_w				11.41
Quadratic equation for two independent variables: $\eta_z = -2.3039 \cdot \alpha_w + 0.0174 \cdot \alpha_w^2 + 0.0837 \cdot n_w + 101.1434$				

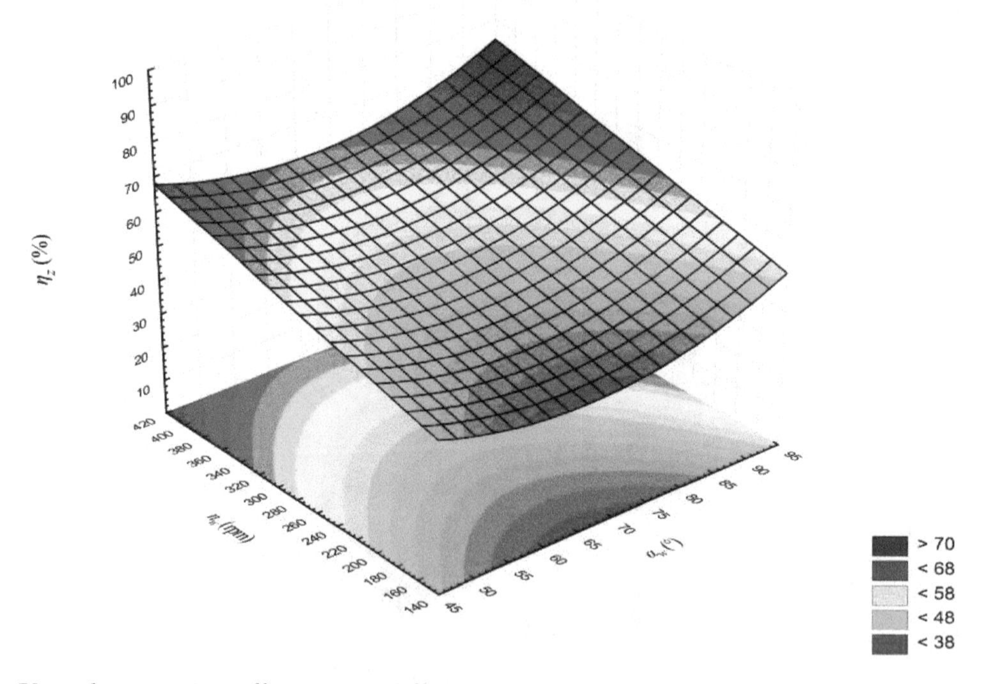

Figure 5. Kernel separation efficiency at different rotor blade angles α_w and rotational speeds n_w in a wire mesh cylinder with 4 × 4 mm openings.

Husk separation efficiency was measured at different rotor blade angles and rotational speeds; the results were processed statistically and are presented in Table 3. In the wire mesh cylinder with 4 × 4 mm openings, the correlation coefficient between husk separation efficiency and rotor blade angle was determined at 0.67, and the correlation coefficient between husk separation efficiency and rotational speed was determined at 0.54. The multiple regression equation describing husk separation efficiency was characterized by a very good fit to empirical data. The value of the coefficient of determination for the tested mesh cylinder was 0.89.

Table 3. A correlation analysis of husk separation efficiency during threshing in a wire mesh cylinder with 4 × 4 mm openings.

General Data: Correlation Coefficients are Significant at $\alpha < 0.05$ N = 25				
No.	**Variable**	**Mean**	**Standard Deviation**	**Coefficient of Variation (%)**
1.	Rotor blade angle α_w (°)	70.00	14.43	20.62
2.	Rotational speed n_w (rpm)	280.00	86.60	30.93
3.	Husk separation efficiency η_p (%)	42.42	9.52	22.45
Correlation Matrix				
		α_w	n_w	η_p
	α_w	1	0	0.673
	n_w		1	0.541
	η_p			1
Variable	**F-Statistic**	**Coefficient of Determination R²**	**Standard Error of the Estimate**	***t*-Statistic**
Free term				5.85
α_w	54.86	0.89	3.42	−4.95
α_w^2				5.36
$\alpha_w \cdot n_w \cdot$				7.20
Quadratic equation for two independent variables: $\eta_p = -2.8529 \cdot \alpha_w + 0.0219 \cdot \alpha_w^2 + 0.0008 \cdot \alpha_w \cdot n_w + 114.3675$				

The quadratic equation describing husk separation efficiency at different rotor blade angles and rotational speeds is presented graphically in Figure 6. In the wire mesh cylinder with 4 × 4 mm openings, husks were most effectively separated at the rotational speed of 400 rpm and the rotor blade angle of 90°.

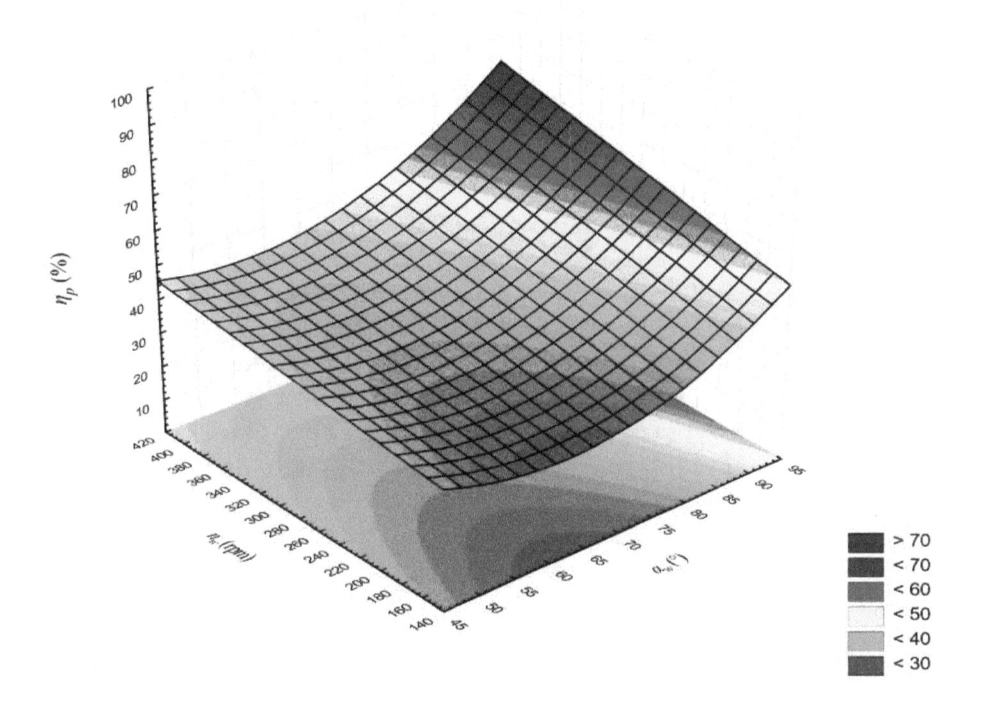

Figure 6. Husk separation efficiency at different rotor blade angles α_w and rotational speeds n_w in a wire mesh cylinder with 4 × 4 mm openings.

4. Conclusions

In the proposed spelt dehuller with an adjustable blade angle, where spelt kernels were separated by the mechanical impact of friction in a wire mesh cylinder with 4 × 4 mm openings, the highest proportion of damaged kernels U_z (13.77%) was noted at a rotational speed of 400 rpm and a rotor blade angle of 80°. Kernel separation efficiency η_z was highest (68.72%) at a rotational speed of 400 rpm and a rotor blade angle of 50°. Husk separation efficiency η_p was highest (60.53%) at a rotational speed of 400 rpm and a rotor blade angle of 90°.

The proportion of damaged kernels, kernel separation efficiency, and husk separation efficiency in the proposed spelt dehuller can be described by linear equations or stepwise regression for a second-order polynomial model with the elimination of non-significant variables, where the rotor blade angle and rotational speed are the independent variables. The multiple regression equations for the separation process in a wire mesh cylinder with 4 × 4 mm openings were characterized by good and very good fit to empirical data. The values of the coefficient of determination R^2 ranged from approximately 0.70 to 0.90.

It can be concluded that the dehulling efficiency of spelt is significantly influenced by both tested variables, i.e., the rotational speed of the impeller n_w and the rotor blade angle α_w.

Grain processing is particularly important in the case of spelt, which has high nutritional value and can be used in the food industry, thus contributing to the development of sustainable agriculture and the extensive use of agricultural resources.

Author Contributions: E.K. and D.J.C. conceived and designed the experiments; E.K. performed the experiments; A.A., D.J.C., S.K., and Z.K. contributed to the literature study; E.K., D.J.C., and S.K. analyzed the data; A.A., E.K., and Z.K. wrote the paper; A.A., D.J.C., and S.K. critically revised the manuscript.

References

1. Cegielska, A.; Gromulska, W. Różnorodność produktów z orkiszu (Diverse spelt products). *Przegląd Zbożowo-Młynarski* **2008**, *5*, 30–31. (In Polish)
2. Majewska, K.; Dąbkowska, E.; Żuk-Gołaszewska, K.; Tyburski, J. Baking quality of flour obtained from grain of chosen spelt varieties (*Triticum spelta* L.). *Żywność Nauka Technol. Jakość* **2007**, *2*, 60–71.
3. Capouchová, I. Technological quality of spelt (*Triticum spelta* L.) from ecological growing system. *Sci. Agric. Biochem.* **2001**, *32*, 307–322.
4. Kohajdova, Z.; Karovicova, J. Nutritional Value and Baking Applications of Spelt Wheat. *Acta Sci. Pol. Technol. Aliment.* **2008**, *7*, 5–14.
5. Wieser, H. Comparative investigations of gluten proteins from different wheat species. III. N–terminal amino acid sequences of a–gliadins potentially toxic for celiac patients. *Eur. Food Res. Technol.* **2001**, *213*, 183–186.
6. Abdel-Aal, E.S.M.; Hucl, P. Amino acid composition and in vitro protein digestibility of selected ancient wheats and their end products. *J. Food Comp. Anal.* **2002**, *15*, 737–747. [CrossRef]
7. Babalski, M.; Przybylak, Z.; Przybylak, K. *Uzdrawiające Ziarna Zbóż (Cereal Grains with Healing Properties)*; Eko Media: Bydgoszcz, Poland, 2013; pp. 1–191. ISBN 9788363537128. (In Polish)
8. Tyburski, J.; Babalski, M. *Uprawa Pszenicy Orkisz (Spelt Cultivation)*; Centrum Doradztwa Rolniczego w Brwinowie, Oddział w Radomiu: Radom, Poland, 2006; pp. 1–25. ISBN 8360185263. (In Polish)
9. Frączek, J.; Reguła, T. Method of evaluation of susceptibility of spelt grains to mechanical damages during the threshing process. *Inżynieria Rol.* **2010**, *4*, 51–58.
10. Choszcz, D.J.; Konopka, S.; Zalewska, K. Characteristics of physical properties of selected varieties of spelt. *Inżynieria Rol.* **2010**, *4*, 23–28.
11. Budzyński, W. (Ed.) *Pszenice—Zwyczajna, Orkisz, Twarda: Uprawa i Zastosowanie (Common Wheat, Spelt and Durum Wheat: Cultivation and Applications)*; Powszechne Wydawnictwo Rolnicze i Leśne: Poznań, Poland, 2012; pp. 1–328. ISBN 9788309011354. (In Polish)

12. Tudor, N. Farmer Built Spelt Dehuller. USDA Sustainable Agriculture Research and Education Report. 2012. Available online: http://mysare.sare.org/MySare/ProjectReport.aspx?do=viewRept&pn=FNE11--731&t=1&y=2014 (accessed on 18 November 2015).
13. Kolankowska, E.; Choszcz, D. Urządzenie do Usuwania Plew z Ziarna Orkiszu (Device for Removing Chaff from Spelt Grain). Patent PL408757, 29 November 2019. (In Polish).
14. Kolankowska, E.; Choszcz, D. Urządzenie do Usuwania Plew z Ziarna Orkiszu (Device for Removing Chaff from Spelt Grain). Utility Model Application No W123522, 31 May 2017. (In Polish).
15. Kolankowska, E. Doskonalenie Procesu Usuwania Plew z Ziarna Orkiszu (Optimization of the Dehulling Process for Spelt Kernels). Ph.D. Thesis, Uniwersytet Warmińsko-Mazurki w Olsztynie, Wydział Nauk Technicznych, Olsztyn, Poland, 2019. (In Polish).
16. Grochowicz, J. *Maszyny do Czyszczenia i Sortowania Nasion (Seed Cleaning and Sorting Machines)*; Wydawnictwo Akademii Rolniczej: Lublin, Poland, 1994; pp. 1–326. ISBN 839016129X. (In Polish)
17. Kolankowska, E.; Choszcz, D. Separator powietrzny (Air Separator). Utility Model Application No W124984, 31 December 2018. (In Polish).
18. Stanisz, A. *Przystępny kurs Statystyki w Oparciu o Program STATISTICA PL na Przykładach z Medycyny. Tom 1. Statystyki Podstawowe (Accessible Course in Statistics Based on the STATISTICA PL Software on Examples from Medicine. Tome 1. Basic Statistics)*; StatSoft Polska: Kraków, Poland, 2007; pp. 1–532. ISBN 8388724185. (In Polish)

13

Optimization of Extraction Conditions for the Antioxidant Potential of Different Pumpkin Varieties (*Cucurbita maxima*)

Bartosz Kulczyński [1], Anna Gramza-Michałowska [1,*] and Jolanta B. Królczyk [2]

[1] Department of Gastronomy Sciences and Functional Foods, Faculty of Food Science and Nutrition, Poznań University of Life Sciences, 31 Wojska Polskiego Str., 60–624 Poznań, Poland; bartosz.kulczynski@up.poznan.pl

[2] Department of Manufacturing Engineering and Production Automation, Faculty of Mechanical Engineering, Opole University of Technology, ul. Prószkowska 76, 45–758 Opole, Poland; j.krolczyk@po.opole.pl

* Correspondence: anna.gramza@up.poznan.pl

Abstract: Antioxidants are a wide group of chemical compounds characterized by high bioactivity. They affect human health by inhibiting the activity of reactive oxygen species. Thus, they limit their harmful effect and reduce the risk of many diseases, including cardiovascular diseases, cancers, and neurodegenerative diseases. Antioxidants are also widely used in the food industry. They prevent the occurrence of unfavourable changes in food products during storage. They inhibit fat oxidation and limit the loss of colour. For this reason, they are often added to meat products. Many diet components exhibit an antioxidative activity. A high antioxidative capacity is attributed to fruit, vegetables, spices, herbs, tea, and red wine. So far, the antioxidative properties of various plant materials have been tested. However, the antioxidative activity of some products has not been thoroughly investigated yet. To date, there have been only a few studies on the antioxidative activity of the pumpkin, including pumpkin seeds, flowers, and leaves, but not the pulp. The main focus of our experiment was to optimize the extraction so as to increase the antioxidative activity of the pumpkin pulp. Variable extraction conditions were used for this purpose, i.e., the type and concentration of the solvent, as well as the time and temperature of the process. In addition, the experiment involved a comparative analysis of the antioxidative potential of 14 pumpkin cultivars of the *Cucurbita maxima* species. The study showed considerable diversification of the antioxidative activity of different pumpkin cultivars.

Keywords: pumpkin; *Cucurbita maxima*; antioxidative activity; Oxygen Radical Absorbance Capacity (ORAC); cluster analysis

---

1. Introduction

Food consumption is directly linked to maintaining health, well-being, and preventing hunger. In this context, proper nutrition is undoubtedly a key aspect of sustainable food, and thus also environmental sustainability. Plant products are the only alternative for the consumers to alter current meat products consumption and more sustainable way of living towards reducing negative impact on the environment. The application of plant matrices enriched with mineral components, instead of relevant animal products, allows the strive to develop practical alternatives that often change existing processes and products not corresponding with a sustainable production. Therefore, sustainable food production in an environmentally acceptable manner to meet the increasing demands of a growing population is an inevitable challenge for agricultural production. Antioxidants are natural molecules found in living organisms which prevent oxidative stress. These compounds are characterized by the ability to scavenge (neutralize) reactive oxygen species (ROS), including hydroperoxide radicals, superoxide anion radicals, singlet oxygen, hydrogen peroxide, and hydroxyl radicals [1,2]. The following compounds are usually listed as strong antioxidants: Carotenoids (e.g., beta-carotene, lycopene, astaxanthin), tocopherols (e.g., alpha-tocopherol, gamma-tocopherol), flavonoids (e.g., anthocyanins, flavanols, flavonones, isoflavones), phenolic acids (hydroxybenzoic acids, hydroxycinnamic acids), stilbenes, some vitamins (vitamin C), coenzyme Q10, sulphur compounds (e.g., allicin), mineral components (e.g., selenium, zinc) [1,3–5]. Products of plant origin (vegetables, fruits, mainly berries, herbs, spices, juices, wine, tea, some cereals, and grains) are rich sources of antioxidants. However, they can also be found in meat products [6]. The antioxidative activity of the compounds we consume every day is important for our health [1,7–10]. Antioxidants inhibit free radicals and thus they reduce the risk of various diseases of affluence, including cardiovascular diseases (atherosclerosis, hypertension, heart attack, stroke), diabetes, cancers, neurodegenerative diseases (e.g., Alzheimer's disease, Parkinson's disease), and osteoporosis [11–15]. Antioxidants may soothe inflammations and viral infections [16,17]. They are said to prevent age-related eye diseases [18,19]. Antioxidants are not only important to maintain good health, but they are also widely used in food technology. For example, these compounds inhibit oxidation and limit the degradation of phytosterols [20] and fats [21]. In this way, they may, for example, prevent the spoilage of meat products [22]. Antioxidants also preserve the right colour of food products [23,24]. It should be noted, however, that their activity depends on many factors, such as chemical structure and profile of antioxidants, various interactions (synergistic, antagonistic effect), and also the multidimensional characteristics of the food matrix [25].

The pumpkin (*Cucurbita* L.) is a plant material containing antioxidants, e.g., carotenoids, tocopherols, phenolic acids, and flavonols. It is commonly grown in Europe, Asia, South America, North America, and Africa [26]. It is estimated that the annual pumpkin production exceeds 20 million tonnes [27,28]. It is a valuable dietary component because its pulp is rich in carotenoids [29–31] and its seeds are a source of unsaturated fatty acids [32]. Pumpkin flowers and leaves are less popular, but they are also edible [33]. Pumpkin pulp can be consumed both raw and after being processed, e.g., cooked, or as compotes, jams, purees, and juices [34]. The most common pumpkin species are: *Cucurbita maxima*, *Cucurbita pepo*, and *Cucurbita moschata*. Each of these species has numerous varieties [34,35]. So far there have been few studies on the antioxidative activity of the pumpkin pulp. As a result, scientific publications do not provide a comparison of the antioxidative properties of the pulp of different pumpkin varieties. Most of the available studies only determine the antioxidative potential of pumpkin seeds and oils [36–38]. Therefore, it is difficult to point out the trends in the contribution of specific compounds and the impact of various factors to the total antioxidant potential.

Thus far, there have not been many studies comparing the cultivars-dependent antioxidant capacity of pumpkin pulp. Additionally, reference publications lack data on the correlation analysis between the antioxidant activity tests in *Cucurbita* maxima cultivars extracts and its cluster analysis. The gap in the current literature to which this study is addressed includes an assessment of the antioxidant capacity of selected pumpkin cultivars, determined by selected radicals scavenging assays. The aim of this study was to compare the antioxidative activity of the pulp of various pumpkin varieties of the *Cucurbita maxima* species. The study also assesses the influence of various pumpkin pulp extraction conditions on its antioxidative potential.

2. Materials and Methods

2.1. Chemicals and Reagents

Gallic acid, sodium carbonate, Folin and Ciocalteu's phenol reagent, (±)-6-hydroxyl-2,5,7,8-tetramethylchromane-2-carboxylic acid (Trolox), 2,2′-azino-bis(3-ethylbenzothiazoline-6-sulfonic acid) diammonium salt (ABTS), potassium persulphate, 2,2-diphenyl-1-picrylhydrazyl (DPPH), sodium acetate trihydrate, acetic acid, 2,4,6-tris(2-pyridyl)-s-triazine (TPTZ), iron(III) chloride hexahydrate, hydrochloric acid, iron (II) sulphate heptahydrate, iron (II) chloride tetrahydrate, 3-(2-pyridyl)-

5,6-diphenyl-1,2,4-triazine-4′,4″-disulfonic acid sodium salt (Ferrozine), ethylenediaminetetraacetic acid (EDTA), sodium phosphate monobasic dehydrate, sodium phosphate monobasic dehydrate, potassium phosphate dibasic, 2,2′-azobis(2-methylpropionamidine) dihydrochloride (AAPH), and fluorescein sodium salt were purchased from Sigma-Aldrich (Darmstadt, Germany).

2.2. *Sample Collection*

The pulp of 14 pumpkin cultivars of the *Cucurbita* maxima species ('Buttercup', 'Golden Hubbard', 'Galeux d'Eysines', 'Melonowa Żółta', 'Hokkaido', 'Jumbo Pink Banana', 'Marina di Chiggia', 'Flat White Boer Ford', 'Jarrahdale', 'Blue Kuri', 'Green Hubbard', 'Gomez', 'Shokichi Shiro', 'Porcelain Doll') was used as the research material. All the cultivars were purchased from 'Dolina Mogilnicy' Organic Farming Products Cooperative (Wolkowo, Poland). While the plants were being grown, they were irrigated, weeded, and the soil was loosened. The pumpkins were harvested in October 2018 and transported to a laboratory, where they were immediately cleaned. The experimental unit consisted of two randomly chosen pumpkins from each variety. All chemical analyses for each pumpkin were performed in triplicate. The edible pulp was cut into pieces, freeze-dried, and then subjected to analysis. The dried pulp was stored at room temperature, without access to oxygen and light.

2.3. *Extract Preparation*

Preliminary studies involved solvents most commonly used for the preparation of plant extracts (acetone, ethyl acetate, ethanol, methanol, water). The extraction was carried out for various concentrations of selected solvents (20%, 40%, 60%, 80%, 100%) and the extraction time (0.5, 1, 2, 3, 4 h). Tests were carried out in three temperature ranges, which are used in plant components extraction as the most effective (30, 50, and 70 °C). Variants with the highest DPPH radical scavenging values were selected for further stages of the study. As a result, the above factors were limited to the previously mentioned nine solvents, two times and three extraction temperatures. In order to optimize the extraction pumpkin extracts were prepared by weighing 5 g of freeze-dried pumpkin and dissolving them in 50 mL of a solvent (water, methanol, 80% water–methanol solution, ethanol, 80% water–ethanol solution, acetone, 80% water–acetone solution, ethyl acetate, 80% ethyl acetate, and water solution). Next, the whole was shaken in a water bath (SWB 22N) at 30, 50, or 70 °C for 1 or 2 h. The extracts were centrifuged (1500 rpm, 10 min) and filtered through paper. Next, their antioxidative activity was assessed. The extraction of all samples was triplicated. The most favourable extraction conditions were selected on the basis of the optimization results: 2 h and 70 °C, maintaining the ratio between the amount weighed and the volume of the solvent used (1:10). Extracts for the pumpkin cultivars were prepared according to these conditions to find differences in their antioxidative activity, which included determination of the ABTS and DPPH radical scavenging assay, oxygen radical absorbance capacity (ORAC), the ferric reducing antioxidant capacity assay (FRAP), and the iron chelating activity assay.

2.4. *Share of Individual Plant Elements*

The pumpkin was carefully peeled in order to determine the content of individual parts of the fruit. Then, the pulp was cleared of seeds. Each part of the pumpkin was weighed with an accuracy of three decimal digits. The results were used to calculate the percentage content of the skin, seeds, and pulp. Each sample was measured in triplicate.

2.5. *Pumpkin Flesh Colour*

The colour of the pumpkin flesh was determined with a colorimeter (Chroma Meter CR-410, Konica Minolta Sensing Inc., Osaka, Japan). The test consisted in measuring the colour of the sample in reflected light and calculating the L * a * b value (L—color brightness; a—color in the range from green to red; b—color from blue to yellow). Before the measurement a homogeneous pulp was prepared from the pumpkin flesh. The analysis was triplicated.

2.6. Moisture Content

The water content was measured by drying 1 ± 0.001 g of fresh pumpkin, weighed, and then the sample was dried at 105 °C for 3 h. Next, the samples were cooled to room temperature and their weight was checked. The samples were then placed in a dryer again for 30 min. The samples were cooled again and weighed. The procedure was repeated until the sample weight between two measurements differed by more than 0.004 g. Each sample was measured in triplicate.

2.7. Active Acidity

The potentiometric method was used to measure acidity (pH) with a pH meter (CP-401). The result of measurement of the potential difference between the indicator and comparative electrodes was recorded. Each sample was measured in triplicate.

2.8. DPPH Radical Scavenging Activity Assay

The analysis was made in accordance with the methodology invented by Brand-Williams et al. (1995) [39]. An amount of 0.01 g of 2,2-diphenyl-1-picrylhydrazyl (DPPH) was weighed and transferred into a 25-mL volumetric flask together with the solvent (80:20 methanol/water (*v*/*v*). Next, the flask was filled up to the marking level. A calibration curve for Trolox (Tx) was also prepared. Assay: An amount of 100 μL of the extract was collected and 2.0 mL of the solvent, as well as 250 μL of the DPPH reagent were added. The whole was shaken on a Vortex at ambient temperature and left in darkness for 20 min. Next, the absorbance was measured at a wavelength $\lambda = 517$ nm (Meterek SP 830). The extraction reagent + DPPH solution was used as a control sample. The assay was triplicated. The results were given as mg Trolox equivalents (Tx) per 100 g of dry mass using the calibration curve $y = 81.2991x - 2.4922$ with a confidence coefficient $R^2 = 0.9963$, and a relative standard deviation of residuals of 2.3%, slope 2.8450, and intercept 1.7469.

2.9. ABTS Radical Scavenging Assay

The analysis was conducted in accordance with the methodology invented by Re et al. (1999) [40]. An amount of 0.192 g of 2,2′-azino-bis(3-ethylbenzothiazoline-6-sulfonic acid) diammonium salt (ABTS) was weighed on an analytical balance with an accuracy of 0.001 g. Next, the weighed ABTS was transferred into a 50-mL volumetric flask. The rest was filled with deionized water up to the marking level. An amount of 0.0166 g of potassium persulphate ($K_2S_2O_8$) was weighed and transferred into a 25-mL volumetric flask. The rest was filled with deionized water up to the marking level. The ABTS and $K_2S_2O_8$ solution was mixed at a 1:0.5 ratio. The mixture was stored at room temperature in darkness for 16 h. Next, the mixture was diluted with a solvent to obtain an absorbance of 0.700 at a wavelength of $\lambda = 734$ nm. A calibration curve for Trolox was also prepared. Assay: An amount of 30 μL of the extract was collected into a tube and 3 mL of the ABTS reagent was added. The whole was mixed by shaking on a Vortex. After 6 min the absorbance was measured at a wavelength of $\lambda = 734$ nm (Meterek SP 830). The extraction reagent + ABTS solution was used as a control sample. The assay was triplicated. The results were given as mg Trolox equivalents (Tx) per 100 g of dry mass using the calibration curve $y = 174.5970 - 0.2115$ with a confidence coefficient $R^2 = 0.9941$, and a relative standard deviation of residuals of 2.5%, slope 9.5155, and intercept 2.6059.

2.10. Total Polyphenol Content

The total polyphenol content was measured with the method invented by Sanchez-Moreno et al. (1998) [41]. Preparation of saturated sodium carbonate solution: An amount of 10.6 g of sodium carbonate was weighed on an analytical balance and placed in a beaker. Then, 100 mL of deionized water was added and the whole was mixed on a magnetic stirrer. A calibration curve was prepared using gallic acid (GAE) as a standard. The Folin-Ciocalteau reagent (FCR) was diluted with deionized water at a 1:1 ratio. Assay: An amount of 125 μL of the FCR reagent, 2 mL of deionized water, 125 μL of the sample, and 250 μL of the saturated sodium carbonate solution were collected into a test tube. The whole was mixed thoroughly on a Vortex and left at ambient temperature for 25 min. Next, the absorbance was measured at $\lambda = 725$ nm (Meterek SP 830). The assay was triplicated. The results were given as mg Gallic Acid equivalents (GAE) per 100 g of dry mass using the calibration curve $y = 0.0017x + 0.0071$ with a confidence coefficient $R^2 = 0.9943$, and a relative standard deviation of residuals of 2.4%, slope 0.0001, and intercept 0.0124.

2.11. Ferric Reducing Antioxidant Power (FRAP) Assay

The assay consists of measuring the increase in absorbance of the FRAP reagent, which takes place after incubation with the active ingredients contained in the plant extract due to the reduction of Fe(III) ions. It can be monitored by measuring variation in absorbance at a wavelength of 593 nm. The assay was based on the methodology invented by Benzie and Strain (1996) [42]. Preparation of acetic buffer (300 mM; pH 3.6): 3.1 g of sodium acetate trihydrate was weighed and combined with 16 mL of glacial acetic acid. The whole was placed in a 1 L flask and the rest was filled with deionized water up to the marking level. TPTZ (2,4,6-tris(2-pyridyl)-s-triazine) (10 mM in 40 mM HCl): 1.46 mL of concentrated HCl was collected into a 1 L volumetric flask, which was filled with deionized water up to the marking level. Next, 0.031 g of TPTZ was weighed and dissolved in 10 mL of 40 mM HCl in a water bath at 50 °C. FeC13 x $6H_2O$ (20 mM): 0.054 g of FeClx x $6H_2O$ was weighed and dissolved in 10 mL of deionized water. Before use the three reagents were mixed at a 10:1:1 ratio. Assay: An amount of 100 μL of the sample and 3 mL of the FRAP reagent (heated to 37 °C) was collected into a test tube. The whole was mixed on a Vortex and incubated at 37 °C for 4 min. Next, the absorbance was measured at a wavelength of $\lambda = 593$ nm (Meterek SP 830). A calibration curve was prepared using $FeSO_4$x $7H_2O$. The assay was triplicated. The results were given as mM FE(II) per 100 g of dry mass using the calibration curve $y = 0.6884x - 0.0003$ with a confidence coefficient R2 = 0.9977, and a relative standard deviation of residuals of 1.8%, slope 0.0166, and intercept 0.0100.

2.12. Iron Chelating Activity Assay

The method invented by Decker and Welch (1990) [43] was used to assay the ability of compounds to bind Fe(II) ions. Iron chloride tetrahydrate (2 mM): 0.0398 g of iron chloride tetrahydrate was weighed and transferred quantitatively into a 100-mL volumetric flask, which was filled with deionized water up to the marking level. Ferrozine (5 mM): 0.123 g of ferrozine was weighed and transferred quantitatively into a 50-mL volumetric flask, which was filled with deionized water up to the marking level. Assay: 1 mL of the extract was collected into a test tube. Next, 3.7 mL of deionized water was added. The whole was mixed on the Vortex. Next, 0.1 mL of iron chloride (2 mM) and 0.2 mL of ferrozine (5 mM) were added. The whole was mixed again and incubated at ambient temperature for 20 min. Next, the absorbance was measured at $\lambda = 562$ nm (Meterek SP 830). A calibration curve was also prepared using ethylenediaminetetraacetic acid (EDTA) disodium salt. The assay was triplicated. The results were given as ppm EDTA per 100 g of dry mass using the calibration curve $y = 1.4053x - 2.8349$ with a confidence coefficient $R^2 = 0.9981$, and a relative standard deviation of residuals of 1.2%, slope 0.0308, and intercept 1.4825.

2.13. Oxygen Radical Absorbance Capacity (ORAC) Assay

The ORAC method based on the methodology invented by Ou et al. (2001) [44] was used to determine the antioxidative capacity. A solution of 42 nM of fluorescein and 153 nM of AAPH as well as 0.075 M phosphate buffer (pH 7.4) were prepared. The extract was dissolved in 75 mM of phosphate buffer. The reaction mixture was prepared in a quartz cuvette to which 0.04 μM of disodium fluorescein in 0.075 M phosphate buffer was added. Next, the extract was added to the mixture, which was kept at 37 °C without access to light. Assay: The chemical reaction was initiated by adding 153 nM of the AAPH solution, which was a source of peroxide radicals. Fluorescence was measured with a Hitachi F-2700 spectrofluorometer at an excitation wavelength $\lambda = 493$ nm and emission wavelength $\lambda = 515$ nm. The first measurement was made immediately after adding the AAPH solution. Then, the fluorescence of the samples was measured every 5 min (f1, f2 ... ...) for 45 min. A standard curve was prepared for the Trolox solution. The assay was triplicated. The results were given as μM of Trolox equivalent per 1 g of dry mass using the calibration curve y = 5.5280x - 3.5001 with a confidence coefficient $R^2 = 0.9943$, and a relative standard deviation of residuals of 2.9%, slope 0.2088, and intercept 2.2231.

2.14. Statistical Analysis

The results were analyzed statistically (STATISTICA 13.1 software, StatSoft Inc., Kraków, Poland). An analysis of variance was made to detect statistically significant differences. A multiple comparison analysis was made using post-hoc LSD tests. The significance level was assumed at $p = 0.05$. Pearson's linear correlation coefficients ($p = 0.05$; $p = 0.01$; $p = 0.001$) were calculated for the antioxidative activity of the samples obtained in various tests. The Ward method was used for a hierarchical cluster analysis, by means of which the pumpkin cultivars were grouped according to their antioxidative activity.

3. Results

3.1. General Characteristics of Pumpkin Cultivars

The pumpkin cultivars were characterized in general (Table 1). The visual assessment showed that the pulp of all the pumpkin cultivars was yellow or orange. The colorimetric measurement method showed that the brightness of the pumpkin pulp ranged from 48.38 to 64.43 (parameter L). There were differences in the pumpkin juice pH. The 'Gomez' (pH = 5.09), 'Buttercup' (pH = 5.16), and 'Golden Hubbard' (pH = 5.16) cultivars were characterized by the highest acidity. The lowest acidity was noted in the 'Hokkaido' (pH = 6.13) and 'Marina di Chiggia' (pH = 6.10) cultivars. The cultivars differed in the water content in the pulp. The highest content was found in the 'Porcelain Doll' (94.85%), 'Galeux d'Eysines' (94.78%), and 'Melonowa Żółta' (93.25%) cultivars, whereas the lowest content was found in the 'Blue Kuri' (78.76%) and 'Marina di Chiggia' (81.17%) cultivars. The cultivars differed in the content of skin, seeds, and pulp. The following cultivars had the highest percentage content of pulp: 'Jumbo Pink Banana' (87.11%), 'Galeux d'Eysines' (85.09%), 'Buttercup' (84.03%), 'Jarrahdale' (83.94%), 'Flat White Boer' (83.29%), and 'Melonowa Żółta' (83.21%). The smallest content of the pulp was found in the following varieties: 'Golden Hubbard' (63.29%), 'Green Hubbard' (63.47%), 'Porcelain Doll' (69.24%), 'Marina di Chiggia' (70.09%), and 'Hokkaido' (70.59%).

3.2. Optimization of Extraction of Pumpkin Antioxidative Components

3.2.1. DPPH Radical Scavenging Activity

The methanol–aqueous (80%) and aqueous extracts exhibited the highest antioxidative activity against the DPPH radical regardless of the extraction time and temperature (Table 2). The extracts prepared with acetone, acetone and water (80%), ethyl acetate and methanol exhibited the lowest ability to inactivate the DPPH radical. The analysis of the influence of the extraction time on the antioxidative activity of the samples showed that the samples extracted for 2 h exhibited a higher antioxidative activity only for the ethyl acetate (30 and 50 °C), methanol (30 °C), and 80% ethyl acetate (50 and 70 °C) extracts. The test also showed that the temperature of the process affected the DPPH radical scavenging ability. As the temperature increased, so did the antioxidative activity. The exceptions were the 80% ethanol extract (the activity decreased at 1 h and 50 °C, but it increased after 2 h) and the methanol extracts, whose activity decreased at 70 °C.

Table 1. General characteristics of used pumpkin cultivars.

Pumpkin Cultivars	Shape	Flesh Colour				pH of Pulp Juice	Moisture Content in the Pulp (%)	Share of Individual Pumpkin Plant Elements		
		Description	L	*a*	*b*			Skin Content (%)	Seeds Content (%)	Pulp Content (%)
Hokkaido	round, slightly elongated at the tail	orange	53.59 ± 1.97b	28.17 ± 0.74g	47.57 ± 1.23ab	6.13 ± 0.02h	88.47 ± 0.59e	17.82 ± 2.06de	11.46 ± 2.38e	70.59 ± 3.97b
Blue Kuri	round	yellow	48.38 ± 2.93a	21.29 ± 1.40de	43.23 ± 1.25a	5.94 ± 0.02f	78.76 ± 1.90a	18.14 ± 2.08e	8.08 ± 1.78cd	75.63 ± 2.46bc
Buttercup	round, slightly flattened	yellow	56.58 ± 4.13bc	27.76 ± 1.96g	48.62 ± 1.91bc	5.16 ± 0.03b	83.93 ± 0.84c	14.08 ± 1.89c	3.44 ± 0.86b	84.03 ± 2.99cd
Gomez	round	orange	56.43 ± 3.07bc	18.82 ± 1.78cd	50.76 ± 0.88c	5.09 ± 0.02a	91.33 ± 0.72f	18.73 ± 2.41e	5.82 ± 1.40c	79 ± 3.71c
Shokichi Shiro	round	orange	53.56 ± 2.59b	13.21 ± 0.87b	45.25 ± 1.80ab	5.72 ± 0.04e	88.79 ± 1.49e	20.7 ± 2.97ef	6.67 ± 1.76c	72.74 ± 2.15b
Jumbo pink banana	elongated	orange	55.78 ± 2.05b	25.81 ± 1.09f	49.96 ± 2.04c	5.49 ± 0.02d	90.61 ± 1.33f	10.39 ± 2.54a	4.2 ± 1.58bc	87.11 ± 3.13d
Golden hubbard	round, slightly elongated at the tail	orange	61.96 ± 2.65de	25.14 ± 1.95f	56.03 ± 2.79d	5.16 ± 0.02b	90.85 ± 1.03f	32.41 ± 3.45h	5.06 ± 1.02c	63.39 ± 3.10a
Flat White Boer Ford	round, clearly flattened	orange	55.01 ± 2.11b	25.25 ± 1.65f	45.89 ± 0.85ab	5.22 ± 0.03c	90.1 ± 0.83f	15.52 ± 2.04cd	3.72 ± 0.50b	83.29 ± 3.39cd
Jarrahdale	round, flattened, irregular	orange	63.59 ± 2.85f	20.02 ± 0.80d	58.13 ± 1.7de	6.06 ± 0.03g	85.78 ± 0.69d	12.53 ± 2.43ab	5.82 ± 0.96c	83.94 ± 4.27cd
Porcelain Doll	round, slightly flattened	orange	58.28 ± 1.55c	17.78 ± 1.80c	55.29 ± 2.49d	5.51 ± 0.03d	94.85 ± 0.78h	22.78 ± 2.84f	9.7 ± 1.31d	69.24 ± 3.61b
Galeux d Eysines	flattened	orange	55.47 ± 2.08bc	22.73 ± 1.01e	49.67 ± 1.21c	5.95 ± 0.04f	94.78 ± 0.45h	11.44 ± 2.13a	5.93 ± 1.07c	85.09 ± 2.71cd
Green hubbard	elongated	orange	59.38 ± 2.54cd	19.11 ± 0.57cd	60.94 ± 1.81e	5.69 ± 0.06e	88.26 ± 1.92e	32.54 ± 4.75h	2.79 ± 0.92a	63.47 ± 3.82a
Marina di Chiggia	round	yellow	64.43 ± 2.55f	8.36 ± 0.45a	60.07 ± 1.45e	6.10 ± 0.02h	81.17 ± 1.29b	26.7 ± 2.43fg	4.23 ± 0.79bc	70.09 ± 4.21b
Melonowa Żółta	round	orange	60.48 ± 1.97	20.2 ± 2.48d	55.97 ± 1.84d	5.93 ± 0.02f	93.25 ± 1.32g	14.37 ± 2.83c	2.45 ± 0.48a	83.21 ± 3.54e

L: Color brightness; *a*: Color in the range from green to red (positive values indicate the proportion of red, and negative values—green); *b*: Color from blue to yellow (positive values indicate the proportion of yellow, and negative values—blue); A–h: Means in the same column followed by the same letters shown in superscript do not significantly differ ($p < 0.05$) in terms of analyzed variables.

Table 2. Effect of extraction parameters on the antioxidants properties of pumpkin extracts determined by DPPH assay (mg Tx/100 g dm).

Solvent	Extraction Time and Temperature					
	1 h			2 h		
	30 °C	50 °C	70 °C	30 °C	50 °C	70 °C
Acetone	74.61 ± 1.25bA	90.94 ± 0.81bB	n.a.	82.37 ± 0.98bA	91.29 ± 0.96bB	n.a.
Acetone–water (80%)	88.81 ± 0.99cA	88.42 ± 1.08bA	n.a.	97.46 ± 1.24cA	111.51 ± 1.31bB	n.a.
Ethyl acetate	63.58 ± 123aA	89.98 ± 1.21bB	116.15 ± 1.54bC	59.57 ± 0.65aA	88.25 ± 1.1aB	133.59 ± 1.76bC
Ethyl acetate–water (80%)	106.93 ± 1.32dA	112.3 ± 1.46cB	161.67 ± 1.76cC	111.15 ± 1.18dA	111 ± 0.97bA	161.67 ± 1.76cB
Ethanol	127.78 ± 2.13fA	133.36 ± 1.13dB	177.16 ± 2.07dC	143.8 ± 1.54eA	156.56 ± 1.29dB	182.51 ± 1.86dC
Ethanol–water (80%)	115.37 ± 1.33eA	79.36 ± 1.61aB	114.93 ± 2.09bA	119.05 ± 1.04dA	132.28 ± 1.17cB	157.52 ± 2.06cC
Methanol	88.44 ± 0.46cA	88.58 ± 1.16bA	60.46 ± 1.60aB	82.29 ± 1.25bA	91.59 ± 1.47bB	84.33 ± 1.90aC
Methanol–water (80%)	136.31 ± 1.70gA	151.26 ± 1.98eB	198.95 ± 2.39fC	165.87 ± 1.38fA	180.41 ± 1.8eB	210.97 ± 2.13eC
Water	151.49 ± 1.33hA	158.37 ± 2.46eA	180.70 ± 3.43eB	168.27 ± 1.20fA	174.45 ± 2.76eB	183.40 ± 3.08dC

A–h: Different letters represent statistically significant differences ($p < 0.05$) between solvents (in column) in the variables under analysis; A–C: Different letters represent statistically significant differences ($p < 0.05$) between temperatures of extraction (separately for 1 and 2 h) in the variables under analysis; n.a.: Not analyzed.

3.2.2. ABTS Radical Scavenging

The aqueous and methanol–aqueous (80%) extracts exhibited the highest ability to scavenge ABTS cation radicals (Table 3). The acetone extract was characterized by the lowest antioxidative activity. The antioxidative activity increased along with the extraction time in most of the samples. The exceptions were: Acetone and 80% ethyl acetate (50 °C). The comparison of the results showed that the extraction temperature affected the efficiency of the extracts in sweeping the ABTS cation radical. As the temperature increased, the antioxidative activity of the samples increased, too. The antioxidative properties decreased only in the 80% acetone (1 h), 80% ethyl acetate (1 h, 70 °C), and 80% ethanol (1 h, 50 °C) extracts.

Table 3. Effect of extraction parameters on the antioxidants properties of pumpkin extracts determined by the ABTS assay (mg Tx/100 g dm).

Solvent	Extraction Time and Temperature					
	1 h			2 h		
	30 °C	50 °C	70 °C	30 °C	50 °C	70 °C
Acetone	24.62 ± 0.32aA	31.51 ± 0.97aB	n.a.	19.06 ± 0.19aA	27.5 ± 0.44aB	n.a.
Acetone–water (80%)	45.9 ± 1.04dA	45.18 ± 0.98cB	n.a.	55.30 ± 1.00dA	62.11 ± 0.35eB	n.a.
Ethyl acetate	24.9 ± 0.98aA	30.83 ± 0.90aB	41.23 ± 1.27bC	33.63 ± 1.02bA	35.76 ± 0.26bB	52.25 ± 1.11aC
Ethyl acetate–water (80%)	35.53 ± 1.23cA	44.57 ± 0.38cB	36.78 ± 0.55aA	40.48 ± 0.66cA	42.33 ± 0.42cB	50.05 ± 0.95aC
Ethanol	28.43 ± 0.50bA	38.21 ± 1.09bB	52.68 ± 1.01cC	37.89 ± 0.99bA	54.8 ± 1.06dB	60.69 ± 1.27bC
Ethanol–water (80%)	62.08 ± 0.48eB	57.24 ± 2.08dA	66.68 ± 0.52dC	58.89 ± 1.07dA	71.81 ± 1.90fB	73.80 ± 1.17cB
Methanol	35.47 ± 0.99cA	49.23 ± 0.71cB	62.26 ± 0.38dC	40.07 ± 0.62cA	58.86 ± 0.73dB	70.65 ± 0.44cC
Methanol–water (80%)	74.07 ± 1.11eA	84.44 ± 0.82eB	104.09 ± 1.02eC	82.27 ± 1.03eA	95.31 ± 0.32gB	110.84 ± 0.57dC
Water	79.30 ± 0.39eA	108.5 ± 1.53fB	108.16 ± 0.65eB	85.16 ± 0.86eA	127.31 ± 1.39hB	124.70 ± 0.59eC

A–i: Different letters represent statistically significant differences ($p < 0.05$) between solvents (in column) in the variables under analysis; A–C: Different letters represent statistically significant differences ($p < 0.05$) between temperatures of extraction (separately for 1 and 2 h) in the variables under analysis; n.a.: Not analyzed.

3.2.3. Total Phenolic Content (TPC)

The highest total phenolic content was found in the aqueous and methanol–aqueous (80%) extracts (Table 4). The lowest concentration of polyphenolic compounds was found in the ethyl acetate, ethyl acetate–aqueous (80%), and acetone extracts. The analysis of the influence of the extraction time on the content of polyphenolic compounds showed that it was higher in the samples extracted for 2 h than in the ones extracted for 1 h, except the acetone (30 and 50 °C), acetone–aqueous (30 °C), and ethyl acetate (70 °C) extracts, where the samples extracted for 1 h had higher content of polyphenols. The extraction temperature also affected the results. The total polyphenolic content in the extracts prepared at 70 °C was greater than in the samples extracted at 30 and 50 °C. The exceptions were 80% acetone (1 h), ethanol–aqueous (80%) (1 and 2 h), and methanol–aqueous (80%) extracts (1 and 2 h), where the content of polyphenolic compounds was lower at 50 °C. However, the highest content of polyphenols in the 80% ethanol–aqueous and 80% methanol–aqueous extracts was found at 70 °C.

3.2.4. Ferric Reducing Antioxidant Power (FRAP)

The methanol–aqueous (80%) and ethanol–aqueous (80%) extracts exhibited the highest ferric ion reducing capacity (Table 5). The analysis of variation in the ferric ion reducing capacity over time showed that the samples extracted for 2 h exhibited greater activity than the ones extracted for 1 h.

Only the ethyl acetate, ethyl acetate (70 °C), ethanol–aqueous (80%) (30 °C), methanol (70 °C), and aqueous extracts were characterized by an inverse dependence. In most cases, the higher extraction temperature increased the ferric ion reducing capacity of the extracts. The higher extraction temperature caused a slight decrease in the antioxidative activity of the acetone extract only.

Table 4. Effect of extraction parameters on the total phenolic content (TPC) of pumpkin extracts (mg GAE/100 g dm).

Solvent.	Extraction Time and Temperature					
	1 h			2 h		
	30 °C	50 °C	70 °C	30 °C	50 °C	70 °C
Acetone	12.29 ± 0.22bA	24.10 ± 0.17cB	n.a.	7.49 ± 0.10aA	11.546 ± 0.10bB	n.a.
Acetone–water (80%)	89.51 ± 1.50dA	64.19 ± 0.71dB	n.a.	67.05 ± 0.96dA	74.51 ± 0.44dB	n.a.
Ethyl acetate	3.09 ± 0.06aA	8.09 ± 0.08aB	22.88 ± 0.11bC	5.2 ± 0.05aA	7.82 ± 0.03aB	18.54 ± 0.15aC
Ethyl acetate–water (80%)	12.82 ± 0.21bA	14.32 ± 0.16bB	18.49 ± 0.29aC	16.84 ± 0.45bA	24.95 ± 0.24cB	30.23 ± 0.21bC
Ethanol	46.27 ± 0.48cA	75.95 ± 1.01eB	130.59 ± 1.53cC	62.00 ± 0.52cA	102.35 ± 0.89fB	164.47 ± 1.88dC
Ethanol–water (80%)	106.52 ± 0.77fA	93.25 ± 0.44fB	132.86 ± 1.61cC	131.63 ± 1.22fA	128.11 ± 1.13gB	186.75 ± 2.05eC
Methanol	49.41 ± 0.46cA	92.88 ± 0.77fB	147.06 ± 1.28dC	58.51 ± 1.10cA	88.52 ± 0.61eB	136.67 ± 1.26cC
Methanol–water (80%)	128.62 ± 0.76gA	101.69 ± 1.12gB	162.51 ± 1.49eC	166.23 ± 0.97gA	160.12 ± 1.20hB	206.47 ± 2.44fC
Water	93.41 ± 1.46eA	155.83 ± 1.31hB	186.47 ± 2.93fC	108.2 ± 1.15eA	171.97 ± 2.15iB	237.94 ± 2.42gC

A–h: Different letters represent statistically significant differences ($p < 0.05$) between solvents (in column) in the variables under analysis; A–C: Different letters represent statistically significant differences ($p < 0.05$) between temperatures of extraction (separately for 1 and 2 h) in the variables under analysis; n.a.: Not analyzed.

Table 5. Effect of extraction parameters on ferric ion reducing antioxidant power (FRAP) of pumpkin extracts (mM Fe(II)/100 g dm).

Solvent	Extraction Time and Temperature					
	1 h			2 h		
	30 °C	50 °C	70 °C	30 °C	50 °C	70 °C
Acetone	71.32 ± 0.98dA	118.96 ± 1.38cB	n.a.	88.21 ± 0.99dA	127.91 ± 2.46dB	n.a.
Acetone–water (80%)	189.12 ± 1.65gA	180.17 ± 3.14eA	n.a.	249.59 ± 2.40iA	240.89 ± 2.28gA	n.a.
Ethyl acetate	28.56 ± 0.33bA	56.69 ± 0.63bB	182.27 ± 1.83cC	23.32 ± 0.23bA	40.82 ± 0.74bB	158.37 ± 1.50cC
Ethyl acetate–water (80%)	36.05 ± 0.67cA	55.57 ± 0.62bB	131.42 ± 2.57bC	43.08 ± 0.41cA	57.71 ± 0.64cB	119.87 ± 2.48bC
Ethanol	97.67 ± 1.52eA	172.86 ± 1.62dB	344.25 ± 4.09eC	128.13 ± 1.09fA	226.36 ± 2.40fB	351.06 ± 2.53eC
Ethanol–water (80%)	221.36 ± 2.03hA	217.73 ± 1.86fA	309.05 ± 2.01dB	171.18 ± 1.12gA	263.49 ± 3.09hB	346.79 ± 4.40eC
Methanol	106.1 ± 1.66fA	181.93 ± 1.81eB	360.74 ± 3.50fC	117.75 ± 2.22eA	222.13 ± 2.13fB	349.59 ± 3.99eC
Methanol–water (80%)	230.73 ± 3.54iA	231.35 ± 2.60gA	320.91 ± 4.65dB	199.39 ± 3.60hA	213.35 ± 3.21eB	290.15 ± 3.48dC
Water	13.15 ± 0.14aA	33.76 ± 0.25aB	85.06 ± 1.08aC	8.67 ± 0.25aA	22.51 ± 0.38aB	48.57 ± 1.33aC

A–i: Different letters represent statistically significant differences ($p < 0.05$) between solvents (in column) in the variables under analysis; A–C: Different letters represent statistically significant differences ($p < 0.05$) between temperatures of extraction (separately for 1 and 2 h) in the variables under analysis; n.a.: Not analyzed.

3.2.5. Iron Chelating Activity

The methanol–aqueous (80%) extracts exhibited the highest Fe(II) ion chelating activity (Table 6). The lowest Fe(II) chelating activity was characteristic of the ethyl acetate–aqueous (80%) and ethyl acetate extracts. As the extraction time increased, so did the Fe(II) ion chelating activity of all the samples except the ethyl acetate–aqueous (80%) extract (50 and 70 °C), and ethanol extract (50 °C). Apart from that, the assay showed that the Fe(II) ion chelating activity increased along with temperature. Only the chelating activity of the ethanol (30 vs. 50 °C), ethanol–aqueous (80%) (30 vs. 50 °C), and ymethanol–aqueous (80%) extracts (30 vs. 50 °C) decreased. However, the chelating activity of these extracts increased again at 70 °C.

Table 6. Effect of extraction parameters on iron chelating activity of pumpkin extracts (ppm EDTA/100 g dm).

Solvent	Extraction Time and Temperature					
	1 h			2 h		
	30 °C	50 °C	70 °C	30 °C	50 °C	70 °C
Acetone	1338.66 ± 23.26dA	1618.62 ± 17.42dB	n.a.	1435.13 ± 23.41dA	1781.61 ± 16.62dB	n.a.
Acetone–water (80%)	1116.18 ± 13.55cA	1452.60 ± 24.50cB	n.a.	1343.64 ± 12.54cA	1527.80 ± 12.67cB	n.a.
Ethyl acetate	385.00 ± 3.93bA	454.31 ± 3.17bB	1563.68 ± 22.42bC	589.78 ± 5.48bA	888.62 ± 6.08bB	1563.68 ± 25.42bC
Ethyl acetate–water (80%)	102.37 ± 5.13aA	174.30 ± 3.18aB	1276.61 ± 18.34aC	438.09 ± 4.22aA	723.49 ± 4.24aB	1236.57 ± 14.16aC
Ethanol	2542.92 ± 28.88fA	1809.22 ± 16.08eB	3023.88 ± 25.61dC	3146.40 ± 31.62fA	2787.12 ± 13.93eB	3486.03 ± 36.51dC
Ethanol–water (80%)	3641.36 ± 32.92hA	2715.41 ± 12.23gB	3864.89 ± 34.29eC	3905.50 ± 35.72hA	3219.77 ± 18.90fB	4144.63 ± 45.86fC
Methanol	3015.23 ± 33.00gA	3279.077.37.34hB	3854.02 ± 46.36eC	3497.55 ± 28.65gA	3579.78 ± 39.10gB	3974.26 ± 27.17eC
Methanol–water (80%)	3897.77 ± 44.08hA	3553.34 ± 28.12iB	4060.22 ± 28.37fC	4202.27 ± 37.15iA	4071.18 ± 39.20hB	4381.1 ± 52.25gC
Water	1796.45 ± 12.16eA	2108.03 ± 16.61fB	2159.47 ± 16.74cC	2009.48 ± 16.53eA	2566.80 ± 24.05eB	2224.65 ± 26.57cC

A–i: Different letters represent statistically significant differences ($p < 0.05$) between solvents (in column) in the variables under analysis; A–C: Different letters represent statistically significant differences ($p < 0.05$) between temperatures of extraction (separately for 1 and 2 h) in the variables under analysis; n.a.: Not analyzed.

3.3. Comparison of the Antioxidative Activity of Selected Extracts of Various Pumpkin Cultivars

3.3.1. DPPH Radical Scavenging Activity

The research showed that the 'Melonowa Żółta' (245.98 and 222.23 mg Tx/100 g dm), 'Hokkaido' (210.97 and 183.40 mg Tx/100 g dm), 'Galeux d'Eysiness' (206.99 mg Tx/100 g dm), and 'Buttercup' cultivars (185.19 mg/Tx/100 g dm) exhibited the highest antioxidative activity against the DPPH radical (Table 7). On the other hand, the aqueous extracts of the following pumpkin cultivars were characterized by the lowest antioxidative potential: 'Gomez' (34.11 mg Tx/100 g dm), 'Shokichi Shiro' (42.22 mg Tx/100 g dm), 'Golden Hubbard' (52.5 mg Tx/100 g dm), and 'Green Hubbard' (54.39 mg Tx/100 g dm). The methanol–aqueous (80%) extracts exhibited greater DPPH radical scavenging activity than the aqueous extracts in all the pumpkin cultivars. The pumpkin pulp exhibited low DPPH radical inhibiting activity (5.8 μmol Tx/g dm). The same experiment showed that other raw materials exhibited much higher antioxidative activity, e.g., artichokes (70.1 μmol Tx/g dm), lettuce (77.2 μmol Tx/g dm), spinach (50.9 μmol Tx/g dm), turmeric (57.6 μmol Tx/g dm). The following raw materials were characterized by lower DPPH radical inactivating ability: Leek (3.2 μmol Tx/g dm), cucumber (2.3 μmol Tx/g dm), celery (3.8 μmol Tx/g dm), carrots (3.5 μmol Tx/g dm), beans (3.8 μmol Tx/g dm) (44). The antioxidative activity of pumpkin seeds and skin (*Cucurbita pepo*) was confirmed in a test with DPPH radical. The test showed that the aqueous extracts (72.36% inhibition) and 70% ethanol extracts (71.0% inhibition) exhibited the highest DPPH radical scavenging activity in the pumpkin skin, whereas the 70% ethanol extract (20.5% inhibition) and 70% methanol extract (18.9% inhibition) exhibited the highest antioxidative activity in the pumpkin seeds. The radicals inhibition of the aqueous extract amounted to 4.12% [37]. Valenzuela et al. researched various pumpkin species seeds and found the highest total polyphenolic content in the *Cucurbita mixta* Pangalo species (275 μmol GAE/g of the extract). There were lower total polyphenolic content levels in the seeds of *Cucurbita maxima* Duchense (212.87 μmol GAE/g of the extract) and *Cucurbita moschata* (Duchense ex Lam.) species (118.79 μmol GAE/g of the extract) [45]. The above research showed that the methanol–aqueous and methanol extracts of pumpkin seeds were characterized by high DPPH radical scavenging activity, and amounted to 69.18% and 86.85%, respectively [46].

Table 7. Antioxidant activity of pumpkin extracts determined by ABTS and DPPH assays.

Pumpkin Cultivars	ABTS (mg Tx/100 g dm)		DPPH (mg Tx/100 g dm)	
	Aqueous–Methanol Extract	Aqueous Extract	Aqueous–Methanol Extract	Aqueous Extract
Hokkaido	110.84 ± 0.57cA	124.70 ± 0.59dB	210.97 ± 2.13hA	183.40 ± 3.08iB
Blue Kuri	85.12 ± 1.73aA	104.78 ± 1.45bB	145.44 ± 1.80fA	67.43 ± 0.88eB
Buttercup	116.37 ± 1.64dA	114.04 ± 1.00cA	185.19 ± 2.01gA	76.58 ± 1.13fB
Gomez	127.36 ± 1.66eA	107.19 ± 1.5bB	86.12 ± 1.90bA	34.11 ± 0.86aB
Shokichi Shiro	99.79 ± 1.71bA	123.75 ± 0.96dB	57.54 ± 0.88aA	42.22 ± 1.15bB
Jumbo pink banana	146.91 ± 0.68fA	173.11 ± 1.37gB	86.72 ± 0.96bA	72.9 ± 0.9fB
Golden hubbard	117.24 ± 1.20dA	152.01 ± 1.46gB	127.5 ± 1.64dA	52.5 ± 0.67dB
Flat White Boer Ford	113.07 ± 1.24cdA	95.92 ± 1.39aB	204.12 ± 2.28gA	96.49 ± 1.54gB
Jarahdale	95.90 ± 1.44bA	104.25 ± 1.28bB	136.90 ± 2.62eA	70.23 ± 1.28fB
Porcelain doll	109.74 ± 1.24cA	138.04 ± 1.47eB	82.42 ± 0.83bA	49.99 ± 1.21cB
Galeux d' Eysines	122.32 ± 1.19eA	110.07 ± 1.34bB	206.99 ± 2.24hA	127.85 ± 1.03hB
Green hubbard	103.87 ± 1.54cA	136.11 ± 1.13eB	101.98 ± 1.21cA	54.39 ± 0.75dB
Marina di Chiggia	118.28 ± 1.12dA	143.52 ± 2.43fB	105.63 ± 1.07cA	73.86 ± 0.85fB
Melonowa Żółta	152.86 ± 1.64fA	187.17 ± 2.55hB	245.98 ± 3.10iA	222.23 ± 3.87jB

A–i: Different letters represent statistically significant differences ($p < 0.05$) between antioxidative activity of pumpkin cultivars; A,B: Different letters represent statistically significant differences ($p < 0.05$) between antioxidative activity of extracts (aqueous–methanol vs. aqueous).

3.3.2. ABTS Radical Scavenging

The ABTS cation radical test showed that the 'Melonowa Żółta' pumpkin cultivar was characterized by the highest antioxidative potential (Table 7). The antioxidative activity of the aqueous extracts (187.17 mg Tx/100 g dm) was higher than that of the methanol–aqueous (80%) extracts (152.86 mg Tx/100 g dm). The lowest ABTS cation radical scavenging activity was noted in the 'Blue Kuri' (methanol–aqueous (80%) extract—85.12 mg Tx/100 g dm), 'Jarahdale' (methanol–aqueous (80%) extract—95.90 mg Tx/100 g dm), and 'Flat White Boer Ford' cultivars (aqueous extract—95.92 mg Tx/100 g dm). In most cases, the aqueous extracts exhibited greater antioxidative activity than the methanol–aqueous (80%) extracts. The research showed that the pumpkin pulp had relatively low ABTS cation radical scavenging ability (11.0 μM Tx/g dm), as compared with other vegetables: Artichoke (39.9 μM Tx/g dm), asparagus (37.5 μM Tx/g dm), broccoli (43.0 μM Tx/g dm), lettuce (85.8 μM Tx/g dm), radishes (61.7 μM Tx/g dm), and turmeric (118.6 μM Tx/g dm) (44). Sing et al. noted that pumpkin pulp (*Cucurbita maxima*) extracted with ethanol and water (50%) exhibited the highest ABTS radical scavenging activity, i.e., 2.04 μM Tx/g. The ABTS radical scavenging ability of the pumpkin extract was slightly lower than that of watermelon (2.24 μM Tx/g) and melon (2.78 μM Tx/g). The authors of the study also found that pumpkin skin extracts were characterized by higher antioxidative potential [47].

3.3.3. Total Phenolic Content (TPC)

The total phenolic content in the pumpkin cultivars was analyzed with the Folin-Ciocalteu method (Table 8). The analysis showed that the content of phenolic compounds in the pumpkin pulp varied depending on the cultivar and the extraction solvent. The highest concentration of total polyphenols was found in the following cultivars: 'Melonowa Żółta' (232.5 and 255.69 mg GAE/100 g dm), 'Hokkaido' (206.47 and 237.94 mg GAE/100 g dm), and 'Gomez' (172.63 and 188.22 mg GAE/100 g dm). The lowest content of phenolic compounds in the methanol–aqueous extracts was found in the 'Blue Kuri' (49.78 mg GAE/100 g dm) and 'Marina di Chiggia' cultivars (49.99 mg GAE/100 g dm). On the other hand, among the aqueous extracts, the lowest total polyphenolic level was found in the 'Blue Kuri' (65.66 mg GAE/100 g dm), 'Shokichi Shiro' (66.64 mg GAE/100 g dm), 'Flat White Boer Ford' (66.01 mg GAE/100 g dm), and 'Jumbo Pink Banana' cultivars (66.40 mg GAE/100 g dm). Apart from that, it is noteworthy that when water was used as the solvent, there was higher content of polyphenols in all the cultivars except for the 'Flat White Boer Ford'. Saavedra et al. conducted a study in which they measured the total polyphenolic content in fresh pumpkin skins and seeds. The highest content was found in aqueous extracts. The total polyphenolic content in the pumpkin skin was 741–1069 mg GAE/100 g dm, whereas in the seeds it was 234–239 mg GAE/100 g dm. The results showed that the polyphenol content in the pumpkin skin and seeds was higher than in its pulp. Only in the 'Melonowa Żółta' cultivar the polyphenol content (255.69 mg/100 g dm) was slightly higher than in the seeds. The Hokkaido cultivar had a similar content of polyphenols (237.94 mg/100 g dm) [37]. Kiat et al. also observed the presence of polyphenols in pumpkin seeds. They noted that the concentration of polyphenols in the methanol–aqueous extract (80%) (72 mg/100 g dm) was higher than in the methanol extract (44 mg/100 g dm) [48]. Bayili et al. showed that the total polyphenolic content in the pumpkin of the *Cucurbita pepo* species amounted to 100.2 mg/100 g fresh weight. The authors observed that the pumpkin contained more polyphenols than some other vegetables, e.g., tomatoes, eggplant, cucumber, and vegetable cabbage. The content of polyphenols was about two times greater in some cultivars, e.g.,: 'Hokkaido', 'Gomez', 'Porcelain Doll', 'Melonowa Żółta', but these results were calculated per dry mass [46]. Singh et al. observed that the content of polyphenolic compounds in pumpkin pulp (extracted with water) amounted to 13.92 mg GAE/100 g fresh weight. There was a higher content of polyphenolic compounds in the methanol–aqueous (19.36–30.69 mg/100 g dm) and ethanol–aqueous solvents (21.45–33.48 mg/100 g dm) [47]. Dar et al. found polyphenolic compounds in various extracts of pumpkin leaves (*Cucurbita pepo*). They observed that the polyphenolic content varied depending on the type of extract. The concentration of polyphenols in the ethanol extract was 40.37 mg/g GAE, in the aqueous extract—40.12 mg/g GAE, butanol extract—92.62 mg/g GAE, ethyl acetate extract—85.12 mg/g

GAE, chloroform extract—21.25 mg/g GAE, and n-hexane extract—12.50 mg/g GAE [49]. Oloyede et al. conducted research on the content of total polyphenols in pumpkin pulp (*Cucurbita pepo* Linn.). They found that the concentration of polyphenolic compounds in pumpkin depended on the degree of ripening. The content of polyphenolic compounds in ripe fruit was 33.5 mg/100 g dm, whereas in unripe fruit it was 10.3 mg/100 g dm. This content was similar to the results observed in methanol–aqueous extracts in the 'Blue Kuri', 'Marina di Chiggia', and 'Jumbo Pink Banana' cultivars [50].

Table 8. Total polyphenols content and oxygen radical absorbance capacity (ORAC) of pumpkin extracts.

Pumpkin Cultivars	Total Phenolic Content (mg GAE/100 g dm)		ORAC (μM Tx/g dm)
	Aqueous–Methanol Extract	Aqueous Extract	
Hokkaido	206.47 ± 2.44hA	237.94 ± 2.42iB	89.97 ± 2.07d
Blue Kuri	49.78 ± 0.76aA	65.66 ± 0.71aB	58.47 ± 0.70b
Buttercup	87.99 ± 1.52eA	115.24 ± 1.85fB	102.08 ± 1.84f
Gomez	172.63 ± 1.06gdA	188.22 ± 2.73hB	99.35 ± 3.05e
Shokichi Shiro	58.01 ± 1.08cA	66.64 ± 1.90aB	108.47 ± 2.32g
Jumbo pink banana	50.23 ± 1.61bA	66.40 ± 0.83aB	86.60 ± 1.19d
Golden hubbard	91.46 ± 1.16eA	101.05 ± 1.13eB	116.38 ± 1.91h
Flat White Boer Ford	70.99 ± 1.03dA	66.01 ± 1.33aB	43.04 ± 1.72a
Jarahdale	56.35 ± 0.90cA	72.61 ± 1.44bB	65.34 ± 1.83c
Porcelain doll	174.53 ± 1.37gA	189.40 ± 0.78hB	98.35 ± 2.36e
Galeux d' Eysines	113.40 ± 1.42fA	95.56 ± 1.36dB	94.87 ± 0.53e
Green hubbard	58.62 ± 1.36cA	126.08 ± 1.72gB	102.90 ± 1.64f
Marina di Chiggia	49.99 ± 1.49aA	77.71 ± 1.35cB	104.76 ± 1.84f
Melonowa Żółta	232.5 ± 2.63iA	255.69 ± 4.29jB	122.73 ± 3.39i

A–j: Different letters represent statistically significant differences ($p < 0.05$) between antioxidative activity (ORAC) and total polyphenolic content of pumpkin cultivars. A, B: Different letters represent statistically significant differences between antioxidative activity (ORAC) and total polyphenolic content ($p < 0.05$) of extracts (aqueous–methanol vs. aqueous).

3.3.4. Ferric Reducing Antioxidant Power (FRAP)

In order to determine the antioxidative potential of the pumpkin cultivars the ability of the aqueous and methanol–aqueous extracts to reduce ferric ions was also investigated. The following cultivars exhibited showed the highest antioxidative activity: 'Buttercup' (592.78 mM Fe(II)/100 g dm), 'Melonowa Żółta' (555.63 mM Fe(II)/100 g dm), 'Galeux d'Eysines' (524.90 mM Fe(II)/100 g dm), 'Flat White Boer Ford' (509.28 mM Fe(II)/100 g dm), and 'Porcelain Doll' (501.19 mM Fe(II)/100 g dm) (Table 9). The lowest ability to reduce the degree of ferric ion oxidation was found in the 'Hokkaido' (48.57 mM Fe(II)/100 g dm), 'Marina di Chiggia' (58.19 mM Fe(II)/100 g dm), and 'Green Hubbard' cultivars (66.34 mM Fe(II)/100 g dm). When 80% methanol was used as the solvent, all the cultivars exhibited higher antioxidative activity. Tiveron et al. observed that pumpkin pulp (*Cucurbita maxima*) was capable of reducing ferric ions at an amount of 19.5 μM Fe2+/g dm. This activity was greater than that of other vegetables, such as: Celery, carrot, cucumber, and leek. On the other hand, it was about 10–15 times lower than that of chicory, broccoli, spinach, and watercress [51]. Fidrianny et al. observed that the pumpkin leaf ethanol extract (*Cucurbita moschata*) exhibited relatively low ferric ion reducing activity, i.e., 1.37%, as compared with 7.39% for ascorbic acid. Ethyl acetate (1.37%) and hexane (0.28%) extracts exhibited lower ferric ion reducing ability [52]. One study showed that the compounds contained in pumpkin pulp (*Cucurbita maxima*) were capable of reducing ferric ions. The highest reducing activity was observed for the ethanol aqueous extract (50%) (3.23 μM Fe(II)/g) and ethanol aqueous extract (50%) (2.66 μM Fe(II)/g). The aqueous extracts exhibited the lowest activity (1.83 μM Fe(II)/g), which was consistent with the results of this study [46].

Table 9. Iron chelating activity and ferric ion reducing antioxidant power (FRAP) of pumpkin extracts.

Pumpkin Cultivars	Iron Chelating Activity (ppm EDTA/100 g dm)		FRAP (mM Fe(II)/100 g dm)	
	Aqueous–Methanol Extract	Aqueous Extract	Aqueous–Methanol Extract	Aqueous Extract
Hokkaido	4381.1 ± 52.25aA	2224.65 ± 26.57aB	290.15 ± 3.48bA	48.57 ± 1.33aB
Blue Kuri	11385.16 ± 56.32gA	7561.89 ± 46.15jB	337.39 ± 1.63cA	123.14 ± 1.73gB
Buttercup	5702.74 ± 59.19bA	2938.91 ± 30.21bB	592.78 ± 2.24jA	283.34 ± 2.43lB
Gomez	12804.62 ± 90.97gA	8125.81 ± 31.32kB	407.05 ± 1.72eA	111.54 ± 1.66fB
Shokichi Shiro	11350.7 ± 100.45gA	5029.95 ± 25.77fB	324.31 ± 1.60cA	84.50 ± 1.84dB
Jumbo pink banana	7499.11 ± 83.49dA	5194.75 ± 45.07fB	386.83 ± 2.43dA	140.94 ± 1.76hB
Golden hubbard	5635.76 ± 78.83bA	3255.99 ± 38.26cB	405.43 ± 2.81eA	95.31 ± 1.12eB
Flat White Boer Ford	7421.45 ± 78.67dA	6131.97 ± 24.33iB	509.28 ± 3.62gA	169.41 ± 1.16iB
Jarahdale	6197.18 ± 50.75cA	5314 ± 26.52gB	478.86 ± 4.04fA	110.17 ± 1.83fB
Porcelain doll	9321.91 ± 81.67fA	5782.54 ± 56.89hB	501.19 ± 5.79gA	202.2 ± 2.13jB
Galeux d' Eysines	6119.04 ± 84.52cA	4302.19 ± 65.47eB	524.90 ± 4.25hA	213.30 ± 2.47kB
Green hubbard	10760.30 ± 99.76gA	7349.57 ± 81.84jB	293.67 ± 2.28bA	66.34 ± 1.27cB
Marina di Chiggia	7753.32 ± 25.21eA	5715.51 ± 69.28hB	266.24 ± 2.97aA	58.19 ± 1.24bB
Melonowa Żółta	5615.31 ± 49.74bA	4095.36 ± 47.03dB	555.63 ± 6.94iA	364.90 ± 6.70mB

A–m: Different letters represent statistically significant differences ($p < 0.05$) between antioxidative activity (chelating properties and FRAP) of pumpkin cultivars. A,B: Different letters represent statistically significant differences ($p < 0.05$) between antioxidative activity of extracts (aqueous–methanol vs. aqueous).

3.3.5. Iron Chelating Activity

The analysis of metal ion binding properties is a method of determining the antioxidative potential of food. For this purpose, the ability to chelate Fe(II) ions by compounds contained in the extracts of various pumpkin cultivars was assayed (Table 9). The test revealed that the methanol–aqueous (80%) extracts of the following cultivars were characterized by the highest antioxidative activity: 'Gomez' (12,804.62 ppm EDTA/100 dm), 'Blue Kuri' (11,385.16 ppm EDTA/100 g dm), 'Shokichi Shiro' (11,350.7 ppm EDTA/100 g dm). By contrast, the aqueous extracts of the following cultivars exhibited the lowest ferric ion chelating ability: 'Hokkaido' (2,224.65 ppm EDTA/100 g dm), 'Buttercup' (2,938.91 ppm EDTA/100 g dm), 'Golden Hubbard' (3,225.99 ppm EDTA/100 g dm), 'Melonowa Żółta' (4,095.36 ppm EDTA/100 g dm), and 'Galeux d'Eysines' (4,302.19 ppm EDTA/100 g dm). As in the FRAP test, the methanol–aqueous (80%) extracts exhibited the highest antioxidative activity.

3.3.6. Oxygen Radical Absorbance Capacity (ORAC)

The antioxidative activity of the pumpkin cultivars was also measured with the oxygen radical absorbance capacity test (ORAC). This is a very high sensitivity method, which is mainly used to measure the activity of hydrophilic antioxidants [53]. It is thought to be one of the most preferable tests for measuring the antioxidative activity [54]. The ORAC test has been frequently used to prepare rankings of food products according to their antioxidative capacity [55]. The method is recommended to quantify the peroxide radical scavenging capacity. The highest antioxidative capacity was found in the following cultivars: 'Melonowa Żółta' (122.73 μM Tx/g dm), 'Jumbo Pink Banana' (117.72 μM Tx/g dm), and 'Golden Hubbard' (108.47 μM Tx/g dm) (Table 8). The following cultivars exhibited the lowest reactive oxygen species scavenging capacity: 'Flat White Boer Ford' (42.38 μM Tx/g dm), 'Blue Kuri' (58.47 μM Tx/g dm), 'Jarahdale' (65.00 μM Tx/g dm), 'Shokichi Shiro' (87.60 μM Tx/g dm), and 'Hokkaido' (89.97 μM Tx/g dm). So far, the ability of pumpkin pulp to absorb oxygen free radicals has not been analyzed. Parry et al. investigated the antioxidative activity (ORAC) of the oil extract from roasted pumpkin seeds. They noted that the product had very low antioxidative potential (1.1 μM Tx/g fat), as compared with parsley seed extract (1097.5 μM Tx/g fat), cardamom extract (941.5 μM Tx/g fat), milk thistle extract (125.2 μM Tx/g fat), and onion extract (17.5 μM Tx/g fat) [56].

3.3.7. Correlation between Antioxidative Properties of *Cucurbita maxima* Cultivars Observed in Different Tests

The results of the antioxidative activity assays were used to analyze correlations between the antioxidative activity exhibited both by the aqueous and aqueous–methanol extracts. As far as the aqueous extracts are concerned, there was a strong positive correlation between the results obtained in the ABTS and oxygen radical absorbance capacity (ORAC) tests ($\varrho = 0.64$; $p < 0.001$), as well as between the total polyphenolic content and the ORAC ($\varrho = 0.59$; $p < 0.001$) and ABTS tests ($\varrho = 0.41$; $p < 0.01$) (Table 10). The results showed a strong negative correlation between the ORAC and FRAP tests ($\varrho = -0.47$; $p < 0.01$), as well as between total polyphenolic content and the chelating activity ($\varrho = -0.47$; $p < 0.01$). Likewise, the analysis of the results for the aqueous–methanol extracts revealed a positive correlation between the total polyphenolic content and the antioxidative activity assayed in the ABTS test ($\varrho = 0.37$; $p < 0.05$) (Table 11). In addition, there was a positive correlation between both FRAP and DPPH tests ($\varrho = 0.45$; $p < 0.05$) and between the FRAP test and the total polyphenolic content ($\varrho = 0.39$; $p < 0.05$). On the other hand, there were negative correlations between the chelation activity and the total polyphenolic content ($\varrho = -0.63$; $p < 0.001$) and the DPPH radical scavenging ability ($\varrho = -0.53$; $p < 0.001$).

Table 10. Results of the correlation analysis between the antioxidant activity tests in *Cucurbita maxima* cultivars aqueous extracts.

	ABTS	DPPH	Total Phenolic Content	Iron Chelating Activity	FRAP	ORAC
ABTS	x	−0.01	0.41 **	−0.37 *	−0.38 *	0.64 ***
DPPH	−0.01	x	0.14	−0.26	0.32	−0.07
Total polyphenols content	0.41 **	0.14	x	−0.47 **	−0.24	0.59 ***
Iron chelating activity	−0.37 *	−0.26	−0.47 **	x	−0.01	−0.35 *
FRAP	−0.38 *	0.32	−0.24	−0.01	x	−0.47 **
ORAC	0.64 ***	−0.07	0.59 ***	−0.35 *	−0.47 **	x

$p < 0.05$ *; $p < 0.01$ **; $p < 0.001$ ***: Determination of statistically significant correlations between tested variables.

Table 11. Results of the correlation analysis between the antioxidant activity tests in *Cucurbita maxima* cultivars aqueous–methanol extracts.

	ABTS	DPPH	Total Phenolic Content	Iron Chelating Activity	FRAP
ABTS	x	−0.04	0.37 *	−0.39 *	−0.06
DPPH	−0.04	x	0.38	−0.53 ***	0.45*
Total Polyphenols Content	0.37 *	0.38	x	−0.63***	0.39*
Iron Chelating Activity	−0.39 *	−0.53 ***	−0.63 ***	x	−0.16
FRAP	−0.06	0.45 *	0.39 *	−0.16	x

$p < 0.05$ *; $p < 0.01$ **; $p < 0.001$ ***: Determination of statistically significant correlations between tested variables.

3.3.8. Cluster Analysis

A cluster analysis was made to isolate groups of the pumpkin cultivars according to their antioxidative activity (Figure 1). The assumption was that the selected groups of cultivars should differ from each other in terms of the most determining variables. The Ward method was used for a hierarchical cluster analysis. Figure 1 shows into which clusters the pumpkin cultivars were categorized.

Two groups of cultivars were distinguished on the basis of the analysis: Cluster 1 ('Buttercup', 'Golden Hubbard', 'Galeux d'Eysines', 'Melonowa Żółta', 'Hokkaido', 'Jumbo Pink Banana', 'Marina Di Chiggia', 'Flat White Boer Ford', 'Jarrahdale'), and cluster 2 ('Blue Kuri', 'Green Hubbard', 'Gomez', 'Shokichi Shiro', 'Porcelain Doll') (Table 12). The Student's t-test for independent samples was applied to check whether there were intergroup differences. The analysis showed that cluster 1 exhibited significantly greater ability to scavenge ABTS (120.49 vs. 104.84 mg Tx/100 g dm) and DPPH radicals

(190.21 vs. 95.24 mg Tx/100 g dm) (in the aqueous–methanol extracts) and DPPH radicals (91.74 vs. 49.49 mg Tx/100 g dm) (in the aqueous extracts). Apart from that, cluster 1 had higher total polyphenolic content (84.46 vs. 62.78 mg GAE/100 g dm). On the other hand, cluster 2 was characterized by greater ferric ion chelating capacity in the aqueous–methanol extracts (6210.87 vs. 11061.61 ppm EDTA/100 g dm) and aqueous extracts (4346.98 vs. 6745.29 ppm EDTA/100 g dm). There were no statistically significant differences between the groups in the oxygen radicals absorbance capacity (ORAC).

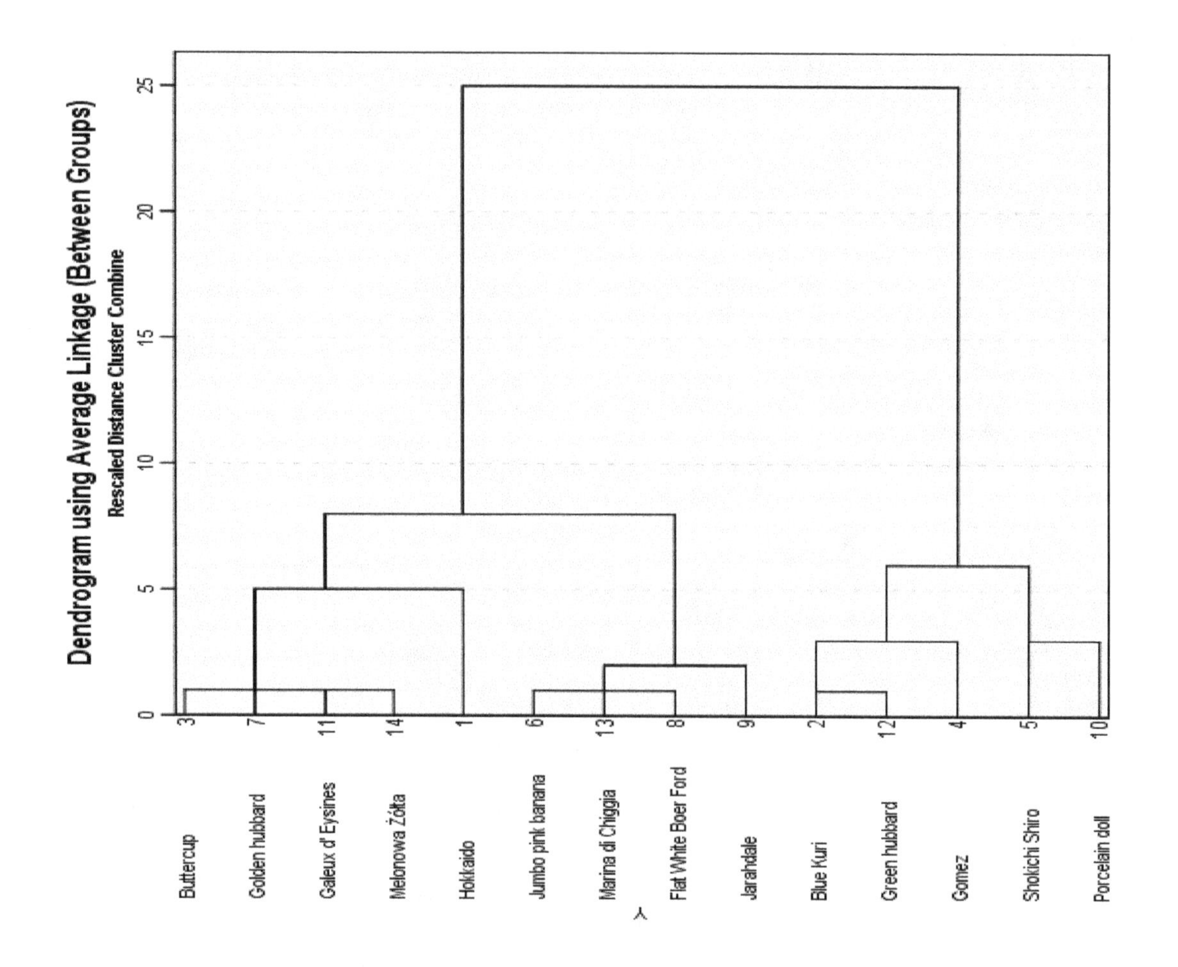

Figure 1. Dendrogram of studied pumpkin variety divisions.

Table 12. Variable levels divided into groups of pumpkin varieties.

		Cluster 1	Cluster 2	p Value
Aqueous–methanol extract	ABTS	120.49 ± 13.72	104.84 ± 14.01	$p < 0.05$
	DPPH	190.21 ± 99.12	95.24 ± 30.17	$p < 0.05$
	Total polyphenols content	84.46 ± 31.73	62.78 ± 9.72	$p < 0.05$
	Iron chelating activity	6210.87 ± 1188.94	11061.61 ± 1127.68	$p < 0.05$
	FRAP	415.28 ± 118.85	372.52 ± 76.37	NS.
Aqueous extract	ABTS	130.6 ± 25.69	122.04 ± 14.5	NS.
	DPPH	91.74 ± 40.25	49.49 ± 11.66	$p < 0.05$
	Total polyphenols content	95.54 ± 29.02	87.29 ± 22.81	NS.
	Iron chelating activity	4346.98 ± 1301.87	6745.29 ± 1191.65	$p < 0.05$
	FRAP	135.33 ± 71.86	117.54 ± 47.87	NS.
	ORAC	91.75 ± 24.88	93.51 ± 19.98	NS.

$p < 0.05$: Determination of statistically significant differences between the groups of varieties; NS: Not significant; dm: Dry mass; the results are expressed as the mean values ± standard deviation of the triplicate samples; the results are expressed in the following units: ABTS, DPPH (mg Tx/100 g dm), iron chelating activity (ppm EDTA/100 g dm), FRAP (mM Fe(II)/100 g dm), total phenolic content (mg GAE/100 g dm), ORAC (μM Tx/g dm).

4. Conclusions

The extraction methods were optimized to test the antioxidative activity of pumpkin pulp. The research showed that the aqueous and aqueous–methanol (80%) extracts exhibited the highest antioxidative potential. The best effects were achieved when the extraction was conducted at 70 °C for 2 h. The acetone and ethyl acetate extracts exhibited low antioxidative activity. The extraction optimization results were used for comparative analysis of the antioxidative potential of 14 pumpkin cultivars of the *Cucurbita maxima* species. To date there has not been such an extensive analysis of the antioxidative properties of the pumpkin pulp of various cultivars. The following tests were conducted to measure the antioxidative activity: ABTS, DPPH, FRAP, ORAC, chelating activity. The total polyphenolic content was also analyzed. The research clearly showed that the pumpkin cultivars under analysis were significantly diversified in their ability to scavenge free radicals and to reduce and chelate ferric ions. They also differed in the total polyphenolic content. The antioxidant activity of pumpkin extracts is generally affected by different variables such as the chemical structure and profile of antioxidants, e.g., carotenoids, tocopherols or phenolic compounds. However, this is the topic of the upcoming publication in which the profile of antioxidant compounds will be characterized. Antioxidants arise as easily extractable compounds, soluble, and as the residue of the extract. That is why it is difficult to identify and categorize the key trends in the contribution of various compounds and the concomitant influence of different factors to the total antioxidant capacity. Hence, as indicated in other studies on the antioxidant activity of the food matrix, consideration should also be given to substances bound in the extract residues. The following pumpkin cultivars exhibited high antioxidative activity: 'Melonowa Żółta', 'Hokkaido', 'Porcelain Doll', and 'Gomez'. The research showed that pumpkin pulp exhibited strong antioxidative properties, which might be significant for human health. The inclusion of pumpkin or pumpkin pulp-based food products into a diet may help protect the human body from the harmful effect of free radicals. It is important to reduce the risk of diseases of affluence. Properly selected powdered pumpkin pulp can be used in the food industry, e.g., as an additive increasing the stability of fat in meat.

Author Contributions: B.K., conceptualization, formal analysis, investigation, writing, review, and editing; A.G.-M., conceptualization, investigation, writing, editing, supervision, and funding acquisition; J.B.K., investigation, editing, and funding acquisition. All authors have read and agreed to the published version of the manuscript.

References

1. Oroian, M.; Escriche, I. Antioxidants: Characterization, natural sources, extraction and analysis. *Food Res. Int.* **2015**, *74*, 10–36. [CrossRef]
2. Nathan, C.; Cunningham-Bussel, A. Beyond oxidative stress: An immunologist's guide to reactive oxygen species. *Nat. Rev. Immunol.* **2013**, *13*, 349–361. [CrossRef]
3. Evans, J.R.; Lawrenson, J.G. Antioxidant vitamin and mineral supplements for preventing age-related macular degeneration. *Cochrane Database Syst. Rev.* **2017**, *2017*, CD000253. [CrossRef] [PubMed]
4. Prasad, K.; Laxdal, V.A.; Raney, B.L. Antioxidant activity of allicin, an active principle in garlic. *Mol. Cell. Biochem.* **1995**, *148*, 183–189. [CrossRef] [PubMed]
5. Pala, R.; Beyaz, F.; Tuzcu, M.; Er, B.; Sahin, N.; Cinar, V.; Sahin, K. The effects of coenzyme Q10 on oxidative stress and heat shock proteins in rats subjected to acute and chronic exercise. *J. Exerc. Nutr. Biochem.* **2018**, *22*, 14–20. [CrossRef] [PubMed]
6. Kulczyński, B.; Gramza-Michałowska, A. Characteristics of Selected Antioxidative and Bioactive Compounds in Meat and Animal Origin Products. *Antioxidants* **2019**, *8*, 335. [CrossRef]
7. Kulczyński, B.; Gramza-Michałowska, A. The importance of selected spices in cardiovascular diseases. *Postepy Hig. Med. Dosw.* **2016**, *70*, 1131–1141. [CrossRef]
8. Gramza-Michałowska, A.; Sidor, A.; Reguła, J.; Kulczyński, B. PCL assay application in superoxide anion-radical scavenging capacity of tea Camellia sinensis extracts. *Acta Sci. Pol. Technol. Aliment.* **2015**, *14*, 331–341. [CrossRef]
9. Salvador, A.C.; Król, E.; Lemos, V.C.; Santos, S.A.; Bento, F.P.; Costa, C.P.; Almeida, A.; Szczepankiewicz, D.;

Kulczyński, B.; Krejpcio, Z.; et al. Effect of Elderberry (*Sambucus nigra* L.) Extract Supplementation in STZ-Induced Diabetic Rats Fed with a High-Fat Diet. *Int. J. Mol. Sci.* **2017**, *18*, 13. [CrossRef]
10. Gramza-Michałowska, A.; Kulczyński, B.; Xindi, Y.; Gumienna, M. Research on the effect of culture time on the kombucha tea beverage's antiradical capacity and sensory value. *Acta Sci. Pol. Technol. Aliment.* **2016**, *15*, 447–457. [CrossRef]
11. Kulczyński, B.; Gramza-Michałowska, A.; Kobus-Cisowska, J.; Kmiecik, D. The role of carotenoids in the prevention and treatment of cardiovascular disease–Current state of knowledge. *J. Funct. Foods* **2017**, *38*, 45–65. [CrossRef]
12. Sidor, A.; Drożdżyńska, A.; Gramza-Michałowska, A. *Black chokeberry* (*Aronia melanocarpa*) and its products as potential health-promoting factors—An overview. *Trends Food Sci. Technol.* **2019**, *89*, 45–60. [CrossRef]
13. Renaud, J.; Martinoli, M.-G. Considerations for the Use of Polyphenols as Therapies in Neurodegenerative Diseases. *Int. J. Mol. Sci.* **2019**, *20*, 1883. [CrossRef] [PubMed]
14. Shahidi, F.; De Camargo, A.C. Tocopherols and Tocotrienols in Common and Emerging Dietary Sources: Occurrence, Applications, and Health Benefits. *Int. J. Mol. Sci.* **2016**, *17*, 1745. [CrossRef] [PubMed]
15. Thayagarajan, A.; Sahu, R.P. Potential Contributions of Antioxidants to Cancer Therapy: Immunomodulation and Radiosensitization. *Integr. Cancer Ther.* **2018**, *17*, 210–216. [CrossRef] [PubMed]
16. Gramza-Michałowska, A.; Sidor, A.; Kulczyński, B. Berries as a potential anti-influenza factor–A review. *J. Funct. Foods* **2017**, *37*, 116–137. [CrossRef]
17. Arulselvan, P.; Fard, M.T.; Tan, W.S.; Gothai, S.; Fakurazi, S.; Norhaizan, M.E.; Kumar, S.S. Role of Antioxidants and Natural Products in Inflammation. *Oxid. Med. Cell. Longev.* **2016**, *2016*, 1–15. [CrossRef]
18. Buscemi, S.; Corleo, D.; Di Pace, F.; Petroni, M.L.; Satriano, A.; Marchesini, G. The Effect of Lutein on Eye and Extra-Eye Health. *Nutrients* **2018**, *10*, 1321. [CrossRef]
19. Braakhuis, A.; Raman, R.; Vaghefi, E. The Association between Dietary Intake of Antioxidants and Ocular Disease. *Diseases* **2017**, *5*, 3. [CrossRef]
20. Kmiecik, D.; Korczak, J.; Rudzińska, M.; Gramza-Michałowska, A.; Hes, M.; Kobus-Cisowska, J. Stabilisation of phytosterols by natural and synthetic antioxidants in high temperature conditions. *Food Chem.* **2015**, *173*, 966–971. [CrossRef]
21. Gramza-Michalowska, A.; Sidor, A.; Hes, M. Herb extract influence on the oxidative stability of selected lipids. *J. Food Biochem.* **2011**, *35*, 1723–1736. [CrossRef]
22. Hęś, M.; Gramza-Michałowska, A. Effect of Plant Extracts on Lipid Oxidation and Changes in Nutritive Value of Protein in Frozen? Stored Meat Products. *J. Food Process. Pres.* **2017**, *41*, e12989. [CrossRef]
23. Ismail, H.; Lee, E.; Ko, K.; Paik, H.; Ahn, D. Effect of Antioxidant Application Methods on the Color, Lipid Oxidation, and Volatiles of Irradiated Ground Beef. *J. Food Sci.* **2009**, *74*, C25–C32. [CrossRef] [PubMed]
24. Liu, F.; Xu, Q.; Dai, R.; Ni, Y. Effects of natural antioxidants on colour stability, lipid oxidation and metmyoglobin reducing activity in raw beef patties. *Acta Sci. Pol. Technol. Aliment.* **2015**, *14*, 37–44. [CrossRef]
25. Durazzo, A.; Lucarini, M. A current shot and re-thinking of antioxidant research strategy. *Br. J. Anal. Chem.* **2018**, *5*, 9–11. [CrossRef]
26. Tanaka, R.; Kikuchi, T.; Nakasuji, S.; Ue, Y.; Shuto, D.; Igarashi, K.; Okada, R.; Yamada, T. A Novel 3a-p-Nitrobenzoylmultiflora-7:9(11)-diene-29-benzoate and Two New Triterpenoids from the Seeds of Zucchini (*Cucurbita pepo* L.). *Molecules* **2013**, *18*, 7448–7459. [CrossRef]
27. Zhou, C.-L.; Mi, L.; Hu, X.-Y.; Zhu, B.-H. Evaluation of three pumpkin species: Correlation with physicochemical, antioxidant properties and classification using SPME-GC–MS and E-nose methods. *J. Food Sci. Technol.* **2017**, *54*, 3118–3131. [CrossRef]
28. Nishimura, M.; Ohkawara, T.; Sato, H.; Takeda, H.; Nishihira, J. Pumpkin Seed Oil Extracted from Cucurbita maxima Improves Urinary Disorder in Human Overactive Bladder. *J. Tradit. Complement. Med.* **2014**, *4*, 72–74. [CrossRef]
29. Kulczyński, B.; Gramza-Michałowska, A. The Profile of Carotenoids and Other Bioactive Molecules in Various Pumpkin Fruits (*Cucurbita maxima* Duchesne) Cultivars. *Molecules* **2019**, *24*, 3212. [CrossRef]
30. Kulczyński, B.; Gramza-Michałowska, A. The Profile of Secondary Metabolites and Other Bioactive Compounds in *Cucurbita pepo* L. and Cucurbita moschata Pumpkin Cultivars. *Molecules* **2019**, *24*, 2945. [CrossRef]
31. Zaccari, F.; Galietta, G. α-Carotene and β-Carotene Content in Raw and Cooked Pulp of Three Mature Stage Winter Squash "Type Butternut". *Foods* **2015**, *4*, 477–486. [CrossRef] [PubMed]

32. Montesano, D.; Blasi, F.; Simonetti, M.S.; Santini, A.; Cossignani, L. Chemical and Nutritional Characterization of Seed Oil from *Cucurbita maxima* L. (var. Berrettina) Pumpkin. *Foods* **2018**, *7*, 30. [CrossRef] [PubMed]
33. Kim, M.Y.; Kim, E.J.; Kim, Y.N.; Choi, C.; Lee, B.H. Comparison of the chemical compositions and nutritive values of various pumpkin (*Cucurbitaceae*) species and parts. *Nutr. Res. Pract.* **2012**, *6*, 21–27. [CrossRef] [PubMed]
34. Paris, H.S.; Daunay, M.-C.; Pitrat, M.; Janick, J. First Known Image of Cucurbita in Europe, 1503–1508. *Ann. Bot.* **2006**, *98*, 41–47. [CrossRef]
35. Lust, T.A.; Paris, H.S. Italian horticultural and culinary records of summer squash (*Cucurbita pepo, Cucurbitaceae*) and emergence of the zucchini in 19th-century Milan. *Ann. Bot.* **2016**, *118*, 53–69. [CrossRef]
36. Nawirska-Olszańska, A.; Kita, A.; Biesiada, A.; Sokół-Łętowska, A.; Kucharska, A.Z. Characteristics of antioxidant activity and composition of pumpkin seed oils in 12 cultivars. *Food Chem.* **2013**, *139*, 155–161. [CrossRef]
37. Saavedra, M.J.; Aires, A.; Dias, C.; Almeida, J.A.; De Vasconcelos, M.C.; Santos, P.; Rosa, E.A. Evaluation of the potential of squash pumpkin by-products (seeds and shell) as sources of antioxidant and bioactive compounds. *J. Food Sci. Technol.* **2015**, *52*, 1008–1015. [CrossRef]
38. Can-Cauich, C.A.; Sauri-Duch, E.; Moo-Huchin, V.M.; Betancur-Ancona, D.; Cuevas-Glory, L.F. Effect of extraction method and specie on the content of bioactive compounds and antioxidant activity of pumpkin oil from Yucatan, Mexico. *LWT-Food Sci. Technol.* **2019**, *285*, 186–193. [CrossRef]
39. Brand-Williams, W.; Cuvelier, M.; Berset, C. Use of a free radical method to evaluate antioxidant activity. *LWT-Food Sci. Technol.* **1995**, *28*, 25–30. [CrossRef]
40. Re, R.; Pellegrini, N.; Proteggente, A.; Pannala, A.; Yang, M.; Rice-Evans, C. Antioxidant activity applying an improved ABTS radical cation decolorization assay. *Free. Radic. Boil. Med.* **1999**, *26*, 1231–1237. [CrossRef]
41. Sánchez-Moreno, J.; Fulgencio, S. A procedure to measure the antiradical efficiency of polyphenols. *J. Sci. Food Agric.* **1998**, *76*, 270–276. [CrossRef]
42. Benzie, I.F.; Strain, J. Ferric reducing/antioxidant power assay: Direct measure of total antioxidant activity of biological fluids and modified version for simultaneous measurement of total antioxidant power and ascorbic acid concentration. *Methods Enzymol.* **1999**, *299*, 15–27. [PubMed]
43. Decker, E.A.; Welch, B. Role of ferritin as a lipid oxidation catalyst in muscle food. *J. Agric. Food Chem.* **1990**, *38*, 674–677. [CrossRef]
44. Ou, B.; Huang, D.; Hampsch-Woodill, M.; Flanagan, J.A.; Deemer, E.K. Analysis of Antioxidant Activities of Common Vegetables Employing Oxygen Radical Absorbance Capacity (ORAC) and Ferric Reducing Antioxidant Power (FRAP) Assays: A Comparative Study. *J. Agric. Food Chem.* **2002**, *50*, 3122–3128. [CrossRef] [PubMed]
45. Kiat, V.V.; Siang, W.K.; Madhavan, P.; Jin, C.J.; Ahmad, M.; Akowuah, G. FT-IR profile and antiradical activity of dehulled kernels of apricot, almond and pumpkin. *Res. J. Pharm. Biol. Chem. Sci.* **2014**, *5*, 112–120.
46. Tiveron, A.P.; Melo, P.S.; Bergamaschi, K.B.; Vieira, T.M.F.S.; Regitano-D'Arce, M.A.B.; Alencar, S.M. Antioxidant Activity of Brazilian Vegetables and Its Relation with Phenolic Composition. *Int. J. Mol. Sci.* **2012**, *13*, 8943–8957. [CrossRef] [PubMed]
47. Singh, J.; Shukla, S.; Singh, V.; Rai, A.K. Phenolic Content and Antioxidant Capacity of Selected Cucurbit Fruits Extracted with Different Solvents. *J. Nutr. Food Sci.* **2016**, *6*, 6. [CrossRef]
48. Bayili, R.G.; Abdoul-Latif, F.; Kone, O.; Diao, M.; Bassole, I.; Dicko, M. Phenolic compounds and antioxidant activities in some fruits and vegetables from Burkina Faso. *Afr. J. Biotechnol.* **2011**, *10*, 62.
49. Valenzuela, G.M.; Soro, A.S.; Tauguinas, A.L.; Gruszycki, M.R.; Cravzov, A.L.; Giménez, M.C.; Wirth, A. Evaluation Polyphenol Content and Antioxidant Activity in Extracts of *Cucurbita* spp. *OALib* **2014**, *1*, 1–6. [CrossRef]
50. Dar, P.; Farman, M.; Dar, A.; Khan, Z.; Munir, R.; Rasheed, A.; Waqas, U. Evaluation of Antioxidant potential and comparative analysis of Antimicrobial activity of Various Extracts of *Cucurbita pepo* L. Leaves. *J. Agric. Sci. Food Technol.* **2017**, *3*, 103–109.
51. Oloyede, F.; Agbaje, G.; Obuotor, E.; Obisesan, I. Nutritional and antioxidant profiles of pumpkin (*Cucurbita pepo* Linn.) immature and mature fruits as influenced by NPK fertilizer. *Food Chem.* **2012**, *135*, 460–463. [CrossRef] [PubMed]
52. Fidrianny, I.; Darmawati, A.; Sukrasno, S. Antioxidant capacities from different polarities extracts of cucurbitaceae leaves using FRAP, DPPH assays and correlation with phenolic, flavonoid, carotenoid content. *Int. J. Pharm. Pharm. Sci.* **2014**, *6*, 858–862.

53. Chensom, S.; Okumura, H.; Mishima, T. Primary Screening of Antioxidant Activity, Total Polyphenol Content, Carotenoid Content, and Nutritional Composition of 13 Edible Flowers from Japan. *Prev. Nutr. Food Sci.* **2019**, *24*, 171–178. [CrossRef] [PubMed]
54. Garrett, A.R.; Murray, B.K.; Robinson, R.A.; O'Neill, K.L. Measuring antioxidant capacity using the ORAC and TOSC assays. *Methods Mol. Biol.* **2010**, *594*, 251–262.
55. Haytowitz, D.B.; Bhagwat, S. USDA Database for the Oxygen Radical Absorbance Capacity (ORAC) of Selected Foods, Release 2. Available online: https://naldc.nal.usda.gov/download/43336/PDF (accessed on 2 May 2010).
56. Parry, J.; Hao, Z.; Luther, M.; Su, L.; Zhou, K.; Yu, L.L. Characterization of cold-pressed onion, parsley, cardamom, mullein, roasted pumpkin, and milk thistle seed oils. *J. Am. Oil Chem. Soc.* **2006**, *83*, 847–854. [CrossRef]

14

Health-Promoting Properties of Fresh and Processed Purple Cauliflower

Joanna Kapusta-Duch [1,*], Anna Szeląg-Sikora [2], Jakub Sikora [2], Marcin Niemiec [3], Zofia Gródek-Szostak [4], Maciej Kuboń [2], Teresa Leszczyńska [1] and Barbara Borczak [1]

1 Department of Human Nutrition, Faculty of Food Technology, University of Agriculture in Krakow, 30-149 Krakow, Poland
2 Institute of Agricultural Engineering and Informatics, University of Agriculture in Krakow, 30-149 Krakow, Poland
3 Department of Agricultural and Environmental Chemistry, University of Agriculture in Krakow, 31-120 Krakow, Poland
4 Department of Economics an Organization of Enterprises, Cracow University of Economics, 31-510 Krakow, Poland
* Correspondence: joannakapustaduch@interia.pl

Abstract: Plant-based foods should be fresh, safe, and natural, with nutritional value and processed in sustainable ways. Among all consumed vegetables, Brassica vegetables are considered to be the most important ones. As they are eaten in large quantities and frequently, they may constitute an important source of nutrients and bioactive compounds in a daily diet. This work is aimed at assessing the effect of technological processing (blanching and traditional cooking in water and in a convection steam oven) as well as the method of frozen storage (in PE-LD zipper bags and vacuum packing) on the content of selected components in purple cauliflower. The material was examined for the content of dry matter, vitamin C, total polyphenols, anthocyanins, thiocyanates, nitrates, and nitrites, as well as antioxidant activity. All technological processes caused significant changes in the contents of examined nutritive and non-nutritive compounds as well as in antioxidant activity or the level of selected chemical pollutions. A trend was also observed towards lower constituents' losses as a result of convection steaming, compared to traditional cooking in water. Moreover, the reduction in the content of examined compounds was smaller in vacuum-packed and frozen-stored vegetables then in those stored in zipper PE-LD bags.

Keywords: nutritional value; brassica vegetables; antioxidative properties; quality of food; nitrates and nitrites; frozen storage; processing of vegetables

---

1. Introduction

Sustainable food consumption is a significant aspect of sustainable development. Throughout the last two decades, Brassica crops were of intense interest to many researchers due to their health benefits [1]. Numerous epidemiological and pharmacological works showed a significant role of a diet abundant in Brassica vegetables, which may protect against many chronic diseases, including type II diabetes, cardiovascular disease, age-related macular degeneration, dementia, immune dysfunction, obesity, and some kinds of cancers [2,3]. Cauliflower (*Brassica oleracea* var. *botrytis*) belongs to the very popular Brassica species and is broadly used as a dish or an ingredient of soups or salads. It is rich in vitamins B_1, B_2, B_3, B_5, B_6, folic acid and C, E, K, as well as omega-3 fatty acids, dietary fiber, potassium, phosphorus, magnesium manganese, and iron. Chemical components contents in Brassica vegetables vary among cultivars, which is especially relevant to cauliflower because, in addition to the large group of white-curded cultivars, breeding techniques have resulted in commercially available genotypes forming green, purple, and orange curds, with enhanced synthesis of chlorophylls, anthocyanins, and carotenoids, respectively. Cauliflower genotypes show differences in the content of bioactive compounds, as well as in the chemical composition [4–6]. In addition, this vegetable contains also a lot of valuable and healthy plant's metabolites, including flavonoids, terpenes, S-methylcysteine sulfoxide, sulfur-containing glucosinolates, coumarins, and other minor compounds. These compounds of cauliflower, and other Brassica vegetables, were found to be effective in the protection of some kinds of cancer as cancer-fighting components [4,7,8]. The number of studies has suggested protective effects of these compounds on human health. Brassica species are commonly present in a diet as additives to meat dishes and other products rich in fat, which constituents favour cell transformation and cancer growth [9]. Therefore, some of Brassica crops are classified as functional foods [7].

Thermal treatment of different types of food products leads to various physical, chemical and biological changes occurring in nutrients as well as non-nutrient compounds. In vegetable processing, cooking is the most commonly used technique; although, the application of high temperature may affect the basic chemical composition of as well as activity of bioactive compounds. Two contrary phenomena may be responsible for changes in the content of bioactive compounds during this process: denaturation of enzymes that are involved in degradation of nutrients and bioactive compounds and softening effect of cooking, which increases the extractability of bioactive compounds. As a result, their amount in cooked products is higher than in the raw material [10]. Therefore, in order to evaluate their accessibility with a diet, more knowledge should be gained about the role and the final concentration of bioactive compounds before and after food processing.

New thermal technologies with higher energy efficiency, less nutrient loss and less environmental impacts are being developed. It is believed that for example the steam blanching is relatively inexpensive and retains more water-soluble ingredients and minerals in comparison to water blanching.

The main drawback to vegetables is that they are perishable and available seasonally. Food technology is focused on the development of such methods of food processing, which will affect its chemical composition to the least extent. In view of this, freezing, which is a simple and fast method, is one the most universal and convenient ways of food preservation. Knowledge about optimization of freezing and storing processes is crucial to preserve beneficial and bioactive compounds of vegetables [6]. In this context, the importance of packaging and its role is huge, especially considering its fundamental role in the product protection against the external conditions as well as mechanical damages. In order to guarantee high product quality in terms of their sensory and nutritional features, selection of suitable packaging materials is required [11].

As for packaging properties, low-density polyethylene (PE-LD) is characterized by low permeability to water vapor and good permeability to gases, especially carbon dioxide [12]. Vacuum packaging, as a static form of hypobaric storage, is widely used in food industry. This method allows oxidative reactions to be reduced effectively in a product, at relatively low cost [13].

Until now, the majority of studies concerned other than examined here *Brassicas* cultivars. In consequence, less information is available about such Brassicas like, for example, colored varieties of cauliflower.

High quality foods obtained with sustainable practices during preharvest need proper postharvest practices, including thermal treatments and storage. The aim of the work was to assess the effect of technological processing (blanching, traditional cooking in water and in a convection steam oven) as well as the method of frozen storage (in PE-LD zipper bags and vacuum packing) on the content of selected components in purple cauliflower.

2. Materials and Methods

2.1. Material

The experimental material was purple cauliflower (*Graffiti* cv.) cultivated at the Producer Cooperative "Traf" in Tropiszów (Poland). The experimental field was located in the northern outskirts of the Krakow. The climate of the region is humid continental (Dfb) according to Köppen's classification. The purple cauliflower was grown in black soil on loess framework with neutral pH. Mineral fertilization was applied according to the fertility of soil and the nutritional requirements of the species and condition treatments (mechanical weed control, diseases, and pests) were carried out during the growing season (depending on soil and weather conditions). Mineral fertilization included 400 kg Polifoski PK (MgS) 15-24-(6-7), 100 kg of Saletrzak NH_4NO_3 + $CaCO_3$ and twice foliar application of Folicare NPK 18:18:18; 5:17:40.

Vegetable samples were prepared for analyses directly after harvest. Properly prepared representative medium samples of vegetables (fresh, blanched, cooked in water, convection steamed and frozen, and stored in different package types for 2 and 4 months), were examined for the content of dry matter, vitamin C, total polyphenols, anthocyanins, thiocyanates, nitrates and nitrites as well as antioxidant activity. In addition, the contents of protein, fat, carbohydrates, ash and fiber were determined in fresh material.

The first step of processing (before thermal treatment) included leaf removing, washing in running water, and dividing the heads into roses 4–6 cm in diameter and 5 cm in length. Then, the vegetables were mixed in order to obtain the representative average laboratory samples (a minimum of three for each analysis performed on the fresh material and the same procedure was on material after cooking). Analyses of fresh vegetables were carried out immediately after the pretreatment.

The blanching of material was carried out in the HENDI steam convection oven (model G715RXSD) for 5 min at 100 °C and then cooking to the consumer's softness in the same oven for 20 min at 100 °C. After blanching, the material was chilled and dried at room temperature for about 20 min. Another part of the fresh material was washed, dried using the filter paper, shredded mechanically, frozen at −22 °C, and, finally, freeze-dried in the Christ Alpha 1–4 apparatus. Afterwards, the material was comminuted in the Knifetec 1095 Sample Mill (Tecator) until reaching a homogenous sample with the possibly smallest diameter of particle. Adequately labelled samples were stored at −22 °C in plastic containers in a Liebherr GTS 3612 chamber freezer (Germany), until analyses of protein, fat, dietary fiber and ash.

Another part of the vegetables was cooked in a stainless steel pot using an electric cooking plate, in unsalted water, and in the initial phase of hydrothermal treatment—without a lid but in accordance with the principle "from farm to fork". The proportion of water to the raw material is 5:1 by weight. The cooking time applied was 15 min. The boiled vegetables were then prepared as described for fresh, blanched and steamed vegetables.

The material blanched in a convection steam oven was divided and packed in two types of packaging: the first batch in the conventional polyethylene (PE-LD) zipper bags (0.915–0.935 g/cm^3 in density and 230 × 320 mm in size); the remainder in the special vacuum bags adapted for this purpose, applying a RAMON vacuum packaging machine (60% vacuum; pressure: 0.10 MPa). Next, hermetically sealed samples were stored at −22 °C in a Liebherr GTS 3612 chamber freezer (Germany). Analyses were carried out on the raw material, the material blanched, cooked and the frozen product. Frozen samples were analyzed after 2 and 4 months of frozen storage. The experimental material, taken from every package (on average: 3 roses differing in diameter—from the smallest up to the largest), was collected and then homogenized using a homogenizer (CAT type X 120) to obtain a mean representative sample.

2.2. *Analytical Methods*

The experimental material was taken from every container (on average: 3 roses differing in diameter—from the smallest up to the largest) and homogenized using a homogenizer (CAT type X 120) in order to obtain a mean representative sample.

At the same time, 70% methanol extracts have been prepared to determine total polyphenols (calculated per chlorogenic acid) by means of the colorimetric measurement of colorful substances occurring as a result of the reaction between phenolic compounds and a Folin–Ciocalteu reagent (Sigma) [14]. The same extract was also used to measure antioxidant activity based on the $ABTS^+$

free radical scavenging ability by a colorimetric assessment of the amount of the $ABTS^+$ free radical solution, which remained unreduced by the antioxidant in the products [15]. In addition, 70% acidified methanolic extracts (5 g of raw vegetables in 80 mL of 70% acidified methanol solution) were prepared [16], which were then used to determine anthocyanin content.

The amount of total phenols in the extracts was measured by a spectrophotometric method at 760 nm on a RayLeigh UV-1800 spectrophotometer, according to the procedure described by Folin–Ciocalteu. The results were calculated using a chlorogenic acid equivalent (CGA) and expressed in milligrams per 100 g of fresh or dry mass, based on a standard curve.

Antioxidant activity was measured by means of colorimetric determination of the quantity of the colored solution of $ABTS^+$ free radical (2,2′-azinobis-(3-ethylbenzothiazoline-6-sulfonic acid)) because of its reduction performed by the antioxidants present in the product examined. Absorbance was measured at 734 nm on a RayLeigh UV-1800 spectrophotometer. The values obtained for every sample were compared to the concentration–response curve of the standard Trolox solution and expressed as micromoles of Trolox equivalent per gram of fresh and dry mass (TEAC).

Anthocyanin content, measured as anthocyanins' absorbance in visible light, was determined in buffer solutions at two pH values: 1 and 4.5. At pH 1 anthocyanins took the form of a red-colored flavic cation, while at pH 4.5 they had changed to the colorless pseudobase. Anthocyanin content was converted to mg of cyanidin-3-glucoside [17].

The dry matter content in the vegetable samples was found in agreement to PN-90/A-75101/03 [18]. The determination principle comprised determining the decrease in mass upon removal of water from the product during thermal drying at the temperature of 105 °C, under normal pressure conditions.

The contents of total ascorbic acid and dehydroascorbic acid were determined using 2,6-dichlorophenoloindophenol in accordance with PN-A-04019:1998 [19]. Extraction of the ascorbic acid was performed using oxalic acid solution. Vitamin C content was expressed as milligrams per 100 g of dry mass (mg/100 g d.m.).

Determination of thiocyanate was based on extraction of the sample with trichloroacetic acid (TCA) and the reaction with ferric ions. Under acidic conditions, blood-red coloration was created due to the formation of $Fe(SCN)^{2+}$ to $Fe(SCN)_6^{3-}$ complexes.

Determinations of nitrates and nitrites were carried out in accordance with the Polish standard PN-92/A-75112 [20]. The colorimetric method was used to determine these contaminants based on nitrite colored reaction with Griess I and II. Previously nitrates must be reduced to nitrites. Nitrate content was established using Griess I (sulfanilamide, Sigma-Aldrich) and Griess II (*n*-(1-Naphtyl)ethylene-diamine dihydrochloride, water solution, Sigma-Aldrich). The principle of this method is to induce in acidic conditions, a color reaction of nitrate(III) with *n*-(1-Naphtyl)ethylene-diamine dihydrochloride.

Raw, freshly prepared, lyophilized samples were also examined for protein content using a Tecator Kjeltec 2200; fat content, using a Tecator Soxtec Avanti 2050; ash content, by means of dry mineralisation in muflon oven at 525 °C in accordance with PN-A-79011-8:1998 [21]; and dietary fiber content, with the enzymatic-gravimetry method using the Tecator Fibertec System E.

Protein content was determined by mineralization of the product in concentrated sulfuric acid (IV) (aqueous mineralization), followed by alkalizing the solution, distillation of the ammonia released, and its qualitative determination [22]. As for fat content, at first fat extraction was performed from the dried material with an organic solvent (petroleum ether), then distilling off the solvent, drying the residue, and determining the mass of the extracted "crude fat" [23]. Analysis of dietary fiber content was conducted according to Polish standard [24] by means of enzymatic and gravimetric methods. Lyophilized samples were subjected to gelatinization with a thermally stable α-amylase, then digested by enzymes involving protease and amyloglucosidase to remove protein and starch present in the sample. Soluble dietary fiber was precipitated by adding ethanol. The sediment was then filtered off, washed in ethanol and acetone, and, after drying, weighed. Half the samples was analyzed for the presence of protein and the remainder incinerated. Total dietary fiber has been calculated as the mass of sediment minus the mass of protein and ash.

In addition, the percentage of carbohydrates has been calculated, as the difference between 100 g of fresh product and the sum of water (g), total fat (g), protein (g), and mineral compounds—ash (g).

2.3. Statistical Analysis

All analyses were conducted in three parallel replications and mean ± standard deviations (SD) were calculated for the values obtained. Significance of differences between mean values of raw, blanched, cooked and frozen stored material was checked by one-way analysis of variance (ANOVA). The significance of differences was estimated with the Duncan test at the critical significance level of $p \leq 0.05$. The Statistica 10.1 (Statistica, Tulsa, OK, USA) program was applied.

3. Results

3.1. Basic Composition and Dietary Fiber

Dry matter content in the vegetable varies depending on the process applied, therefore, to show only an effect of the applied process, all the results presented below have been calculated per the dry matter unit. Fresh vegetable contained 19.8 g dry matter, 2.36 g proteins, 0.14 g fat, 0.53 g ash, 16.3 g total carbohydrates, and 2.36 g dietary fiber per 100 g of fresh vegetable (Table 1).

Table 1. Selected antioxidative and bioactive compounds, antioxidant activity, basic composition, and nitrates and nitrites of raw purple cauliflower.

Component	Unit	Mean
Dry Mass	g/100 g	9.18 ± 0.04
Vitamin C	mg/100 g d.m.	689.54 ± 1.54
Total Polyphenols	mg CGA/100 g d.m.	1376.36 ± 3.85
Antioxidant Activity	µmol Trolox/g d.m.	79.85 ± 0.46
Thiocyanates	(SCN) mg/100 g d.m.	26.25 ± 1.69
Anthocyanins	µmol/g d.m.	78.21 ± 5.85
Total Protein	g/100 g d.m.	25.70 ± 1.56
Fat	g/100 g d.m.	1.55 ± 0.21
Ash	g/100 g d.m.	5.73 ± 0.29
Total Carbohydrates	g/100 g d.m.	66.72 ± 1.16
Dietary Fiber	g/100 g d.m.	25.67 ± 1.63
Nitrates	mg $NaNO_3$/kg d.m.	605.23 ± 23.72
Nitrites	mg $NaNO_2$/kg d.m.	17.97 ± 0.31

Technological treatments, both traditional cooking and cooking in the convection steam oven, resulted in a significant ($p \leq 0.05$) increase in dry matter content by 17.1 and 12.7% respectively, compared to fresh cauliflower. Frozen storage of the analyzed material for 2 and 4 months led to a substantial ($p \leq 0.05$) increase in the dry matter content respectively by 31.1 and 32.2% (conventional, in a PE-LD bag) and 30 and 32.2% (in vacuum) compared to the blanched product (Table 2).

Table 2. Content of selected compounds and antioxidant activity of purple cauliflower depending on the technological processing.

The Kind of Processing	Dry Mass g/100 g	Vitamin C mg/100 g d.m.	Total Polyphenols mg CGA/100 g d.m.	Antioxidant Activity µmol Trolox/g d.m.	Thiocyanates (SCN) mg/100 g d.m.	Anthocyanins µmol/g d.m.	Nitrates mg $NaNO_3$/kg d.m.	Nitrites mg $NaNO_2$/kg d.m.
fresh	$9.18^{a} \pm 0.04$	$689.54^{f} \pm 1.54$	$1376.36^{f} \pm 3.85$	$79.85^{d} \pm 0.46$	$26.25^{d} \pm 1.69$	$78.21^{c} \pm 5.85$	$605.23^{e} \pm 23.72$	$17.97^{c} \pm 0.31$
blanched	$8.85^{a} \pm 0.07$	$683.62^{f} \pm 7.99$	$1358.76^{f} \pm 13.58$	$80.79^{d} \pm 0.80$	$26.89^{d} \pm 0.31$	$80.34^{c} \pm 8.15$	$580.79^{e} \pm 33.56$	$17.29^{bc} \pm 0.48$
cooked	$10.75^{b} \pm 0.35$	$305.30^{d} \pm 0.26$	$493.40^{a} \pm 1.84$	$47.53^{a} \pm 2.76$	$12.27^{a} \pm 1.18$	$28.47^{a} \pm 5.92$	$219.16^{b} \pm 7.10$	$11.26^{a} \pm 1.18$
steamed	$10.35^{b} \pm 0.21$	$362.42^{e} \pm 4.24$	$1143.77^{e} \pm 15.30$	$61.55^{c} \pm 1.78$	$19.13^{bc} \pm 0.96$	$67.73^{bc} \pm 8.20$	$470.53^{d} \pm 4.10$	$14.11^{ab} \pm 1.50$
after 2 months of frozen storage (zipper bags)	$11.60^{c} \pm 0.14$	$245.69^{b} \pm 19.51$	$968.97^{c} \pm 6.10$	$52.76^{b} \pm 0.98$	$19.39^{bc} \pm 1.10$	$52.76^{b} \pm 11.83$	$211.81^{b} \pm 2.80$	$13.62^{a} \pm 1.10$
after 2 months of frozen storage (vacuum)	$11.50^{c} \pm 0.14$	$280.87^{c} \pm 2.46$	$1013.04^{d} \pm 7.38$	$59.39^{c} \pm 0.37$	$20.09^{c} \pm 1.23$	$60.43^{bc} \pm 17.09$	$291.39^{c} \pm 7.26$	$14.61^{abc} \pm 0.86$
after 4 months of frozen storage (zipper bags)	$11.70^{c} \pm 0.14$	$197.43^{a} \pm 1.21$	$899.23^{b} \pm 15.59$	$44.87^{a} \pm 1.21$	$13.08^{a} \pm 1.33$	$43.68^{ab} \pm 8.70$	$148.03^{a} \pm 6.41$	$22.99^{d} \pm 1.45$
after 4 months of frozen storage (vacuum)	$11.70^{c} \pm 0.28$	$232.48^{b} \pm 8.46$	$976.32^{c} \pm 11.72$	$53.08^{b} \pm 2.05$	$16.50^{b} \pm 2.05$	$53.85^{b} \pm 8.46$	$234.02^{b} \pm 8.70$	$33.16^{e} \pm 2.78$

Values are presented as mean value ± standard deviation and expressed in dry matter. Means in columns with different superscript letters in common (a, b, c, d, e) differ significantly ($p \leq 0.05$).

3.2. *Vitamin C*

Studies revealed that fresh product contained 63.2 mg/100 g (689.54 mg per 100 g dry matter) vitamin C (Table 2). Traditional cooking and convection steaming led to significant ($p \leq 0.05$) losses of ascorbic acid of 55.7 and 47.4% respectively, compared to the fresh vegetable. In the frozen stored material, all changes were significant in comparison with the blanched cauliflower ($p \leq 0.05$). Losses of vitamin C in products stored for 2 months were 64.1% and 58.9%, respectively, for the conventionally packed and vacuum packed ones. In the material stored for 4 months, vitamin content decreased by 71.1% (traditional packaging) and by 66.0% (vacuum packaging). At every stage of the study, losses were significantly ($p \leq 0.05$) lower in the vegetables stored in vacuum pouches compared to those kept in conventional l zipper bags.

3.3. *Total Polyphenols*

It has been revealed that 100 g of purple cauliflower had 126.35 mg of total polyphenols, expressed as chlorogenic acid (1376.36 mg/100 g dry matter) (Table 2). In comparison with the fresh vegetable, the applied hydrothermal treatment caused statistically significant ($p \leq 0.05$) losses of these components: by 64.2% in traditional cooking and by 16.9% in convection steaming. Losses of total polyphenols after 2 and 4-months of frozen storage in a traditional way and in vacuum, were respectively 28.7 and 33.8% as well as 25.4 and 28.1% and were statistically significant ($p \leq 0.05$) compared to the blanched material. At every stage of this experiment, they were significantly ($p \leq 0.05$) lower for steam-cooked vegetables and those frozen stored in vacuum.

3.4. *Anthocyanins*

Purple cauliflower contained 7.18 mg anthocyanins per 100 g fresh material (78.21 mg/100 g dry matter). Traditional cooking in water significantly reduced their content (63.6%), compared to fresh vegetables. However, the remaining heat treatments (blanching and convection steaming) had no significant effect on anthocyanin content in this cauliflower. After 2 and 4 months of frozen storage, anthocyanin content compared to blanched vegetables decreased significantly ($p \leq 0.05$) by respectively 34.3 and 45.6% (traditional packaging) and 24.8 and 33% (vacuum packaging). At every stage of the research, these losses were significantly ($p \leq 0.05$) lower in the vacuum packed vegetables than in those stored in PE-LD bags.

3.5. *Antioxidant Activity*

The study revealed that antioxidant activity in purple cauliflower was 7.33 μmol Trolox per 1 g fresh of vegetable matter (79.85 μmol Trolox/g dry vegetable matter). The applied technological treatments (except for blanching) caused significant ($p \leq 0.05$) reductions in antioxidant activity compared to not processed vegetables, by 40.5% for traditional cooking and by 22.9% for convection steaming. Frozen storage of vegetables in PE-LD bags and vacuum pouches for 2 and 4 months resulted in a significant fall in antioxidant activity, compared to blanched vegetables, of 34.6 and 44.5%, and 26.5 and 34.3%, respectively. At all stages of the experiment the reduction of this parameter was significantly ($p \leq 0.05$) lower in steam-boiled vegetables and in those stored in vacuum.

3.6. *Thiocyanates*

The content of thiocyanates (SCN) in purple cauliflower was 2.41 mg/100 g fresh mass (26.25 mg/100 g dry mass). As in the case of the majority of the discussed components as well as antioxidant activity, blanching did not change significantly ($p \leq 0.05$) the content of thiocyanates compared to the fresh material. However, losses were recorded after traditional cooking (53.3%) and convection steaming (27.1%). In turn, 2- and 4-month frozen storage of cauliflower reduced significantly ($p \leq 0.05$) the content of these components compared to blanched material by 27.9 and 51.3% (traditional storage in PE-LD bags) and by 25.3 and 38.6% (vacuum storage), respectively.

3.7. Nitrates and Nitrites

This study showed that purple cauliflower had 55.56 mg of nitrates, expressed as potassium nitrate (KNO_3), per 1000 g fresh mass (605.23 mg/1000 g dry mass). Traditional cooking and convectional steaming led to a significant ($p \leq 0.05$) decrease in these compounds compared to fresh vegetables, by 63.8 and 22.2% respectively. In addition, 2- and 4-month frozen storage in two package types—conventional PE-LD and vacuum packaging—caused a significant reduction in nitrates: by 63.5 and 74.5%, and 49.8 and 59.7%, respectively, compared to blanched vegetables. At every stage of this work, the reduction of this parameter was significantly ($p \leq 0.05$) lower in steam-cooked vegetables and in those frozen stored in vacuum.

The content of nitrites in 1000 g of fresh material was 1.65 mg (17.97 mg/1000 g dry matter). The applied hydrothermal treatment reduced significantly ($p \leq 0.05$) nitrites compared to the fresh material, by 37.3% (traditional cooking) and 21.5% (convection steaming). Two months' frozen storage of blanched cauliflower led to decreases in the content of nitrites in the examined material; however, only the difference recorded for conventional packaging (21.2%) was statistically significant ($p \leq 0.05$). In the case of 4-month stored cauliflower, significant ($p \leq 0.05$) increases in the content of nitrites were recorded in the products packed conventionally (by 33%) and in vacuum (by 91.8%).

4. Discussion

4.1. Basic Composition and Dietary Fiber

This study revealed that protein content (g per fresh and dry mass) in purple cauliflower was close to that found by Rumpel [25] and Kunachowicz et al. [26] and Kahlon et al. [27], according to whom, its values were 2.5 g and 2.4 g/100 g fresh weight of white cauliflower and 27.8 g/100 g dry matter, respectively. Fat content in white cauliflower, determined by the latter author, was 1.7 g/100 g dry matter, which is a higher value compared to that obtained in this work. Rumpel [25] reports that the content of assimilable carbohydrates in white cauliflower is 2.6 g/100 g of fresh vegetable, while Schonhof et al. [28] gives two values—2.27 and 2.56—depending on the cultivar, which differs from our findings. On the other hand, the content of dietary fiber found by Kahlon et al. [27] was similar (62.1 g/100 g dry mass) to our results. Both Rumpel [25] and Kunachowicz et al. [26] determined dietary fiber in white rose cauliflowers at the levels of 2.9 and 2.4 g/100 g fresh weight respectively, which is close to our findings. According to Puupponen-Pimiä et al. [29], the content of this constituent was 30.2 g/100 g dry matter unit. The values reported by Kahlon et al. [27] for ash content (10.1 g/100 g dry matter), considerably exceeding that obtained in this work (5.73 g/100 g dry matter). However, Ali [30] noted a similar content of micro- and macroelements, but in white cauliflower florets.

4.2. Dry Matter

According to the literature [31,32], dry matter content in the fresh white rose of Romanesco cauliflowers ranges from 6.96 to 13.95 g/100 g, which agrees with our results. Cooking in water as well as convection steaming increased its content compared to the fresh material, by 17.1% and 12.7% respectively. Similar increases, of 18.0% (white rose cauliflower) and 12.5% (green rose cauliflower), were also reported by Gębczyński and Kmiecik [33]. The reduced water content in the examined material was probably caused by the release of water from the tissues damaged during high-temperature processes [34].

This work showed that after 2-month frozen storage at −22 °C, the content of dry matter increased, which is congruent with the findings of Gębczyński and Kmiecik [33] and Cebula et al. [35], who noted mean increases of 22.3 and of 10.5% (white rose cultivar) and 14.5% (*Romanesco* cv.), respectively. An increase in dry matter in the frozen and then stored material may probably be explained by the phenomenon of denaturing cell walls at low temperature and resulting release of water [36].

4.3. Vitamin C

In 100 g of fresh cauliflower there was 63.20 mg of vitamin C, which is consistent with the findings of Kaulman et al. [37] and Volden et al. [38], who noted values of 63 mg and 64.2 mg (*Graffiti* cv.), respectively. The obtained results were in agreement also with those of Bhandari and Kwak [39] and Picchi et al. [40], who reported that vitamin C in cauliflower ranged from 396.7–649.7 and 346–638 mg of 100 g dry weight, respectively.

This study showed minimal losses of this component due to blanching, which was carried out using a convection steam oven. Cooking in water reduced vitamin C content by 55.7%, while convection steaming by 47.4%. Lower losses of vitamin C in this process were observed by Filipiak-Florkiewicz [31], Davey et al. [41], and Pellegrini et al. [42], and were 38.2% (white cauliflower) and 36.9% (*Romanesco* variety), 14.1–29.5%, and 50.9%, respectively.

Technological treatments for example, pre-processing or blanching may lead to considerable losses particularly in vitamin C. Their extent depends on the applied temperature, length of the exposure to this temperature, and a degree of the product comminuting [43]. Conventional frozen storage of cauliflower did result in a vitamin C decrease of 64.1 (after two months) and 71.1% (after 4 months). Volden et al. [38] noted losses in this constituent content after 3-month frozen storage, of 22.7% for the cultivar Graffiti and 24.1% for the green cauliflower cultivar *Celio*.

Incedayi and Suna [44] treated one group of cauliflower florets with 1,5% citric acid solution and the another group with 0.5% Ca-ascorbate plus citric acid 1% solution. Afterwards, florets were packed in two MAP gas mixtures and into biaxially oriented polypropylene film (BOPP). After 15 days of storage at 4 °C, vitamin C content decreased by 11.5%. Vacuum packaging led to a significantly lower ($p \leq 0.05$) reduction in the vitamin C content of 62.4%, on average, compared to conventional packing in zipper bags.

The losses in vegetable constituents are caused by both internal factors like, for example, respiration, production and the effect of ethylene, or changes in chemical composition, and external (environmental) ones such as relative humidity of the air, temperature, and gas composition of the surrounding atmosphere. Vacuum packaging reduces the access of oxygen to the material, which protects it against losses, for example, oxidation of vitamin C [45]. Quality of plant raw material is affected by factors connected with cultivars and by the broadly understood agricultural engineering factors [46].

4.4. Total Polyphenols

In most of the literature polyphenols are determined in various cauliflower varieties with white florets and calculated as total polyphenol content per 100 g fresh material. Volden et al. [38] and Pellegrini et al. [42] reported per 100 g of fresh vegetable 59.9 and 146 mg of total polyphenols, respectively. The results stated by Kaulman et al. [37], Bahorun et al. [47], and Kaur and Kapoor [48] are much lower: 63.8 mg, 27.8 mg, and 96 mg/100 g, respectively. Significant differences are most probably the result of the variety of the analyzed cultivars, growing conditions and the results' calculation methods.

Compared to fresh cauliflower, losses in total polyphenol content due to cooking in water were substantially larger than from blanching; slightly lower values were noted by Pellegrini et al. [42] and Filipiak-Florinkiewicz [31] for white varieties: 45.4 and 43.6%, respectively. Our results showed that losses resulting from traditional cooking were substantially larger than those found by Mazzeo et al. [49] and Picchi et al. [40] for varieties with white florets, being 30.8% and 18.9% respectively. As for cooking in a convection steam oven, the obtained results are almost identical with that reported by Pellegrini et al. [42] for white cauliflower (17.7%). Frozen storage of purple cauliflower led to a decline in the level of these compounds. Losses in the material stored for 4 months, noted by Gębczyński and Kmiecik [33], were much lower and were 11.6% (white cauliflower) and 20.4% (green cauliflower). This is consistent with the findings of Puupponen-Pimiä et al. [29], who observed a reduction in the content of these compounds by only 14.3% after 6 months of frozen storage.

Chassagne-Berces et al. [50] also noted that long-lasting storage of raw materials enhances the processes of enzymatic or chemical oxidation of polyphenolic compounds, the extent of which depends on the raw material or medium parameters such as temperature, pH, water activity, time, and oxygen content.

4.5. Anthocyanins

As for anthocyanin content, the results reported by Lo Scalzo et al. [51] and Li et al. [52] were lower (4.21 mg/100 g fresh weight) and higher (201.11. mg/100 g dry matter), respectively, compared to our findings. The same authors [51] noted that blanching reduced the content of these components by 64.6%, which markedly exceeds our results. Anthocyanins are known to be thermolabile compounds. However, due to the high temperature and short time of blanching, losses may be insignificant that explains such results [53]. Lo Scalzo et al. [51] and Palermo et al. [10] found larger losses of these compounds of 80.2 and 80%, respectively, after traditional cooking in water. According to these authors, there were no significant changes in the convection-steamed material, which is in line with our results. As opposed to Volden et al. [38], who did not observe significant changes in the material frozen stored for 3, 6, and 12 months, frozen storage of the blanched material caused significant losses in the content of these constituents.

4.6. Antioxidant Activity

Volden et al. [38] found that the cultivar Graffiti is characterized by strong antioxidant activity (25.7 μmol Trolox equivalent per 1 g of fresh vegetable). Antioxidant activity determined by Beecher et al. [53], Filipiak-Florkiewicz [31], Murcia et al. [54], and Mazzeo et al. [49] in other cauliflower varieties was similar or higher: 6.5, 6.15, 9.9, and 17.3 μmol Trolox equivalent per 1 g of fresh vegetable, respectively. Such differences may result from various processing conditions including, for example, methods applied for the measurement of antioxidant activity, different extractants, or various extraction times. Culinary treatments may contribute to the reduction of antioxidant activity due to the penetration of antioxidants into the solution or their degradation under elevated temperature [55]. In this case, however, there were no changes in antioxidant activity due to blanching. Losses observed by Volden et al. [38] and Filipiak-Florkiewicz [31] were larger, by 28.0% in the cultivars Graffiti and Celio, as well as by 10.4% and 25.4% in white and green cultivars, respectively. Cooking lowered antioxidant activity of purple cauliflower. Mazzeo et al. [49] and Filipiak-Florkiewicz [31] found reductions of 35.3 and 22.5%, respectively, compared to the fresh material (white cultivar) and of 34.7 (white cultivar) and 37.3% (green cultivar).

Similar decreases in antioxidant activity resulting from 3-month frozen storage, of 30.4% and 26.6%, were observed by Volden et al. [38] for the cultivars *Graffiti* and *Celio*. According to Murcia et al. [54], 24-h freezing caused a fall in antioxidant activity of 0.6%; whereas, Drużyńska et al. [56] noted an increase in this parameter after 48 h.

4.7. Thiocynates

Glucosinolates alone are not biologically active; only their enzymatic derivatives are. When the plant tissue is damaged, hydrolysis of glucosinolates takes place by the endogenous enzyme 'myrosinase' (thioglucoside glucohydrolase EC 3:2:3:1) with a release of a range of breakdown products, including isothiocyanates, indoles, nitriles, oxazolidines, and thiocyanates [7]. This study showed that purple cauliflower contained 2.41 mg/100 of thiocyanates (SCN) per g fresh weight. To the best of our knowledge, there are few studies concerning not only the amount of glucosinolates and their profile, but also their breakdown products. According to Kapusta-Duch et al. [7], cooking decreased significantly the concentration of isothiocynates in green and purple cauliflowers, by 11.0 and 42.4%, respectively, in comparison with raw vegetables. Similar results were found in this work.

4.8. Nitrates and Nitrites

Nitrites and indirectly nitrates, when consumed with food in too large quantities, can be harmful to health [57]. Among vegetables, Brassicas are characterized by medium or low degree of accumulation of these compounds. However, due to the consumed mass they may constitute significant proportion in the daily food ration [58]. In comparison with our results, most authors report several times higher or similar contents of nitrates in various cauliflower cultivars. For example, the values reported were 143–354 mg [59], 210.6 mg [60], 171.2 mg [61], or 61 mg/kg fresh mass [62]. As far as the content of nitrites is concerned, the results obtained by researchers, depending on the examined cultivar, were lower or markedly higher compared to our findings. For example, Leszczyńska et al. [62] found 3.49 mg and 47 mg nitrites in white and green cauliflower respectively, while Gajewska et al. [60] stated 0.8 mg of these constituents per kg of fresh vegetable. According to Leszczyńska et al. [62], the reductions in these compounds due to blanching were much greater than that in this work, amounting to 72.2% for white rose cauliflower and 37.9% for the cultivar Romanesco. Cooking was also responsible for the reduction of nitrates, which agrees with the findings of Shimada and Ko [63], who determined losses of these compounds within the range of 14 to 79%. Filipiak-Florkiewicz et al. [61] observed higher losses of nitrates, compared to those obtained in this work, of up to 72.2% in white rose cauliflower. In comparison with blanching, frozen storage lowered the content of nitrates, on average by 61.9%. Leszczyńska et al. [62] noted a decline in these constituents by 78.7% in white rose cauliflower.

Compared to fresh vegetables, no significant change in the content of nitrites was found due to blanching. However, losses reported by Leszczyńska et al. [62] due to this process were significant and amounted to 8.9% on average. Cooking purple cauliflower led to an average 29.4% decrease in nitrite content compared to fresh vegetables. Similar decreases, of 22.9 (white rose cauliflower) and 17% (*Romanesco* cv.) obtained Filipiak-Florkiewicz et al. [61] and Leszczyńska et al. [62].

Storage of the material contributes to an increase of nitrite content in cauliflower as a result of the reduction of nitrates which takes place in tissues due to enzyme or bacteria activity [64]. After 2-months frozen storage, the content of nitrites decreased, on average by 18.3%, while after a 4 months increased significantly, by 62.4% on average, compared to blanched vegetables. A similar upward trend (151%) for the level of nitrites noted Leszczyńska et al. [62] for green rose cauliflower, frozen and stored for 4 months. Filipiak-Florkiewicz et al. [61], who cooked a vegetable previously frozen stored for the identical period of time, also observed a 105.7% increase in the amount of these compounds.

The results obtained indicate that further studies are needed to explain changes in nutritive and non-nutritive components during selected treatments like cooking or freezing not only purple cauliflower, but other colored Brassica vegetables. The results presented by other authors and those obtained in this work are not clear-cut, so the research problem seems to be interesting. Therefore, the research goal should be to optimize hydrothermal processes and to select the best type of packaging in order to make the best possible use of pro-health substances occurring in Brassica vegetables in human nutrition in accordance with the idea of sustainable development.

5. Conclusions

Brassica are among the top 10 economic vegetables in the world and a source of valuable nutrients, but their content is strongly affected by the method of their preparation. This study clearly showed that technological treatments had a significant effect on nutrient and health-promoting compounds in purple cauliflower. The nutritional content in Brassica vegetables have been reported to vary during the growth period due to climatic and agronomical factors, but also the postharvest treatments are valid factors influenced on quality of these vegetables.

All technological processes, i.e., blanching, cooking, and frozen storage, led to significant changes in the content of the examined nutritive and non-nutritive compounds as well as antioxidant activity or the level of selected chemical pollutions. A trend was also observed towards lower losses of constituents due to convection steaming, compared to traditional cooking in water. Moreover, the reduction of examined compounds was smaller in vacuum-packed and frozen-stored vegetables then in those stored in zipper PE-LD bags.

Author Contributions: Conceptualization, J.K.D., T.L. and M.K.; Methodology, B.B.; Software, A.S.S. and Z.S.G.; Validation, A.S.S., J.S. and M.N.; Formal Analysis, J.S.; Data Curation, B.B.; Writing Original Draft Preparation, J.K.D., BB.; Writing Review & Editing, J.K.D., A.S.S., J.S. and M.N.; Visualization, Z.S.G.; Supervision, T.L. and M.K.

References

1. Šamec, D.; Urlić, B.; Salopek-Sondi, B. Kale (*Brassica oleracea* var. *acephala*) as a superfood: Review of the scientific evidence behind the statement. *Crit. Rev. Food Sci. Nutr.* **2018**, 1–12. [CrossRef]
2. Dos Reis, L.C.; De Oliveira, V.R.; Hagen, M.E.; Jabloński, A.; Flôres, S.H.; De Oliveira, R.A. Effect of cooking on the concentration of bioactive compounds in broccoli (*Brassica oleracea* var. Avenger) and cauliflower (*Brassica oleracea* var. Alphina F_1) grown in an organic system. *Food Chem.* **2015**, *172*, 770–777. [CrossRef] [PubMed]
3. Shahidi, F.; Ambigaipalan, P. Phenolics and polyphenolics in foods, beverages and spices: Antioxidant activity and health effects—A review. *J. Funct. Foods.* **2015**, *18*, 820–897. [CrossRef]
4. Ahmed, F.A.; Ali, R.F. Bioactive compounds and antioxidant activity of fresh and processed white cauliflower. *Biomed. Res. Int.* **2013**, *2013*, 367819. [CrossRef] [PubMed]
5. Florkiewicz, A.; Filipiak-Florkiewicz, A.; Topolska, K.; Cieślik, E.; Kostogrys, R.B. The effect of technological processing on the chemical composition of cauliflower. *Ital. J. Food Sci.* **2014**, *26*, 275–281. Available online: https://www.researchgate.net/publication/286116360_The_effect_of_technological_processing_on_the_chemical_composition_of_cauliflower (accessed on 9 August 2018).
6. Kalisz, A.; Sękara, A.; Smoleń, S.; Grabowska, A.; Gil, J.; Komorowska, M.; Kunicki, E. Survey of 17 elements, including rare earth elements, in chilled and non-chilled cauliflower cultivars. *Sci. Rep.* **2019**, *9*, 5416. [CrossRef]
7. Kapusta-Duch, J.; Kusznierewicz, B.; Leszczyńska, T.; Borczak, B. Effect of cooking on the contents of glucosinolates and their degradation products in selected *Brassica* vegetables. *J. Funct. Foods* **2016**, *23*, 412–422. [CrossRef]
8. Manchali, A.; Murthy, K.N.Ch.; Patil, B.S. Crucial facts about health benefits of popular cruciferous vegetables. *J. Funct. Foods.* **2012**, *4*, 94–106. [CrossRef]
9. Ribeiro-Santos, R.; Andrade, M.; Sanches-Silva, A.; De Melo, N.R. Essential Oils for Food Application: Natural Substances with Established Biological Activities. *Food Bioproc. Tech.* **2017**, 1–29. [CrossRef]
10. Palermo, M.; Pellegrini, N.; Fogliano, V. The effect of cooking on the phytochemical content of vegetables. *J. Sci. Food Agric.* **2014**, *94*, 1057–1070. [CrossRef]
11. Ghaani, M.; Cozzolino, C.A.; Castelli, G.; Farris, S. An overview of the intelligent packaging technologies in the food sector. *Trends Food Sci. Technol.* **2016**, *51*, 1–11. [CrossRef]
12. Hussein, Z.; Caleb, O.J.; Opara, U.L. Perforation-mediated modified atmosphere packaging of fresh and minimally processed produce—A review. *Food Pack. Shelf Life.* **2015**, *6*, 7–20. [CrossRef]
13. Chowdhury, S.; Nath, S.; Biswas, S.; Dora, K.C. Effect of vacuum packaging on extension of shelf life of Asian Sea-Bass fillet at 5 ± 1 °C. *J. Exp. Zool. India* **2017**, *20*, 1241–1245.
14. Swain, T.; Hillis, W.E. The phenolic constituents of *Prunus domesticus* (L.). the quantity of analysis of phenolic constituents. *J. Sci. Food Agric.* **1959**, *10*, 63–68. [CrossRef]
15. Re, R.; Pellegrini, N.; Proteggente, A.; Pannala, A.; Yang, M.; Rice-Evans, C. Antioxidant activity applying an improved ABTS radical cation decolorization assay. *Free Radic. Biol. Medic.* **1999**, *26*, 1231–1237. [CrossRef]
16. Pellegrini, N.; Del Rio, D.; Colombi, B.; Bianchi, M.; Brighenti, F. Application of the 2′2-azobis (3-ethylbenzothiazoline-6-sulfonic acid) radical cation assay to flow injection system for the evaluation of antioxidant activity of some pure compounds and bevereges. *J. Agric. Food Chem.* **2003**, *51*, 164–260. [CrossRef] [PubMed]
17. Benvenuti, S.; Pellati, F.; Melegani, M.; Beertelli, D. Polyphenols, Anthocyanins, Ascorbic Acid and Radical Scavenging Activity of Rubus, Ribes and Aronia. *J. Food Sci.* **2004**, *69*, 164–169. [CrossRef]

18. Polish Standard. *PN-90/A-75101/03. Fruit and Vegetable Products. Preparation of Samples for Physico-Chemical Studies. Determination of Dry Matter Content by Gravimetric Method*; Polish Committee for Standardization: Warsaw, Poland, 1990. (In Polish)
19. Polish Standard. *PN-A-04019:1998. Food products—Determination of Vitamin C*; Polish Committee for Standardization: Warsaw, Poland, 1998. (In Polish)
20. Polish Standard. *PN-92/A-75112. Fruit and Vegetable Products. Determination of Nitrates and Nitrites*; Polish Committee for Standardization: Warsaw, Poland, 1992. (In Polish)
21. Polish Standard. *PN-A-79011-8:1998. Dry Food Mixes. Test Methods. Determination of Total Ash And Ash Insoluble In 10 Percent (m/m) Hydrochloric Acid*; Polish Committee for Standardization: Warsaw, Poland, 1998. (In Polish)
22. Polish Standard. *PN-EN ISO 8968-1:2004. Milk—Determination of Nitrogen Content—Part 1: Determination of Nitrogen by the Kjeldahl Method*; Polish Committee for Standardization: Warsaw, Poland, 2004. (In Polish)
23. Polish Standard. *PN-A-79011-4:1998. Dry Food Mixes. Test Methods. Determination of Fat Content*; Polish Committee for Standardization: Warsaw, Poland, 1998. (In Polish)
24. Polish Standard. *PN-A-79011-15:1998P. Dry Food Mixes. Test Methods. Determination of Dietary Fiber Contents*; Polish Committee for Standardization: Warsaw, Poland, 1998. (In Polish)
25. Rumpel, J. *Cultivation of Cauliflower*, 1st ed.; Publisher Hortpress: Warszawa, Poland, 2002; pp. 6–13. (In Polish)
26. Kunachowicz, H.; Nadolna, I.; Iwanow, K.; Przygoda, B. *The Nutritional Value of Selected Foods and Typical Dishes*, 6th ed.; PZWL: Warszawa, Poland, 2014. (In Polish)
27. Kahlon, T.; Chin, M.; Chapman, M. Steam cooking significantly improve in vitro bile acid binding of beets, eggplant, asparagus, carrots, green beans and cauliflower. *Nutr. Res.* **2007**, *27*, 750–755. [CrossRef]
28. Schonhof, I.; Krumbein, A.; Bruckner, B. Genotypic effects on glucosinolates and sensory properties of broccoli and cauliflower. *Nahr. Food* **2004**, *48*, 25–33. [CrossRef]
29. Puupponen-Pimiä, R.; Häkkinen, S.T.; Aarni, M.; Suortti, T.; Lampi, A.-M.; Eurola, M.; Piironen, V.; Nuutila, A.M.; Oksman-Caldentey, K.-M. Blanching and long-term freezing affect various bioactive compounds of vegetables in different ways. *J. Sci. Food Agric.* **2003**, *83*, 1389–1402. [CrossRef]
30. Ali, A.M. Effect of food processing methods on the bioactive compound of cauliflower. *Egypt. J. Agric. Res.* **2015**, *93*, 117–131. Available online: http://www.arc.sci.eg/ejar/UploadFiles/Publications/1043314%D8%A7%D9%84%D8%A8%D8%AD%D8%AB%20%D8%A7%D9%84%D8%AB%D8%A7%D9%86%D9%89%20%D8%AA%D9%83%D9%86%D9%88%D9%84%D9%88%D8%AC%D9%8A%D8%A7.pdf (accessed on 25 August 2018).
31. Filipiak-Florkiewicz, A. *Effect of Hydrothermal Treatment on Selected Health-Promoting Properties of Cauliflower (Brassica oleracea var. Botrytis L.)*; Scientific Papers; University of Agriculture in Krakow: Kraków, Poland, 2011; p. 347. (In Polish)
32. Florkiewicz, A. Sous-vide method as alternative to traditional cooking of cruciferous vegetables in the context of reducing losses of nutrients and dietary fibre. *Ż.N.T.J.* **2018**, *25*, 45–57. [CrossRef]
33. Gębczyński, P.; Kmiecik, W. Effects of traditional and modified technology, in the production of frozen cauliflower, on the contents of selected antioxidative compounds. *Food Chem.* **2007**, *1*, 229–235. [CrossRef]
34. Ramesh, M.N. The performance evaluation of a continuous vegetable cooker. *Int. J. Food Sci. Technol.* **2000**, *2*, 377–384. [CrossRef]
35. Cebula, S.; Kunicki, E.; Kalisz, A. Quality changes in curds of white, green an Romanesco cauliflower during storage. *Pol. J. Food Nutr. Sci.* **2006**, *56*, 155–160.
36. Evans, J. Emerging Refrigeration and Freezing Technologies for Food Preservation. In *Innovation and Future Trends in Food Manufacturing and Supply Chain Technologies*; Woodhead Publishing: Sawston, UK; Cambridge, UK, 2016; pp. 175–201. [CrossRef]
37. Kaulman, A.; Jonville, M.; Schneider, Y.; Hoffmann, L. Carotenoids, polyphenols and micronutrient profiles of *Brassica oleraceae* and plum varieties and their contribution to measures of total antioxidant capacity. *Food Chem.* **2014**, *155*, 240–250. [CrossRef] [PubMed]
38. Volden, J.; Bengtsson, B.G.; Wicklund, T. Glucosinolates, L-ascorbic acid, total phenols, anthocyanins, antioxidant capacities and colour in cauliflower (*Brassica oleracea* L. ssp. *Botrytis*); effect of long-term freezer storage. *Food Chem.* **2009**, *112*, 967–976. [CrossRef]

39. Bhandari, S.R.; Kwak, J.H. Chemical composition and antioxidant activity in different tissues of Brassica vegetables. *Molecules* **2015**, *20*, 1228–1243. [CrossRef]
40. Picchi, V.; Migliori, C.; Scalzo, R.L.; Campanelli, G.; Ferrari, V.; Di Cesare, L.F. Phytochemical content in organic and conventionally grown Italian cauliflower. *Food Chem.* **2012**, *130*, 501–509. [CrossRef]
41. Davey, M.; Montagu, M.; Inze, D.; Sanmartin, M.; Kanellis, A.; Smirnoff, M.; Benzie, J.; Strain, J.; Favell, D.; Fletcher, J. Review: Plant L-ascorbic acid: Chemistry, function, metabolism, bioavailability and effects of processing. *J. Sci. Food Agric.* **2000**, *80*, 825–860. [CrossRef]
42. Pellegrini, N.; Chiavaro, E.; Gardana, C.; Mazzeo, T.; Contino, D.; Gallo, M.; Riso, P.; Fogliano, V.; Porrini, M. Effect of different cooking methods on colour, phytochemical concentration and antioxidant capacity of raw and frozen *Brassica* vegetables. *J. Agric. Food Chem.* **2010**, *58*, 4310–4321. [CrossRef] [PubMed]
43. Abushita, A.A.; Daood, H.G.; Biacs, P.A. Change in carotenoids and antioxidant vitamins in tomato as a function of varietal and technologicals factors. *J. Agric. Food Chem.* **2000**, *48*, 2075–2081. [CrossRef] [PubMed]
44. İncedayi, B.; Suna, S. Effects of modified atmosphere packaging on the quality of minimally processed cauliflower. *Acta Aliment.* **2012**, *41*, 401–413. [CrossRef]
45. Van Ooijen, I.; Fransen, M.L.; Verlegh, P.W.J.; Smit, E.G. Atypical food packaging affects the persuasive impact of product claims. *Food Qual. Prefer.* **2016**, *48*, 33–40. [CrossRef]
46. Szeląg-Sikora, A.; Niemiec, M.; Sikora, J.; Chowaniak, M. Possibilities of designating swards of grasses and small-seed legumes from selected organic farms in Poland for feed. In Proceedings of the IX International Scientific Symposium "Farm Machinery and Processes Management in Sustainable Agriculture, Lublin, Poland, 22–24 November 2017; pp. 365–370. [CrossRef]
47. Bahorun, T.; Luximon-Ramma, A.; Crozier, A.; Aruoma, O. Total phenol, flavonoid, proanthocyanidin and vitamin C levels and antioxidant activities of Mauritian vegetables. *J. Sci. Food Agric.* **2004**, *4*, 1553–1561. [CrossRef]
48. Kaur, C.; Kapoor, H.C. Anti-oxidant activity and total phenolic content of some Asian vegetables. *Int. J. Food Sci. Tech.* **2002**, *37*, 153–161. [CrossRef]
49. Mazzeo, T.; N'Dri, D.; Chiavaro, E.; Visconti, A.; Fogliano, V.; Pellegrini, N. Effect of two cooking procedures on phytochemical compounds, total antioxidant, capacity and colour of selected frozen vegetables. *Food Chem.* **2011**, *28*, 617–633. [CrossRef]
50. Chassagne-Berces, S.; Fonseca, F.; Citeau, M.; Marin, M. Freezing protocol effect on quality properties of fruit tissue according to the fruit, the variety and the stage of maturity. *LWT Food Sci. Technol.* **2010**, *43*, 1441–1449. [CrossRef]
51. Lo Scalzo, R.; Genna, A.; Branca, F.; Chedin, M.; Chassaigne, H. Anthocyanin composition of cauliflower (*Brassica oleracea* L. var. *botrytis*) and cabbage (*B. oleracea* L. var. *capitata*) and its stability in relation to thermal treatments. *Food Chem.* **2008**, *107*, 136–144. [CrossRef]
52. Li, H.; Deng, Z.; Zhu, H.; Hu, C.; Liu, R.; Young, J.; Tsao, R. Highly pigmented vegetables: Anthocyanin composition and their role in antioxidant activities. *Food Res. Int.* **2012**, *46*, 250–259. [CrossRef]
53. Beecher, G.R.; Gebhardt, S.E.; Haytowitz, D.B.; Holden, J.M.; Wu, X.; Prior, R.L. Lipophilic and hydrophilic antioxidant capacities of common foods in the United States. *J. Agric. Food Chem.* **2004**, *52*, 4026–4037. [CrossRef]
54. Murcia, M.A.; Lopez-Ayerra, B.; Garcia-Carmona, F. Effect of processing and different blanching times on broccoli: Proximate composition and fatty acids. *LWT Food Sci. Technol.* **2009**, *32*, 238–243. [CrossRef]
55. Cartea, M.E.; Francisco, M.; Soengas, P.; Velasco, P. Phenolic Compounds in *Brassica* Vegetables. *Molecules* **2011**, *16*, 251–280. [CrossRef] [PubMed]
56. Drużyńska, B.; Stępień, K.; Piecyk, M. The influence of cooking and freezing on contents of bioactive components and their antioxidant activity in broccoli. *Bromat. Chemia Toksykol.* **2009**, *42*, 169–176. Available online: http://ptfarm.pl/pub/File/bromatologia_2009/bromatologia_2_2009/Bromat%202,2009%20s.%20169-176.pdf (accessed on 24 July 2010). (In Polish).
57. Sindelar, J.J.; Milkowski, A.L. Human safety controversies surrounding nitrate and nitrite in the diet. *Nitric Oxide* **2012**, *26*, 259–266. [CrossRef] [PubMed]
58. Bahadoran, Z.; Mirmiran, P.; Jeddi, S.; Azizi, F.; Ghasemi, A.; Hadaegh, F. Nitrate and nitrite content of vegetables, fruits, grains, legumes, dairy products, meats and processed meats. *J. Food Comp. Anal.* **2016**, *51*, 93–105. [CrossRef]

59. Samantaria, P. Nitrate in vegetables: Toxity, content, intake and EC regulation. *J. Sci. Food and Agric.* **2006**, *86*, 10–17. [CrossRef]
60. Gajewska, M.; Czajkowska, A.; Bartodziejska, B. The content of nitrates and nitrites in selected vegetables on detail sale in Lodz region. *Ochr Śr. Zasobów Nat.* **2009**, *40*, 388–395. (In Polish)
61. Filipiak-Florkiewicz, A.; Cieślik, E.; Florkiewicz, A. Effect of technological processing on contents of nitrates and nitrites in cauliflower. *Żyw. Człow. Metab.* **2007**, *34*, 1197–1201. Available online: http://agro.icm.edu.pl/agro/element/bwmeta1.element.agro-article-e3ebf1f1-dd46-4dad-8ed5-c9d0526fceac?q=bwmeta1.element.agro-number-fcf445e1-46f5-442d-a933-e0d38017adce;75&qt=CHILDREN-STATELESS (accessed on 18 April 2008). (In Polish).
62. Leszczyńska, T.; Filipiak-Florkiewicz, A.; Cieślik, E.; Sikora, E.; Pisulewski, P. Effects of some processing methods on nitrate and nitrite changes in cruciferous vegetables. *J. Food Comp. Anal.* **2009**, *22*, 315–321. [CrossRef]
63. Shimada, Y.; Ko, S. Nitrate in vegetables. *Chugoku Gakuen J.* **2004**, *3*, 7–10. Available online: https://pdfs.semanticscholar.org/5cbe/9ce8f74a9c1c334008cd295ccd5a1553b9da.pdf (accessed on 15 May 2005).
64. Ranasinghe, R.A.S.N.; Marapana, R.A.U.J. Nitrate and nitrite content of vegetables: A review. *J. Pharmacogn. Phytochem.* **2018**, *7*, 322–328. Available online: https://www.researchgate.net/publication/326533977_Nitrate_and_nitrite_content_of_vegetables_A_review_RASN_Ranasinghe_and_RAUJ_Marapana (accessed on 13 January 2019).

15

Compaction Process as a Concept of Press-Cake Production from Organic Waste

Paweł Sobczak [1], Kazimierz Zawiślak [1], Agnieszka Starek [2,*], Wioletta Żukiewicz-Sobczak [3], Agnieszka Sagan [2], Beata Zdybel [2] and Dariusz Andrejko [2]

1. Department of Food Engineering and Machines, University of Life Sciences in Lublin, Akademicka 13, 20-612 Lublin, Poland; pawel.sobczak@up.lublin.pl (P.S.); kazimierz.zawislak@up.lublin.pl (K.Z.)
2. Department of Biological Bases of Food and Feed Technologies, University of Life Sciences in Lublin, Akademicka 13, 20-612 Lublin, Poland; agnieszka.sagan@up.lublin.pl (A.S.); beata.zdybel@up.lublin.pl (B.Z.); dariusz.andrejko@up.lublin.pl (D.A.)
3. Pope John Paul II State School of Higher Education in Biala Podlaska, Sidorska 95/97, 21-500 Biala Podlaska, Poland; wiola.zukiewiczsobczak@gmail.com

* Correspondence: agnieszka.starek@up.lublin.pl

Abstract: As a result of agri-food production large amounts of organic waste are created in the form of press cakes. Until now, they were mainly used as animal fodder and also utilized for biofuels production. No other way usage has been found yet. A large quantity of these by-products is usually discarded in open areas, which leads to potentially serious environmental problems. The rich chemical composition of these waste products makes it possible to use them for producing other food products valuable for consumers. Based on the test results obtained, it can be stated that moisture content of press cakes is varied and depends on the input material. However, appropriately composed mixtures of various waste products and a properly conducted compaction process allows for obtaining a new product with functional properties. In addition, application of honey powder and starch tablet coating creates a product of resistant to compression and cutting. Results seem to have commercial importance, as they demonstrate that properly processed by-products can be used in food preparations as dietary supplements.

Keywords: press cakes; compaction; disposal; sustainable development; modern products

---

1. Introduction

In the food industry, during processing of raw materials, waste products are generated. Their substantial amounts accumulated in a short time causes serious problems for processing plants, primarily due to the instability of these by-products, including the microbiological one. Therefore, they should be used as an intermediate for further processing. In general, around the world waste management is carried out to convert into useful components as much waste products as possible without endangering the natural environment [1–5].

A particularly large amount of waste is produced during cold-press edible oil extraction. A common method of using press cakes from rapeseed, flaxseed, and sunflower extraction, is their utilisation for animal nutrition as a high protein vegetable fodder or as a component for enriching the protein content in the production of maize silage, which is intended for high-yielding cows. The use of press cakes reduces the cost of animal nutrition products [6–8]. They can also be successfully used as organic fertilizer rich in minerals (Ca, Mg, P, Fe, and Mn) and organic compounds, as well as for energy purposes [9–12] after a compaction process that turns them into pellets [13,14]. Press cakes can be used for manufacturing formulated products such as functional foods. Compaction could be a particularly important method for processing press cakes in the areas where there is a surplus of them, there is no possibility for their fast processing into fodder, and transport over long distances is not economically viable [15].

Production of coconut oil also entails generation of waste. The research done by Ramachandran et al. [16,17] confirms that these press cakes are characterised not only by a clearly sweet taste, but also high content of amino acids (mainly arginine). They are also used as a potential raw material in bioprocesses, as they are an excellent substrate for growing microorganisms (substrates for the production of phytase based on the use of Rhizopus spp. strain).

Avocado oil, pressed from fruit pulp, is nowadays used as a dietary supplement reinforcing body resistance and improving skin appearance. Moreover, it is used for production of cosmetic goods [18]. At the same time avocado pit, until lately considered inedible, has recently been the subject of scientific interest. As reported by Soong et al. [19] and Wang et al. [20] this hard part of the fruit contains many more polyphenols, exhibiting intense antioxidant activity, than the pulp itself. According to the ORAC scale for determining the free radical absorbance capacity of antioxidants, the pit of the fruit has high antioxidant potential, equal to 428.8 μmolTE/g for the Hass variety. For comparison, the same quantity of pulp only has a potential of 11.6 μmolTE/g. Avocado extracts, thanks to their polyphenol content, expressed as gallic acid equivalents, are already used in cosmetic care [21], as well as a means to delay the oxidation of sunflower oil [22]. Moreover, many studies confirm [23–26] that avocado pit has cytotoxic properties, inter alia, against breast, lung, colon, and prostate cancer cells.

Individual types of waste are increasingly used in various industries [27,28]. In the scientific literature, however, there are no reports on their comprehensive use to obtain a valuable products intended for human consumption. In addition, there are difficulties in the proper preparation of this organic waste (ensuring adequate moisture content, disintegration and chemical composition) and their appropriate concentration [14,15]. Therefore, the aim of the study is to determine the potential use of organic waste and combine it in such a way as to obtain a new, stable product with pro-health properties.

2. Materials and Methods

2.1. Research Material

Materials used for the study consisted of flax, sunflower, pumpkin, and coconut press cakes, and powder avocado pit created during cold-press oil production. In order to obtain a uniform product all raw materials were mixed and subjected to compaction in a chamber equipped with a piston (Figure 1), where they were formed into a tablet. Before blending the compound press cakes were ground to a uniform size. The percentage share of particular components constituting the blend is shown in Table 1. The press cakes were sampled immediately after cold-press oil extraction using a screw press (model DUO Farmet, Česká Skalice, Czech Republic), while the avocado pit was cut into smaller pieces, dried in an SLN 15 STD type laboratory drying oven, manufactured by the POL-EKO Company (Wodzislaw Śląski, Poland), at a temperature of 40 °C for 1.5 hours, and then ground in a mill so that powder was obtained.

Table 1. Percentage share of individual raw materials in the compound.

Type of Raw Material	Compound 1	Compound 2
Flax press cake	30	29
Sunflower press cake	20	20
Pumpkin press cake	30	29
Avocado pit	5	5
Coconut press cake	15	15
Honey powder	0	2

Shares of particular components were selected having in mind the possibility of obtaining value products in form of a stable tablet. To Compound no. 2 a 2% of dried honey was added. The purpose of this admixture was to obtain a stable tablet form by combining particles of honey with the remaining raw materials.

Then the tablets were divided into two groups and one of them was coated with a thin layer of starch solution. For this purpose a 40% solution of potato starch was prepared which was heated to the temperature of 40 °C in order to obtain lower viscosity. Surface of the tablets was coated using a spray nozzle. Afterwards, the tablets were left to dry at room conditions for 24 hours.

2.2. *Measurement of Chemical and Physical Properties of Post-Production Waste*

Fat content was determined in the obtained press cakes as well as in the fragmented and dried avocado pit using an automatic Soxhlet apparatus (Tecator Soxtec System HT 1043 extraction unit, Gemini, Apeldoorn, Sweden). Total protein content measurements were carried out according to the Kjeldahl method using a Kjeltec 8400 automatic distiller (Foss, Foss Anatytical AB, Höganäs, Sweden). The total protein content was calculated using a 6.25 conversion factor. Determination of the ash content in the press cakes was done in a muffle furnace according to PN-EN ISO 18122:2016-01 [29] standard. Analysis of mineral composition included the determination of the amount of: calcium, copper, iron, potassium, magnesium, and zinc by using ICP OES spectrometer (SpectroBlue, SPECTRO Analytical Instruments GmbH, Kleve, Germany). Analytical curves were prepared by dilutions of VHG SM68-1-500 Element Multi Standard 1 in 5% HNO_3. Moisture content of the raw materials was determined using a laboratory drying oven with forced air circulation. Samples of raw materials were placed in the oven chamber and then dried at the temperature of 105 °C until constant weight was achieved, in accordance with PN-EN-ISO 18134-3:2015-11 [30] standard.

2.3. *Methodology of the Tablet Forming Process*

The tablet forming process was carried out with a ZO20/TN2S machine (Zwick.Roell AG, BT1-FR0.5TN.D14, Ulm, Germany) using the designed attachment shown in Figure 1.

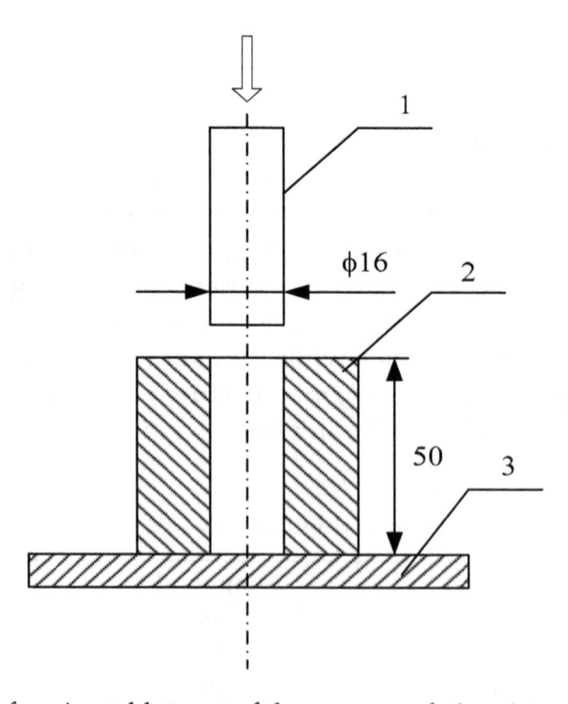

Figure 1. Workstation for forming tablets out of the compound: 1—piston, 2—feed hopper, 3—base.

The pressing force exerted by the piston was constant, equal to 20 kN. The head pressing the piston, allowed for the formation of a single agglomerate with the weight of 2 g. Compaction work and specific density of the obtained tablets was designated by means of testXpert software, which is used to operate the Zwick devices.

2.4. *Methodology for Assessing Strength of the Tablets*

The tablets (Figure 2) were analysed by examining the cutting force and hardness. The stress tests were carried out using a Micro Stable Pro TA.XT Plus system (Stable Micro Systems Ltd, Surrey, United Kingdom).

Figure 2. Tablets obtained out of the compound.

The method of placing a single tablet for the cutting force test and hardness test is depicted in Figure 3. In the cutting test a knife with the cutting edge angle of 45° was used. The test was run until the tablet was cut into half. The hardness test was made using a flat head and recording the compression force until reaching half the diameter of the tablet. The tests were performed in 10 repetitions.

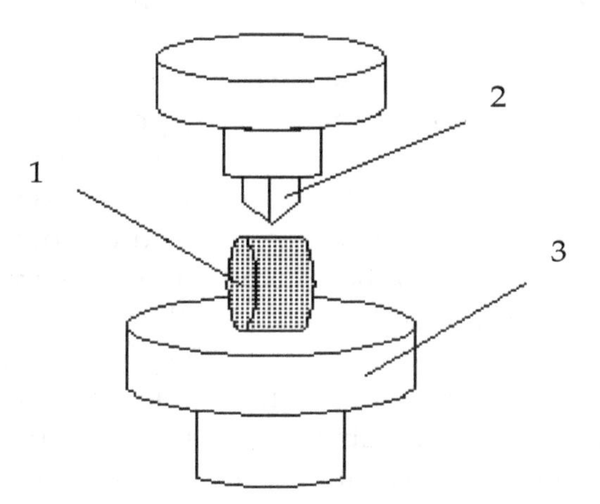

Figure 3. Method of placing the sample for the cutting force test: 1—sample, 2—cutting knife, 3—measuring table.

2.5. *Statistical Analysis*

Results were statistically analysed with Statistica 10.0 software (Statistica 10, StatSoft Inc., Tulsa, OK., U.S.A.), using one-way analysis of variance. The significance of differences between the mean values was tested using Tukey's procedure.

3. Results and Discussion

3.1. Characterization of Chemical Properties of Post-Production Waste

The results of carried out experiments are shown in Tables 2 and 3.

Table 2. Chemical composition of the press cakes and obtained compounds.

Discriminant	By-products					Compound	
	Flax Press Cake	Sunflower Press Cake	Pumpkin Press Cake	Coconut Press Cake	Avocado Pit	1	2
Protein [%]	32.47 ± 0.11	37.65 ± 0.36	53.98 ± 0.07	17.40 ± 0.41	4.78±0.06	36.2± 0.26	36.1± 0.23
Fat [%]	11.52 ± 0.07	31.42 ± 0.25	14.23 ± 0.05	15.02 ± 0.04	1.30±0.21	18.1± 0.04	17.1± 0.06
Ash [%]	5.56 ± 0.02	4.69 ± 0.01	8.09 ± 0.01	4.13 ± 0.01	2.30±0.01	5.75± 0.01	5.70± 0.01

Means of three determinations ± SD (standard deviation).

Table 3. Content of selected elements in the press cake.

By-Products	Content of Selected Elements [ppm]					
	Ca	**Cu**	**Fe**	**K**	**Mg**	**Zn**
Flax press cake	3489.355	11.808	135.916	8067.798	4035.281	69.089
Sunflower press cake	1421.465	23.737	66.131	<0.005	3708.333	111.364
Pumpkin press cake	602.551	6.378	134.694	<0.005	6707.143	155.102
Coconut press cake	394.628	20.661	75.000	10017.355	2021.694	32.645
Avocado pit	913.545	7.977	31.909	7143.500	759.023	8.568

By-products resulting from processing of pumpkin seeds had the highest protein content of 53.98%. In the flax and sunflower press cakes the value of this parameter ranged between 32.39 and 37.91%, while in the case of coconut press cakes its quantity was about half the amount. The avocado pit was characterised by the lowest protein content, equal to 4.78%.

Sunflower press cakes had the highest fat content of 31.415%. The coconut, pumpkin, and flax press cakes were characterised by a significantly lower fat content, i.e. between 15.02 and 11.52%. Ground avocado pit contained a negligible amount of fat equal to 1.30%.

The ash content in the analysed samples varied between 2.30% (for the avocado pit) and 8.09% (for the pumpkin seed press cakes).

As it can be seen in Table 3, the calcium content varied very significantly. Flax press cake contained the highest amount of this element, i.e. 3489 ppm. In the by-products of coconut oil extraction process, calcium content was about 74% lower. Copper content in the analysed production by-products was less varied and fell within the range from 6.378 ppm (for the pumpkin press cake) to 23.737 ppm (for the sunflower press cake). Iron content in the waste originating from flax and pumpkin processing was at a similar levels. The analysis showed that avocado pit was characterised by the lowest amount of this element. Potassium content was also determined in the tested by-products. Coconut press cake had the highest amount of this element, equal to 10017.355 ppm. Concentration of potassium in the waste resulting from flax seed oil extraction and in the avocado pit was lower respectively by approx. 20% and 29%. In the remaining by-products subject to the study amount of potassium was below limit of quantification. Magnesium content in the analysed samples varied within the range between 759.023 and 6707.143 ppm. The lowest content of this element was measured for the avocado pit. Pumpkin press cake was characterized by a several times higher magnesium content. For the entire studied material the highest zinc content was measured for the pumpkin press cake and sunflower press cake. In turn, the lowest amount of this element was recorded in the case of avocado pit.

The chemical composition of the press cakes can be highly variable, depending on the quality of seeds, method of oil extraction, storage parameters, and so on. As confirmed by the studies done by other authors these oil pressing by-products are characterized by high nutritional value. For example, the by-product obtained in the course of oil extraction from dried pumpkin seeds has a high amount of valuable protein and hydrolysates produced from it have antioxidant and functional properties [31]. Research conducted by Salgado et al. [32] shows that sunflower press cakes are also a source of protein with high solubility in water, good physicochemical properties, and high antioxidant activity. The data presented by Ramachandran et al. [17] demonstrate that coconut press cake contains a high level of residual oil consisting of saturated short-chain fatty acids. Moreover, it is characterised by a high content of protein and crude fibre. Literature on chemical composition of sesame press cake [33,34], rapeseed press cake [35], Camelina sativa, and flax press cake [36] also provides a lot of information about their high nutritional value, and confirms that in future by-products will be an important resource for use as a food ingredient for direct human consumption.

3.2. Analysis of the Compaction Process

Before the compaction process moisture content of individual components as well as that of the compound was measured (Figure 4).

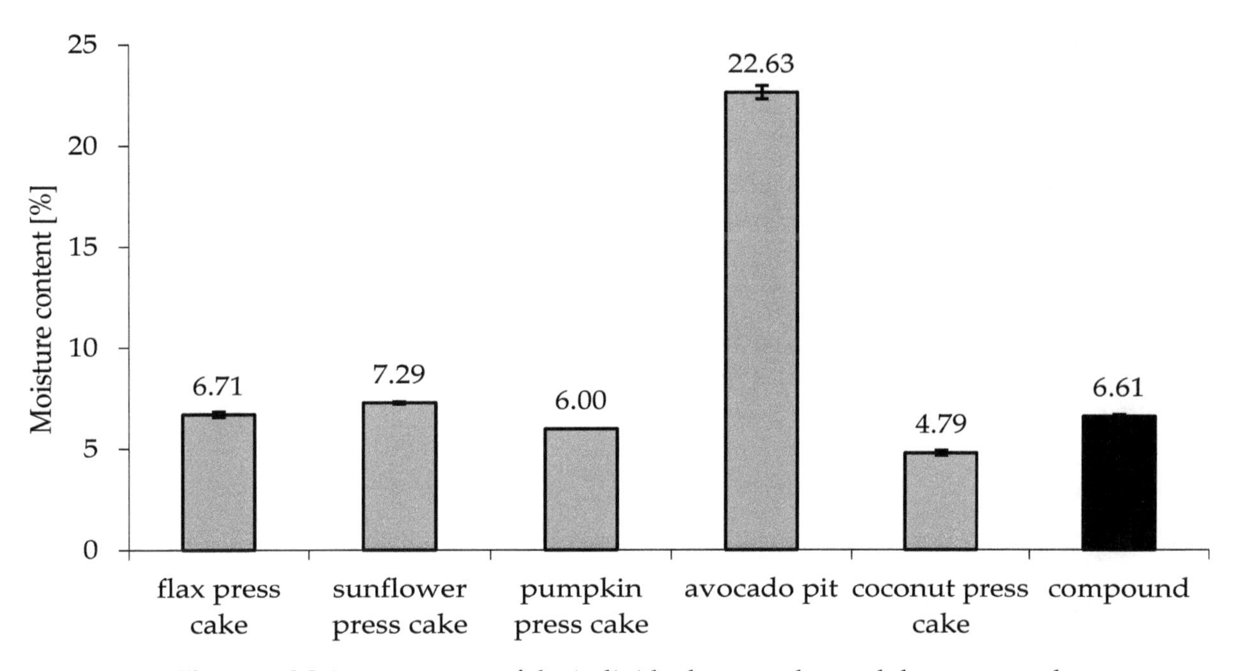

Figure 4. Moisture content of the individual press cakes and the compound.

Avocado pit press cake had the highest moisture content (22.63%). For the remaining raw materials moisture content varied within the range between 4.79% (for the coconut press cake) and 7.29% (for the sunflower press cake). The obtained compound had the moister content of 6.61%.

Moisture content of press cakes is varied and depends on the input material. As reported by research moisture content for oil press cakes ranges from 5 to 9% [13]. Wojdalski et al. [15] conducted research on the compaction process of apple pomace resulting from juice production, for which, at the temperature of 20 °C and relative humidity of the air equal to 31.7%, the total compaction energy in the case of unfragmented apple pomace ranged from 66.60 to 150.00 J/g, while for pellets made from fragmented pomace this parameter was between 34.79 J/g and 149.95 J/g. Specific density of the obtained tablets was respectively within the range of 1114.0-1166.3 and 1114.0-1168.1 g/dm^3. In the case of examined here compounds a significantly smaller compaction work, i.e. of approx. 9.7 J/g, was necessary to obtain tablets. This is due to high fat content, which for compound 1 was equal to 18.1%, while for compound 2 to 17.1% (Figure 5).

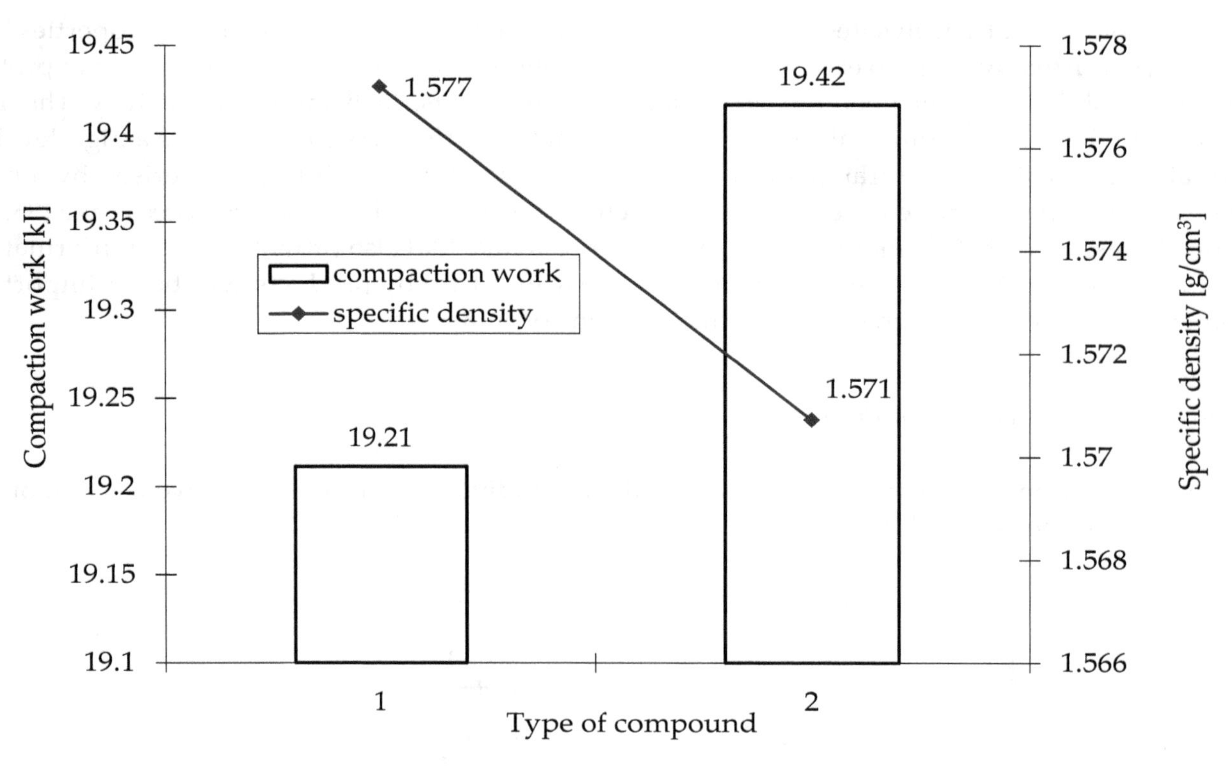

Figure 5. Compaction work and specific density obtained after forming tablets out of the compound.

3.3. Evaluation of the Obtained Tablets

In accordance with the accepted methodology the obtained tablets were subject to stress tests in order to determine the impact of adding honey and coating technology. Figures 6 and 7 and Table 4 show results of respectively the cutting test and hardness test.

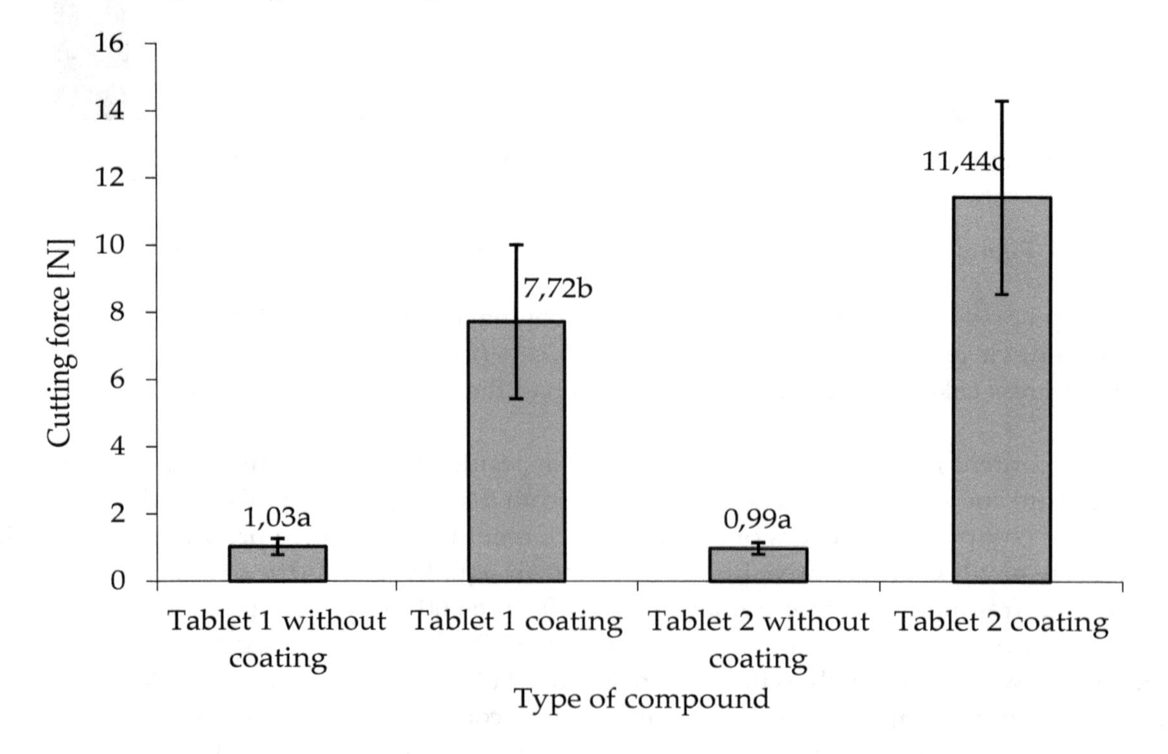

Figure 6. Maximum cutting force for individual tablets.

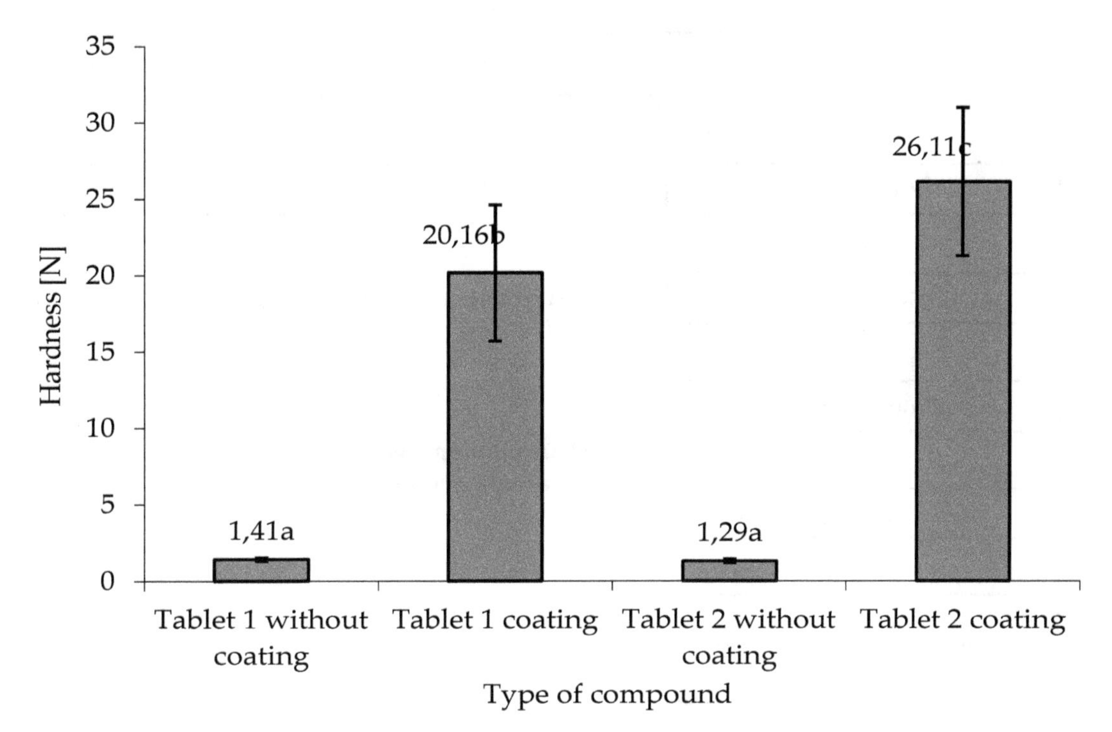

Figure 7. Hardness of the obtained tablets.

Table 4. Analysis of variance (ANOVA).

Source Variables	SS	df	MS	F Value	Probability
Compaction work [J/g]	0.1652	1	0.1652	3.578845	0.063046
Specific density [g·dm^{-3}]	659.469	1	659.469	1.038803	0.311938
Cutting force [N]	570,115	3	190.038	61.1161	0
Hardness [N]	3598.742	3	1199.581	117.406	0

As confirmed by the study tablets with starch coating are characterized by greater resistance to cutting. The maximum cutting force was recorded for tablet 2, i.e. the one with honey powder admixture. Temperature increase caused by friction force resulting from compression transforms powdered honey into a viscous liquid which reinforces produced tablet. In order to obtain high quality agglomerate it is necessary to combine this raw material with other raw materials that provide appropriate strength properties [37,38].

A similar relationship was obtained during the tablet hardness tests. The tablets with addition of honey powder and coated with starch were characterised by highest hardness (26.11N). Statistical analysis of the results confirmed the significance of the differences between the tablets with honey powder admixture and those without it. This greater hardness of the tablets with honey results from its bonding with the other components constituting the compound, as well as with the applied starch coating. Tablets without coating were characterised by low hardness of 1.3–1.4 N, regardless of the admixture of honey powder. Studies on the impact of coating on the properties of final product were done for many sectors of the food industry. As demonstrated by the research done by Fijałkowska et al. [39] after applying a coating of thermally hardened starch on dried apple a slight increase in the hardness of dried apple was recorded as compared with the control sample [40]. Another advantage is extending the freshness of a variety of fruits by applying a coating and thus controlling the water content in the fruits, as well as their colour and firmness [41].

The presented diagram (Figure 8) illustrates the technologies of producing a new product that can be used in food preparations as dietary supplements.

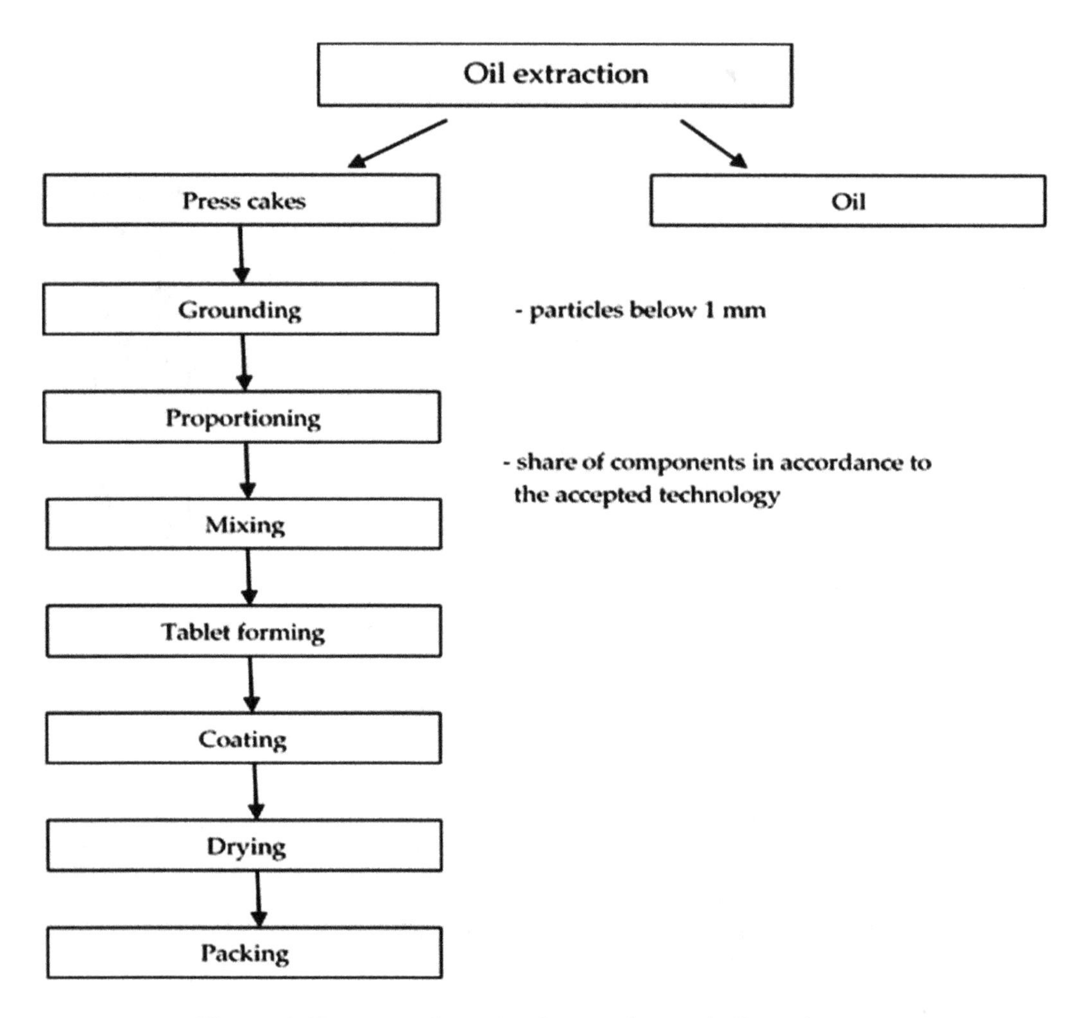

Figure 8. Prospects for using by-products of oil production.

4. Conclusions

The potential of some organic wastes is not always fully appreciated. They can be useful basic raw materials which can be sources of valuable, previously unknown or overlooked ingredients. The studies carried out suggest the possibility for using post-production press cakes in the manufacture of food products immediately after oil pressing (without the need of transport of post-production press cakes). The products obtained have a high protein content and significant amounts of minerals. An appropriate compaction process of the ingredients and application of a coating also allows for obtaining tablet of appropriate quality. Products with starch (40% solution of potato starch) coating and enriched with 2% honey powder are characterized by greater resistance to cutting and better hardness.

The proposal for organic waste compaction process resulting from cold-press oil extraction presented here can significantly increase profitability of processing raw materials. Moreover, manufacture of new products having the characteristics of pro-health food will also enhance the food market and make it more attractive. Through the admixture of various flavor coatings (honey, chocolate, caramel) they could become a product for direct consumption.

Author Contributions: Conceptualization, P.S., K.Z., and D.A.; methodology, P.S., and D.A.; formal analysis, A.S. (Agnieszka Sagan), P.S.; investigation, P.S., A.S. (Agnieszka Starek), B.Z. and W.Ż.-S.; data curation, P.S., A.S. (Agnieszka Starek); writing—original draft preparation, A.S. (Agnieszka Starek), P.S.; writing—review and editing, A.S. (Agnieszka Starek), A.S. (Agnieszka Sagan); visualization, P.S., A.S. (Agnieszka Starek); supervision, P.S. All authors have read and agreed to the published version of the manuscript.

Acknowledgments: The samples of chemical analysis have been examined at the Regional Research Center for Environment, Agricultural and Innovative Technologies, Pope John II State School of Higher Education in Biała Podlaska.

References

1. Darlington, R.; Staikos, T.; Rahimifard, S. Analytical methods for waste minimization in the convenience food industry. *Waste Manag.* **2009**, *4*, 1274–1281. [CrossRef]
2. Garcia-Garcia, G.; Woolley, E.; Rahimifard, S. Identification and analysis of attributes for industrial food waste management modelling. *Sustainability* **2019**, *11*, 2445. [CrossRef]
3. Martin, M.; Danielsson, L. Environmental implications of dynamic policies on food consumption and waste handling in the European union. *Sustainability* **2016**, *8*, 282. [CrossRef]
4. Shin, S.G.; Han, G.; Lim, J.; Lee, C.; Hwang, S.A. Comprehensive microbial insight into two-stage anaerobic digestion of food waste-recycling wastewater. *Water Res.* **2010**, *44*, 4838–4849. [CrossRef] [PubMed]
5. Strotmann, C.; Göbel, C.; Friedrich, S.; Kreyenschmidt, J.; Ritter, G.; Teitscheid, P.A. Participatory approach to minimizing food waste in the food industry—A manual for managers. *Sustainability* **2017**, *9*, 66. [CrossRef]
6. Ahern, N.A.; Nuttelman, B.L.; Klopfenstein, T.J.; MacDonald, J.C.; Erickson, G.E.; Watson, A.K. Comparison of wet and dry distillers grains plus solubles to corn as an energy source in forage-based diets. *Prof. Anim. Sci.* **2016**, *32*, 758–767. [CrossRef]
7. Kaczmarek, P.; Korniewicz, D.; Lipiński, K.; Mazur, M. Chemical composition of rapeseed products and their use in pig nutrition. *Pol. J. Nat. Sci.* **2016**, *4*, 545–562.
8. Świerczewska, E.; Mroczek, J.; Niemiec, J.; Słowiński, M.; Jurczak, M.; Siennicka, A.; Kawka, P. Broiler chick performance and meat quality depending on the type of fat in feed mixtures. *J. Anim. Feed Sci.* **1997**, *6*, 379–389. [CrossRef]
9. Appels, L.; Lauwers, J.; Degrève, J.; Helsen, L.; Lievens, B.; Willems, K.; Dewil, R. Anaerobic digestion in global bio-energy production: Potential and research challenges. *Renew. Sustain. Energy Rev.* **2011**, *15*, 4295–4301. [CrossRef]
10. Maćkowiak, C.; Igras, J. Chemical composition of sewage sludge and food industry wastes of fertilisation value. *Inż. Ekol.* **2005**, *10*, 70–77.
11. Singh, R.P.; Ibrahim, M.H.; Esa, N.; Iliyana, M.S. Composting of waste from palm oil mill: A sustainable waste management practice. *Rev. Environ. Sci. Biotechnol.* **2010**, *9*, 331–344. [CrossRef]
12. Yaakob, Z.; Mohammad, M.; Alherbawi, M.; Alam, Z.; Sopian, K. Overview of the production of biodiesel from waste cooking oil. *Renew. Sustain. Energy Rev.* **2013**, *18*, 184–193. [CrossRef]
13. Kraszkiewicz, A.; Kachel-Jakubowska, M.; Niedziółka, I.; Zaklika, B.; Zawiślak, K.; Nadulski, R.; Mruk, R. Impact of various kinds of straw and other raw materials on physical characteristics of pellets. *Annu. Set Environ. Prot.* **2017**, *19*, 270–287.
14. Zawiślak, K.; Sobczak, P.; Kraszkiewicz, A.; Niedziółka, I.; Parafiniuk, S.; Kuna-Broniowska, I.; Tanaś, W.; Żukiewicz-Sobczak, W.; Obidziński, S. The use of lignocellulosic waste in the production of pellets for energy purposes. *Renew. Energy* **2020**, *145*, 997–1003. [CrossRef]
15. Wojdalski, J.; Grochowicz, J.; Ekielski, A.; Radecka, K.; Stępniak, S.; Orłowski, A.; Kosmala, G. Production and Properties of Apple Pomace Pellets and their Suitability for Energy Generation Purposes. *Annu. Set Environ. Prot.* **2016**, *18*, 89–111.
16. Ramachandran, S.; Roopesh, K.; Nampoothiri, K.M.; Szakacs, G.; Pandey, A. Mixed substrate fermentation for the production of phytase by *Rhizopus* spp. using oilcakes as substrates. *Process Biochem.* **2005**, *40*, 1749–1754. [CrossRef]
17. Ramachandran, S.; Singh, S.K.; Larroche, C.; Soccol, C.R.; Pandey, A. Oil cakes and their biotechnological applications—A review. *Bioresour. Technol.* **2007**, *98*, 2000–2009. [CrossRef]
18. Flores, M.; Saravia, C.; Vergara, C.E.; Avila, F.; Valdés, H.; Ortiz-Viedma, J. Avocado Oil: Characteristics, Properties, and Applications. *Molecules* **2019**, *24*, 2172. [CrossRef]
19. Soong, Y.Y.; Barlow, P.J. Antioxidant activity and phenolic content of selected fruit seeds. *Food Chem.* **2004**, *88*, 411–417. [CrossRef]
20. Wang, W.; Bostic, T.R.; Gu, L. Antioxidant capacities, procyanidins and pigments in avocados of different strains and cultivars. *Food Chem.* **2010**, *122*, 1193–1198. [CrossRef]
21. Alex, S.; Nogent, L.; Caroline, B.; Sophie, L.B.; Philippe, M. Avocado Flesh and/or Skin Extract Rich in

Polyphenols and Cosmetic, Dermatological and Nutraceutical Compositions Comprising Same. United States Patent Application No. 16/117, 511, 30 August 2018.
22. Segovia, F.; Hidalgo, G.; Villasante, J.; Ramis, X.; Almajano, M. Avocado seed: A comparative study of antioxidant content and capacity in protecting oil models from oxidation. *Molecules* **2018**, *23*, 2421. [CrossRef]
23. Alkhalf, M.I.; Alansari, W.S.; Ibrahim, E.A.; ELhalwagy, M.E. Anti-oxidant, anti-inflammatory and anti-cancer activities of avocado (*Persea americana*) fruit and seed extract. *J. King Saud Univ. Sci.* **2018**, *31*, 1358–1362. [CrossRef]
24. Dabas, D.; Elias, R.J.; Ziegler, G.R.; Lambert, J.D. In Vitro Antioxidant and Cancer Inhibitory Activity of a Colored Avocado Seed Extract. *Int. J. Food Sci.* **2019**. [CrossRef]
25. Lara-Marquez, M.; Spagnuolo, P.A.; Salgado-Garciglia, R.; Ochoa-Zarzosa, A.; Lopez-Meza, J.E. Cytotoxic Mechanism of Long-chain Lipids Extracted from Mexican Native Avocado Seed (*Persea americana var. drymifolia*) on Colon Cancer Cells. *FASEB J.* **2018**, *32*, 804–832.
26. Vo, T.S.; Le, P.U. Free radical scavenging and anti-proliferative activities of avocado (*Persea americana* Mill.) seed extract. *Asian Pac. J. Trop. Med.* **2019**, *9*, 91.
27. Berbel, J.; Posadillo, A. Review and analysis of alternatives for the valorisation of agro-industrial olive oil by-products. *Sustainability* **2018**, *10*, 237. [CrossRef]
28. Zafeiriou, E.; Arabatzis, G.; Karanikola, P.; Tampakis, S.; Tsiantikoudis, S. Agricultural commodities and crude oil prices: An empirical investigation of their relationship. *Sustainability* **2018**, *10*, 1199. [CrossRef]
29. *Biopaliwa Stałe—Oznaczanie Zawartości Popiołu*; Polish Committee for Standardization: Warsaw, Poland, 2016; PN-EN ISO 18122:2016-01.
30. *Biopaliwa Stałe—Oznaczanie Zawartości Wilgoci—Metoda Suszarkowa—Część 3: Wilgoć w Próbce do Analizy Ogólnej*; Polish Committee for Standardization: Warsaw, Poland, 2015; PN-EN-ISO 18134-3:2015-11.
31. Popović, L.; Peričin, D.; Vaštag, Ž.; Popović, S.; Krimer, V.; Torbica, A. Antioxidative and functional properties of pumpkin oil cake globulin hydrolysates. *J. Am. Oil Chem. Soc.* **2013**, *90*, 1157–1165. [CrossRef]
32. Salgado, P.R.; Molina Ortiz, S.E.; Petruccelli, S.; Mauri, A.N. Sunflower protein concentrates and isolates prepared from oil cakes have high water solubility and antioxidant capacity. *J. Am. Oil Chem. Soc.* **2011**, *88*, 351–360. [CrossRef]
33. Mohdaly, A.A.; Smetanska, I.; Ramadan, M.F.; Sarhan, M.A.; Mahmoud, A. Antioxidant potential of sesame (*Sesamum indicum*) cake extract in stabilization of sunflower and soybean oils. *Ind. Crop. Prod.* **2011**, *34*, 952–959. [CrossRef]
34. Yasothai, R. Chemical composition of sesame oil cake—Review. *Int. J. Sci. Environ. Technol.* **2014**, *3*, 827–835.
35. Amarowicz, R.; Fornal, J.; Karamac, M. Effect of seed moisture on phenolic acids in rapeseed oil cake. *Grasas Aceites* **1995**, *46*, 354–356. [CrossRef]
36. Terpinc, P.; Čeh, B.; Ulrih, N.P.; Abramovič, H. Studies of the correlation between antioxidant properties and the total phenolic content of different oil cake extracts. *Ind. Crop. Prod.* **2012**, *39*, 210–217. [CrossRef]
37. Miranda, T.; Arranz, J.I.; Montero, I.; Román, S.; Rojas, C.V.; Nogales, S. Characterization and combustion of olive pomace and forest residue pellets. *Fuel Process. Technol.* **2012**, *103*, 91–96. [CrossRef]
38. Obidziński, S. Analysis of usability of potato pulp as solid fuel. *Fuel Process. Technol.* **2012**, *1*, 67–74. [CrossRef]
39. Fijałkowska, A.; Witrowa-Rajchert, D.; Weroński, A. The influence of raw material coating on drying process and reconstitution properties of dried product. *Zesz. Probl. Postęp. Nauk Roln.* **2012**, *571*, 39–47.
40. Krochta, J.M.; De Mulder-Johnston, C. Edible and biodegradable polymer films: Challenges and opportunities. *Food Technol.* **1997**, *51*, 61–74.
41. García, M.A.; Martino, M.N.; Zaritzky, N.E. Plasticized starch-based coatings to improve strawberry (Fragaria×ananassa) quality and stability. *J. Agric. Food Chem.* **1998**, *46*, 3758–3767. [CrossRef]

16

Application of Non-Parametric Bootstrap Confidence Intervals for Evaluation of the Expected Value of the Droplet Stain Diameter Following the Spraying Process

Andrzej Bochniak [1], Paweł Artur Kluza [1,*], Izabela Kuna-Broniowska [1] and Milan Koszel [2,*]

1 Department of Applied Mathematics and Computer Science, University of Life Sciences in Lublin, 20-950 Lublin, Poland; andrzej.bochniak@up.lublin.pl (A.B.); izabela.kuna@up.lublin.pl (I.K.-B.)
2 Department of Machinery Exploitation and Management of Production Processes, University of Life Sciences in Lublin, 20-950 Lublin, Poland
* Correspondence: pawel.kluza@up.lublin.pl (P.A.K.); milan.koszel@up.lublin.pl (M.K.)

Abstract: In the era of sustainable agriculture, the issue of proper and precise implementation of agrotechnical operations, without harmful effects on the natural environment, begins to play an important role. Statistical tools also become important, for example, when assessing the malfunction of plant cultivation equipment. The study presents a comparison of six nonparametric bootstrap methods used for construction of confidence intervals for the expected value of an average diameter of droplet stains following the spraying process. The simulation tests were carried out based on experiment with nozzle sprayer Lechler 110-03 using two spray nozzles: a new one and an old one. It was assumed that the distribution of the droplet stain size was consistent with the lognormal distribution. The paper considers the influence of the sample size, mean value and standard deviation of the droplet stain diameter on the interval range as well as on the estimated coverage probabilities of the confidence intervals. It was shown that in general these methods can be applied for this purpose. For the double bootstrap method and the studentized method, the empirical confidence levels of the constructed intervals turned out to be less distinct than the assumed level but the lengths of these intervals were greater than the lengths of intervals obtained using the other four methods.

Keywords: bootstrap methods; confidence intervals; lognormal distribution; sprayer; droplet diameters; sustainable agriculture

---

1. Introduction

Sustainable agriculture aims to promote a sustainable farming system that relies on the rational use of natural resources, which allows reducing the negative impact of agriculture on the environment and prevents the loss of organic matter in the soil. It is part of a larger movement towards sustainable development, which suggests that natural resources are limited, recognizes the constraints of economic growth and encourages equality in resource allocation. Sustainable agricultural development has been proposed as one of the alternatives to reverse the spiral of resource degradation and poverty. The goal of sustainable agriculture is to protect natural resources [1,2]. Sustainable development is one of the principles on which the European Union (EU) policy should be based, including the EU's common agricultural policy. The common agricultural policy is the main institutional structure and the main factor of reforms affecting the development of sustainable agriculture [3].

Spraying is the basic and the most widely used treatment in plant protection but improperly preformed or defective technical equipment may degrade the quality of agricultural raw materials and pose a threat to humans and the environment. Recently, plant protection has been severely criticized for creating environmental and human hazards. The use of pesticides, in addition to unquestionable effectiveness, has serious disadvantages. Creation of resistant biotypes, which are difficult to combat later, can be another possible threat. In addition, chemicals change crop biochemistry, pollute the environment and often destroy natural enemies who would limit the host population [4].

However, the use of herbicides, despite many controversies, is one of the basic methods of protecting plants against weeds in modern agriculture. The use of plant protection products (generally known as pesticides or agrochemicals) makes it possible to increase food production by destroying weeds and pests that attack crops. As a result, productivity per hectare increases and also reduces losses during transport and storage of food [5]. In addition, the use of pesticides gives the farmer the opportunity to intervene quickly when the crop is directly threatened by diseases and pests. This is especially important in the case of mass and sudden outbreaks of pests. Some chemicals, for example, seed mortars, have a preventive effect. In turn, granulated pesticides introduced into the soil and taken up by plant roots operate for the whole period or a significant part of the growing season [6].

The effect of herbicides depends on both the technical and technological processes that make up the spraying procedure, as well as the behaviour of the active substance on the sprayed surface of the plants and inside them. These processes include the production of spray liquid droplets (atomization), transfer from the atomizer to the plant surface, droplet retention on plants, formation of sediments on the leaf surface after evaporation of water from the spray liquid, up-taking of the active substance (absorption) and its translocation and metabolism [7].

A number of technological, technical and climatic factors influence the quality of the sprayer work. The most important of them include the type of machine, choice of nozzles, spray parameters, temperature and humidity as well as compliance with the instructions of producers of plant agents [8]. It should be noted that the nozzle wear degree has a decisive influence on spray quality. Application of excessive numbers of droplets on the protected surface causes their merging, which worsens the quality of spraying. Gravity makes droplets on the surface of plants penetrate the soil and reach groundwater [5,9]. Besides, if nozzles generate very small droplets, they are drifted away by wind or the liquid evaporates before falling on protected plants.

In spraying operations, it is important that as much of the spray liquid as possible settles on the surface of the plant. This is the basic condition for achieving high effectiveness of the procedure, as well as reducing losses that burden the environment. Liquid settlement on plants depends mainly on the development phase of the plant, crop density, dose, type of usable liquid and operation of the spraying apparatus [10]. It was stated that in the total spray liquid expenditure balance (during wheat spraying) 52.4% were losses in the air (as a result of drift and evaporation) and as a result of falling on the soil, while 47.6% sprayed liquid stopped on the plants [11]. These are significant losses that can cause environmental pollution.

A uniform distribution of the spray liquid has a significant influence on the environmental impact of agrochemicals. Uneven application means that a more larger dose of the agent is applied over a large field area than it results from actual needs. It is not only a waste but it also carries the risk of the accumulation of residues of plant protection agents in agricultural raw materials [12].

For these reasons, statistical analysis of the droplet stains after the spraying process is important for adjustment of the operating parameters of the sprayer. One of the statistical methods that can be used is the analysis of the droplet stain distribution, especially point or interval estimation of the parameters of their distribution. Naturally occurring droplets are usually irregular in shape and different in size. This resulted in the introduction of different mean droplet diameter definitions along with statistical description of the droplet size distribution (linear, surface and volume (Sauter)).

The conducted research regarding the estimation of droplet size and quantity involves the division into raindrops and droplets following the spraying process, where various types of probability distribution can be applied to describe each of the two cases presented above. As regards empirical data obtained for raindrops, the following distributions are used most frequently—exponential [13], gamma [14–16], Gauss [17], lognormal [18,19], Weibull [20,21] and Poisson [22]. The gamma distribution is used to describe the droplet size distribution following the spraying process [23]. Equivalently, the following distributions are also used for the same type of data—upper truncated lognormal [24,25], Rosin–Rammler and Nukiyama–Tanasawa [26] and Weibull [18,27]. The lognormal distributions with the upper truncation [15,23] are applicable for the distribution of droplet diameters because the measured droplet diameters are in a certain interval ranging from zero to a definite upper limit, which in an extreme case can be a maximum measured droplet size.

A number of parametric and non-parametric methods have been designed for interval estimation of the expected value of a random variable with a lognormal distribution. A few such procedures were described and compared in Reference [28], including the Cox method for a large sample size presented in Reference [29], a conservative method described in Reference [30] and a parametric bootstrap method [31]. Based on numerical results, it was demonstrated [28] that all these procedures, with the exception of the Cox method, are too conservative or too liberal disregarding the sample size. Therefore, the methods of construction confidence intervals presented above are not satisfactory for small sample sizes. This problem was solved in Reference [32], where the method was developed for construction of exact confidence intervals for small sample sizes using the idea of a generalized p-value and generalized confidence intervals.

The bootstrap method was designed by Bradley Efron [23,24,26,33]. The group of bootstrap methods includes parametric and non-parametric methods, including inter alia—the basic, percentile, bias-corrected, bias-corrected and accelerated, studentized and double bootstrap methods. These methods differ as regards the selection of appropriate quantiles of the distribution of the estimators from bootstrap samples. Bootstrap is a relatively new method but in the computer era it is frequently used for obtaining confidence intervals and test hypotheses in situations when the sample sizes are small, the distribution of estimators is unknown or when the fulfilment of relevant assumptions is doubtful. The statistical research carried out in Reference [27], showed that the information on the asymmetry of the probability distribution of a random variable or initial estimation of the asymmetry coefficient can be helpful while determining the sample size both in bootstrap tests and classical significance tests.

The aim of this study was to investigating the possibility of using various bootstrap methods to generate confidence intervals to assess the average droplet trace diameter, which can be characterized by an asymmetric probability distribution set close to the lognormal one. The simulations were limited to the ranges of droplet trace sizes obtained in experimental experiments using the Lechler 110-03 sprayer. The usefulness of particular methods is indicated depending on the size of the random sample and the different degree of variability of the obtained droplet trace sizes. The tests carried out are aimed at determining the methods of assessing the degree of wear of the sprayer nozzle.

2. Materials and Methods

2.1. Material

The analysis carried out in the research concerns the estimation of the mean value of droplet stain diameters following the spraying process carried out in an experimental study on a Lechler 110-03 spray nozzle type. The nominal flow rate for this type of the spray nozzle is $1.17\ \mathrm{L \cdot min^{-1}}$. The spray boom shift during spraying was $9\ \mathrm{km \cdot h^{-1}}$ ($2.5\ \mathrm{m \cdot s^{-1}}$) and the pressure was 5 bar (0.5 MPa). After the droplets applied on each piece of foil dried, scans of 5×5 cm images were obtained. The scanner resolution during scanning was 300 dpi. The size of droplet stains and the number of droplets was calculated using the computer software Image Pro+ by Media Cybernetics. The new spray nozzle and older one were examined for comparison in a research experiment. The older nozzle was obtained as a result of laboratory wearing using Kaolin KOM water slime from SURMIN-KAOLIN. The slime was produced by adding 9.1 kg of kaolin to 150 L of water [34]. After reaching 10% wearing, the liquid outflow rate was measured at 1.29 $L \cdot min^{-1}$. In Figure 1 examples of scans obtained from the spraying process using new spray nozzle (a) and older one (b) are shown.

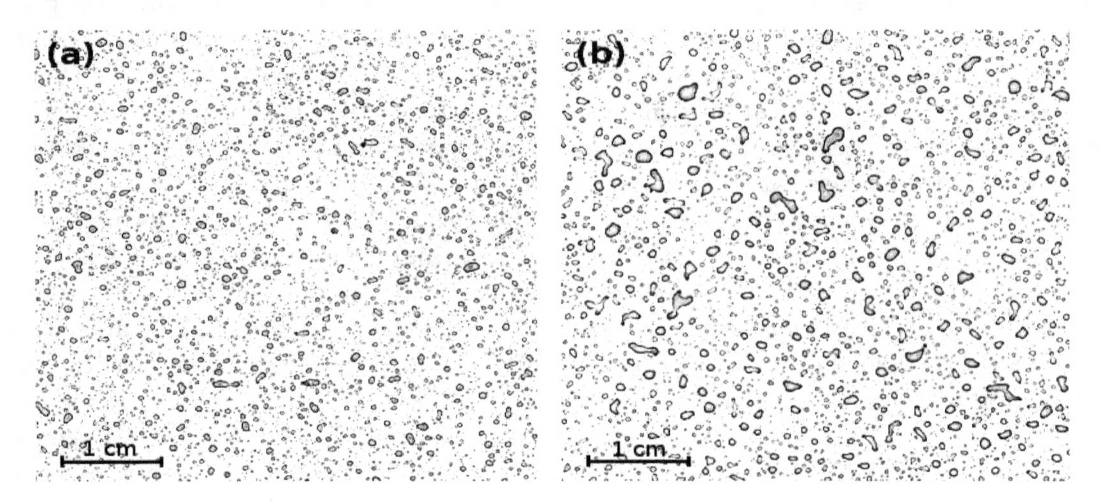

Figure 1. Example scans of droplet stains from: (**a**) new spray nozzle (**b**) spray nozzle in longer use.

2.2. Mathematical Background

The density function $f(x)$ of the lognormal distribution for the diameter size of the splashed droplets following the spraying process is given by:

$$f(x) = \frac{1}{x\sigma\sqrt{2\pi}} exp\left(-\frac{(log(x)-\mu)^2}{2\sigma^2}\right), \tag{1}$$

where:

$x \in (0, +\infty)$ – droplet diameter [μm],
$\mu \in < 0, +\infty)$ – expected value of variable ln(X), location parameter,
$\sigma > 0$ – standard deviation, scale parameter,
$log(\cdot)$ – natural logarithm.
$Y = log(X) \sim N(\mu, \sigma)$.

The parameters of the lognormal distribution turn out to be the functions of both parameters μ and σ^2 and this dependence poses difficulties in obtaining exact and optimal tests and confidence intervals. In particular, the expected value of lognormal random variable X depends on μ and σ^2:

$$E(X) = E(exp(Y)) = exp\left(\mu + \frac{1}{2}\sigma^2\right). \tag{2}$$

The overlapping of droplet stains following the spraying process is an additional difficulty in using these stains to estimate the distribution of the spraying parameters. The poor accuracy in point estimation of the mean value and standard deviation has an impact on the interval estimation of these parameters. Taking into account the above difficulties, six non-parametric bootstrap methods were used to construct the confidence intervals—basic, percentile, bias-corrected, bias-corrected and accelerated, studentized and double bootstrap methods.

During the research work, the sensitivity of the lengths and coverage of intervals constructed using the discussed bootstrap methods were compared, given the changes in the sample sizes and changes in the values of standard deviation. The consistency of empirical coverages was also compared with the assumed confidence level.

Let $\mathbf{x} = (x_1, x_2, \ldots, x_n)$ denote an n-element random sample, whereas $R(x, F)$ is some statistics determined on the sample space with distribution F, θ is the unknown parameter and σ is the standard deviation of this parameter. The bootstrap sample is an n-element realization of the variable X,

$\mathbf{x}^* = (x_1^*, x_2^*, \ldots, x_n^*)$ and its elements are selected at random with replacement from sample x. In order to obtain bootstrap samples, independent B-fold ($B \geq 1000$) generation of n-element data sequences $\mathbf{x}^* = (x_1^*, x_2^*, \ldots, x_n^*)$ is carried out. These data sequences are used to determine the estimator $\hat{\theta}$ of the θ parameter. As a result, B-element sequence of estimators $\hat{\theta}_j^* = R\left(x_j^*\right), j = 1, 2, \ldots, n$ is obtained. Out of the sequence of estimators $\hat{\theta}_j^*$, the quantiles of a fixed rank are calculated, which are then used to obtain the confidence interval for parameter $\theta = R(x)$.

In the simplest basic method (B), the $(1-\alpha)\%$ confidence interval for parameter θ is obtained according to the formula:

$$\left(2\hat{\theta} - \hat{\theta}^*_{(1-\frac{\alpha}{2})}, 2\hat{\theta} - \hat{\theta}^*_{(\frac{\alpha}{2})}\right), \tag{3}$$

where:

$\hat{\theta}^*_{(\alpha)}$ denotes α^{th} quantile from the sequence of values $\hat{\theta}_j^*$,

$\hat{\theta}$ denotes the estimator from the original sample $\mathbf{x} = (x_1, x_2, \ldots, x_n)$.

The percentile (P), bias-corrected (BC) and bias-corrected and accelerated (BCa) methods confidence intervals do not use estimator $\hat{\theta}$ but are calculated exclusively based on quantiles of a relevant rank from the sequence of estimators:

$$\left(\hat{\theta}^*_{(\alpha_L)}, \hat{\theta}^*_{(\alpha_U)}\right), \tag{4}$$

where the ranks α_L and α_U of appropriate percentiles can be calculated according obtained sequence of estimators $\hat{\theta}_j^* = R\left(x_j^*\right), j = 1, 2, \ldots, n$ depending on selected method.

The bootstrap method with bias correction [35] contributes to improvement of the precision of the percentile method in the case when the median of estimator values for all bootstrap samples is not equal to the value of the estimator for the original sample. This method adds two steps after estimation of the distribution of the considered parameter for bootstrap samples:

- calculation of the fraction k of bootstrap samples for which the estimator value of the parameter is lower than the estimator calculated on the basis of the original sample,
- calculation of interval bounds regarding the calculated fraction of samples in order to correct the bias.

Similar to the previous method, the bias-corrected and accelerated bootstrap method (BCa) [24,35] takes into account the fact that estimators of the parameter for bootstrap samples can be biased and the impact of individual values in a sample on the value of the estimator is considered as well. When the distribution is skewed, some adjustment needs to be made and BCa method gives such a possibility.

The ranks α_L and α_U of percentiles in BCa method depend on two constants a and z_0 called the acceleration and bias correction coefficient. The standard error of $\hat{\theta}$ is stabilized by coefficient a and z_0 measures the median bias of $\hat{\theta}^*$. This results in faster convergence of the estimator confidence limit into its exact value than in the case of the quantile method. Thus, the probability of coverage for the BCa method is closer to the nominal probability than in the percentile method.

There are different ways to determine the value of coefficient a. One of them uses the jackknife method [24], which explores the impact of n individual elements of a sample by using n special sample size $n-1$. Each of the n new samples is obtained through deleting each time only one element from the original sample. Based on such samples, n estimators of the θ parameter are calculated.

In the studentized method, sometimes referred to as the bootstrap-t method, the appropriate statistic quantiles are used in the following form:

$$t^* = \frac{\hat{\theta}^* - \hat{\theta}}{\hat{\sigma}\left(\hat{\theta}^*\right)}, \tag{5}$$

where $\hat{\sigma}\left(\hat{\theta}^*\right)$ is the estimator of the standard error of the estimator $\hat{\theta}^*$ determined based on the bootstrap sample $x^* = (x_1^*, x_2^*, \ldots, x_n^*)$.

The double bootstrap method is a modification of the previous method [36], where the standard error $\sigma\left(\hat{\theta}^*\right)$ of $\hat{\theta}^*$ is estimated on bases of additional B_1 bootstrap samples $\mathbf{x}^{**} = (x_1^{**}, x_2^{**}, \ldots, x_n^{**})$ generated from the $\mathbf{x}^*$ sample, where $\bar{\theta}^{**} = \frac{1}{B_1}\sum_{i=1}^{B_1}\hat{\theta}_i^{**}$.

This section outlines the six bootstrap methods, which were used in this study. More details and appropriate mathematical formulas of individual methods can be found in numerous studies available in the cited literature.

2.3. Computer Simulations

On the basis of the experimental tests, the ranges of the average diameters of droplet traces obtained in the spraying process using the tested nozzles were estimated. The determined ranges of means as well as standard deviations obtained for the droplet stains were used as the basis for conducting computer simulations using the six bootstrap methods described in the previous section. The simulations were carried out taking into account different sample sizes (10–600 elements). For each combination of two values of the analyzed mean and standard deviation, ten thousand samples were generated consisting of equal elements each, which were used for generating bootstrap samples. For each random sample x, the total of $B = 2000$ bootstrap samples were generated, based on which the relevant percentiles were estimated. In the case of a double bootstrap, the standard deviation was estimated based on $B_1 = 250$ samples. For each generated sample, all six types of confidence intervals were calculated and the assumed confidence level was $\alpha = 0.05$. Computer simulations were carried out using software MATLAB R2014a by employing our own code and embedded procedures on the computer with the operating system Windows 10 Professional equipped with an Intel Xeon E5-2609 2.40 GHz processor.

3. Results

The analysis took into account the mean values and standard deviation values estimated on the basis of the measurements performed with the Lechler 110-03 sprayer. The first step involved testing the consistency of the distribution of size of droplets stain diameters from the measurements with the lognormal distribution and determining its parameters.

The considered sample scans, which were previously presented on Figure 1, resulted in a highly asymmetrical distribution of droplet stain diameters. The distribution of droplet diameters obtained while spraying using a new and old nozzle are presented in Table 1. The drops traces were divided into 5 size classes—0–150 μm, 150–250 μm, 250–350 μm, 350–450 μm and above 450 μm. The estimates were obtained using computer software Image Pro+ by Media Cybernetics. The table includes the number of objects assigned to classes, the percentage of objects, the percentage of surface covered and the average drop values and standard deviations in each class. The data compiled in Table 1 show a strongly asymmetrical distribution.

In addition, to determine the approximate parameters of the droplet size distribution, a division into a larger number of size classes was made (Figure 2). For both nozzles (new and old), using the MatLab function *fitnlm*, parameters of log-normal distribution were determined, which are best fitted to the obtained histograms for selected droplet size classes. For the new nozzle, the average is $\mu_x = 160.3$ μm and the standard deviation $\sigma_x = 65.8$ μm, while for the older nozzle the average is $\mu_x = 186.7$ μm and the standard deviation $\sigma_x = 81.2$ μm.

Due to the overlapping of droplet stains following the spraying process and the difficulties in construction of the confidence intervals for the expected value of the lognormal distribution, we chose the bootstrap methods [24]. As already mentioned, the confidence intervals, constructed using the discussed bootstrap methods were compared in terms of the sensitivity of their lengths and probabilities of the expected value coverage to changes in sample sizes and in values of standard deviation. The consistency of empirical coverage probabilities with the assumed confidence level was also compared.

Table 1. Droplet traces size distribution in an example datasets.

Stain Class	No. Objects	% Objects	% Area	Diameter Mean	Diameter Std
New nozzle					
<150 μm	511	62.01	43.96	123.89	19.19
150–250 μm	242	29.37	36.72	184.29	22.61
250–350 μm	61	7.40	15.95	293.17	52.33
350–450 μm	9	1.09	3.17	382.51	69.80
>450 μm	1	0.12	0.19	471.00	0.00
Older nozzle					
<150 μm	323	51.43	37.49	132.34	14.37
150–250 μm	204	32.48	35.71	196.42	15.69
250–350 μm	80	12.74	19.09	295.28	24.28
350–450 μm	12	1.91	3.69	412.07	39.19
>450 μm	9	1.43	4.01	521.74	47.16

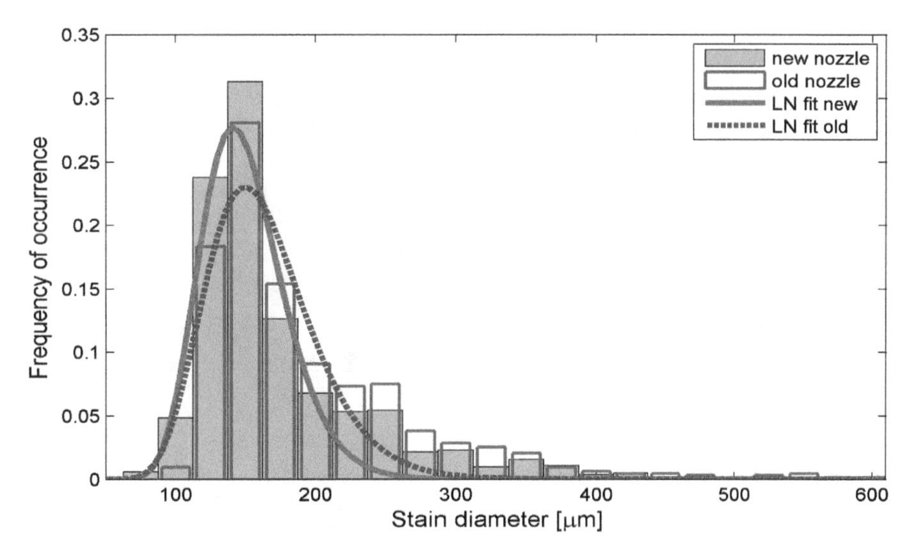

Figure 2. Histograms of the droplet stain diameters from the experiment and fitted lognormal distribution for new and older nozzles.

The analysis took into account the mean values and standard deviation values estimated based on the measurements performed with the Lechler 110-03 sprayer. In order to generate samples from the lognormal distribution for the population mean, the range from 90 μm to 210 μm was taken with a step of ten units, whereas for the standard deviation the range was from 30 μm to 180 μm with a step of ten units. For each combination of two values of the analyzed mean and standard deviation, ten thousand samples were generated consisting of thirty elements each, which were random samples used for generating bootstrap samples. For each random sample x, the total of $B = 2000$ bootstrap samples were generated, based on which the relevant percentiles were estimated. In the case of a double bootstrap, the standard deviation was estimated based on $B_1 = 250$ samples. The values of B and B_1 were determined according to Reference [36]. For each generated sample, all six types of confidence intervals were calculated and the assumed confidence level was $\alpha = 0.05$.

The largest differences between the results obtained by different methods occurred for small samples. Figure 3 presents numerical results of the coverage of the calculated confidence intervals for the assumed 95% confidence levels using the six bootstrap methods with division to the selected lengths of the mean value and standard deviation value (μ_x, σ_x) of variable X. In a majority of cases, all the considered methods give the coverage probabilities lower than the assumed level of 95% for sample sizes of 30. The confidence intervals constructed with the studentized and double bootstrap methods give the coverage probabilities not greater than 94% for each considered value of the mean; noteworthy, the variation of the feature may be high and, in a particular case, the variation coefficient may even be equal to 80%. The other four methods allow obtaining coverage probabilities not greater than 93%; however, only for the variation coefficient not higher than 34% (Figure 3).

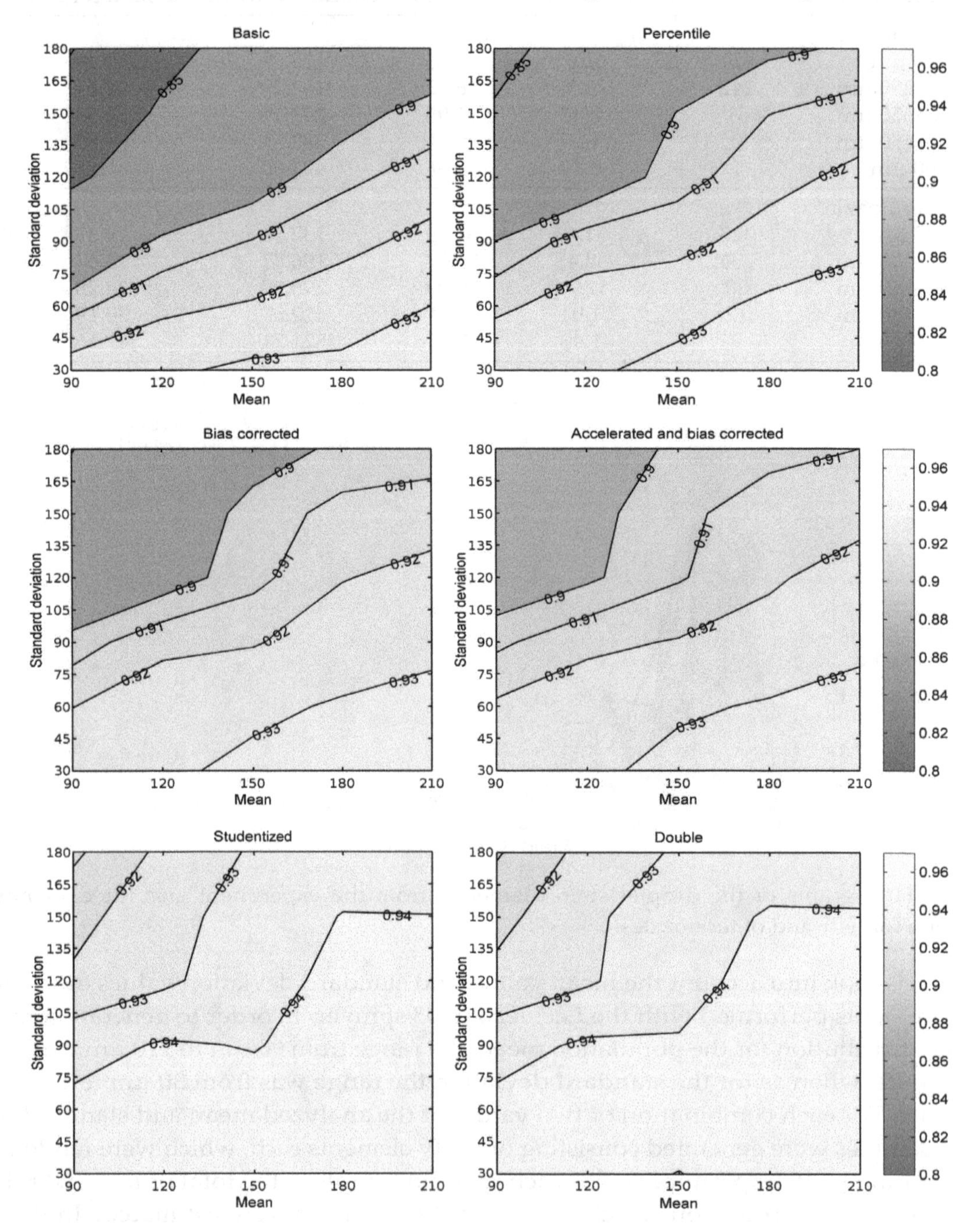

Figure 3. Coverage probabilities of the mean value by confidence intervals constructed by the six types of bootstrap methods ($n = 30$).

In order to study the impact of the sample size on the lengths of confidence intervals for the examined bootstrap methods, the sample sizes considered in the simulations were 10, 20, 30, 50, 80, 100, 150, 200, 400 and 600. Since the interval lengths for larger sample sizes overlap, the figure presents only the lengths for the smaller sample sizes (up to 100 elements). The samples were generated from a population for which the mean value was 160 μm and the standard deviation ranged from 40 to

120 μm. For each of the sizes, five thousand samples were generated and the length of the confidence interval calculated with the six studied methods was estimated based on each sample. The lengths of the intervals obtained were averaged for each method separately. From the graph presented in Figure 4 (for the standard deviation of 80 μm, close to the sample obtained from the experiment), it is evident that the intervals for the studentized and double bootstrap methods are wider, compared with the basic, percentile, bias correction and accelerated methods. Concurrently, it can be noted that the lengths of the confidence intervals for the four methods mentioned above are fairly equal (Figure 4).

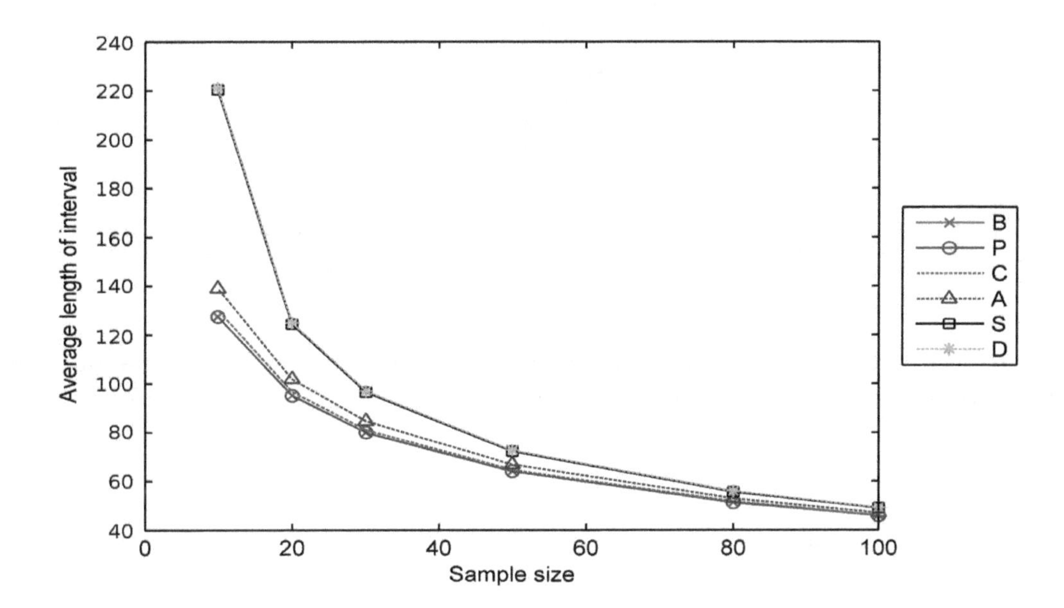

Figure 4. Changes in the length of confidence intervals constructed using nonparametric bootstrap methods depending on the sample size; $\mu_x = 160$ μm, $\sigma_x = 65$ μm (B—basic, P—percentile, C—bias corrected, A—accelerated, S—studentized, D—double bootstrap).

The four methods—namely, basic, percentile, bias correction and accelerated methods—give intervals characterized by similar length but different from the results obtained using the studentized and double bootstrap methods, which are mutually comparable (Hall, 1992). Hence, the graph shows the comparison of the lengths of the confidence intervals only for two methods—the percentile and double bootstrap methods (Figure 4). For small sample sizes and large standard deviation, the lengths of the intervals obtained using the double bootstrap method are even approximately 3 times longer than those obtained using the percentile method. For larger sample sizes, these differences are less visible.

In the study of the influence of the sample size on the coverage probabilities of confidence intervals, obtained in particular for the smaller sample sizes, the follwing sample sizes were included in the simulation—10, 20, 30, 50, 80, 100, 150, 200, 400 and 600 (Figure 5). For each sample size, five thousand samples were generated from the same population as previously (mean value 160 μm and standard deviation from 40 to 120 μm) and the percentage of covering of the assumed value of 160 was checked.

The intervals obtained using the analyzed methods give a similar covering of the true value of the population mean only for samples with a size of at least 100–200 elements. For smaller sample sizes, we obtained a clear division of the discussed methods into three groups. Given the interval lenghts discussed above, the covering is closest to the assumed confidence level for the studentized and double bootstrap intervals, followed by the bias correction, percentile, accelerated and basic methods. According to the considered criterion, the worst covering is provided by the basic method (Figure 6).

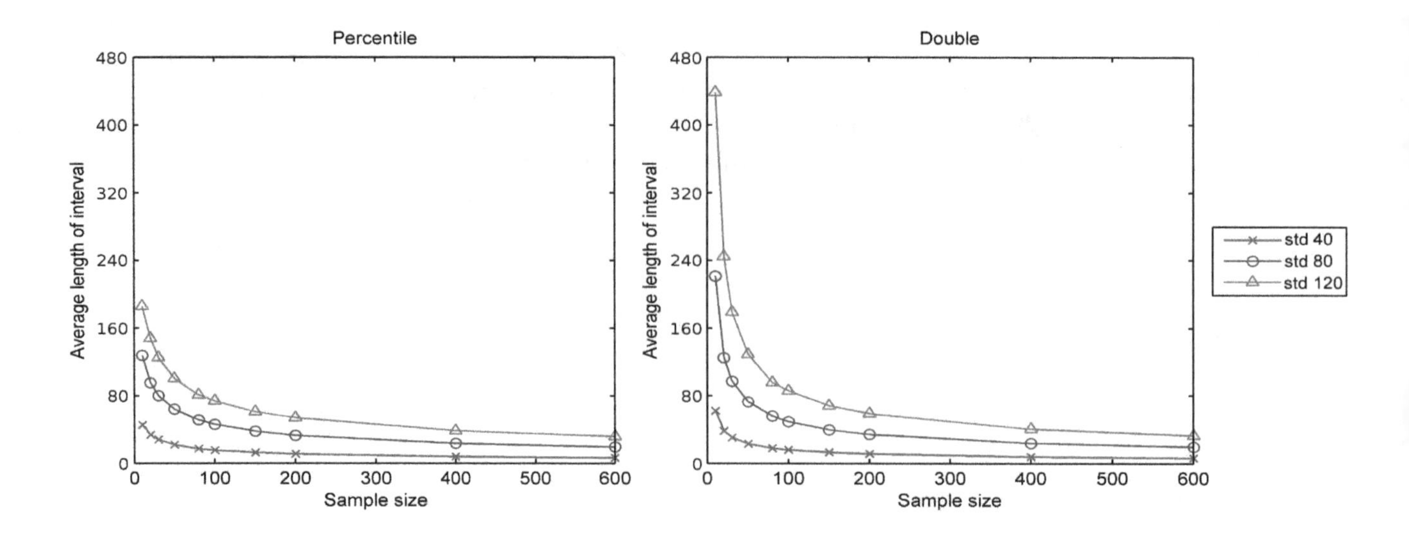

Figure 5. Changes in the lengths of confidence intervals constructed using nonparametric bootstrap methods (percentile and double bootstrap) depending on the sample size and different values of standard deviation; $\mu_x = 160$ µm.

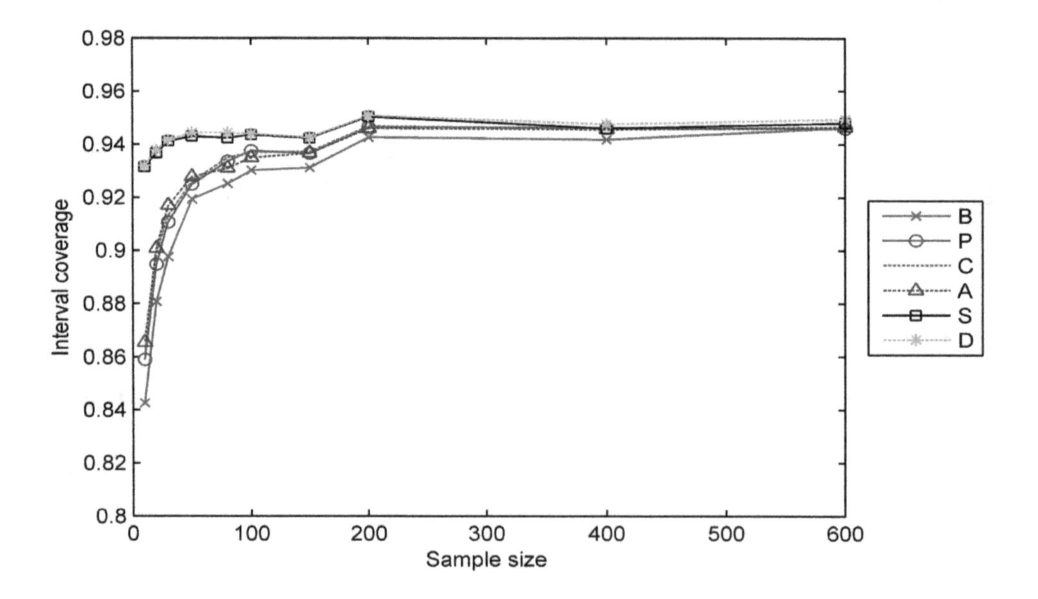

Figure 6. Changes in the empirical confidence level of intervals obtained using the nonparametric bootstrap methods depending on the sample size; $\mu_x = 160$ µm, $\sigma_x = 65$ µm (B—basic, P—percentile, C—bias corrected, A—accelerated, S—studentized, D—double bootstrap).

The graph presented in Figure 7 and Table 2 show the influence of the standard deviation on the coverage of confidence intervals. As in the case of the comparison of the interval lengths, the graph shows a comparison of only the percentile method and the double bootstrap method (Figure 7, Table 2). The difference between the coverages of the intervals obtained using these methods are visible mainly for small sample sizes.

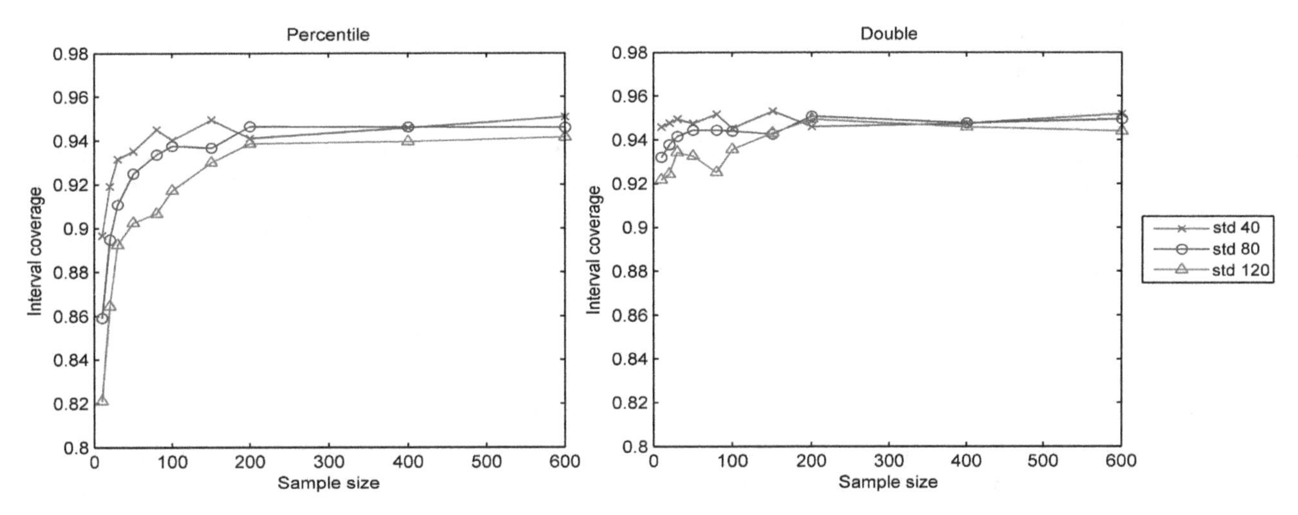

Figure 7. Changes in the empirical confidence level of intervals constructed using nonparametric bootstrap methods (percentile and double bootstrap) depending on the sample size and different values of standard deviation; $\mu_x = 160$ µm.

Table 2. Coverage probabilities of the expected value by confidence intervals constructed using nonparametric bootstrap methods depending the sample size and different values of standard deviation, $\mu_x = 160$ µm.

Sample Size	Standard Deviation	B	P	BC	BCa	S	D
10	40	0.897	0.897	0.895	0.895	0.947	0.946
	80	0.843	0.859	0.861	0.865	0.932	0.932
	120	0.789	0.821	0.828	0.847	0.921	0.922
20	40	0.918	0.919	0.919	0.917	0.946	0.948
	80	0.881	0.895	0.899	0.901	0.937	0.938
	120	0.838	0.864	0.870	0.882	0.924	0.924
30	40	0.928	0.932	0.933	0.934	0.950	0.949
	80	0.898	0.911	0.913	0.917	0.941	0.942
	120	0.869	0.892	0.897	0.905	0.934	0.934
50	40	0.935	0.935	0.936	0.939	0.947	0.947
	80	0.919	0.925	0.926	0.928	0.943	0.944
	120	0.883	0.902	0.907	0.914	0.932	0.933
80	40	0.944	0.945	0.946	0.946	0.950	0.952
	80	0.925	0.934	0.935	0.931	0.942	0.944
	120	0.895	0.907	0.908	0.910	0.926	0.925
100	40	0.943	0.940	0.941	0.941	0.944	0.945
	80	0.930	0.938	0.937	0.935	0.944	0.944
	120	0.905	0.917	0.919	0.920	0.934	0.936
150	40	0.948	0.949	0.948	0.948	0.952	0.953
	80	0.931	0.937	0.937	0.937	0.942	0.943
	120	0.921	0.930	0.932	0.932	0.942	0.943
200	40	0.941	0.941	0.940	0.941	0.945	0.946
	80	0.943	0.946	0.947	0.946	0.950	0.951
	120	0.928	0.939	0.941	0.939	0.948	0.949
400	40	0.947	0.946	0.946	0.945	0.947	0.947
	80	0.942	0.946	0.946	0.946	0.946	0.948
	120	0.936	0.940	0.939	0.941	0.944	0.946
600	40	0.949	0.951	0.949	0.949	0.952	0.952
	80	0.946	0.946	0.946	0.946	0.948	0.949
	120	0.940	0.942	0.941	0.939	0.942	0.944

4. Discussion

Agricultural ecosystems have been designed to provide food, fibre and fuel. Ecosystem services such as nutrient recirculation, pollination and water flow are often referred to as endangered resources in modern societies. Agricultural systems can therefore be seen as heterogeneous but nested socio-ecological systems requiring an institutional environment [3].

Man-made climate change has created major challenges to achieve sustainable agriculture by depleting ecological and natural resources. Existing climate change has an impact on water resources. Some countries are already experiencing severe water scarcity or are reaching their limits [37]. The most promising strategy for achieving sustainable development is to replace hazardous agrochemicals with environmentally friendly preparations from symbiotic microorganisms that could provide protection against biotic and abiotic substances [38].

The main conflict in the pursuit of sustainable development is between economic maximization of growth and environmental protection [39]. On a global scale, agriculture is one of the main sources of degradation of ecosystem services. Also in Europe, most of today's agriculture has a severe negative impact on ecosystems, for example contributing to global carbon cycle disorders, loss of biodiversity and increasing eutrophication. The transition to more sustainable agriculture is necessary to avoid further degradation of ecosystems [3]. The presented assessments of various authors regarding the impact of pesticide use on the natural environment, agricultural production and human health indicate that an important element is proper plant protection technique and compliance with spraying parameters. These parameters can have a significant impact on the quality of the plant protection treatment and the state of the environment.

To reduce the harmful effects of plant protection agents on the environment, we should aim to reduce herbicide doses. In Reference [40], authors based on their own research, say that the dose can be reduced from 12.5% to 50%. In their view, this does not significantly reduce crop yield, while maintaining the required herbicidal effectiveness. The use of herbicides in reduced doses brings the best results in spring cereals, because weeds in these crops are in the early stages of development and thus are easier to eliminate. Very good results can be obtained in plantations with low weed intensity and in the case of controlling species sensitive to a given herbicide. The weaker effect of reduced doses is obtained in winter cereals and especially barley. This fact is caused by the fact that weeds are more advanced in development. Producers of plant protection products are also trying to reduce the content of the active substance, contributing to the reduction of production costs and demonstrating their concern for the state of the environment [8].

Another way to protect the environment is to reduce the number of treatments through the combined use of agrochemicals. Appropriate mixtures of plant protection products give the possibility of wider protection of cereals and affect the increase of yields. The effectiveness of mixtures often does not differ from the effectiveness of preparations used separately and in some cases is even greater. It should be remembered that only well-tested and tested combinations of preparations can be used together [41].

It was therefore considered that it would be more reasonable to systematically implement an integrated model for controlling harmful organisms than to use only chemical methods. The idea of sustainable development combines economic growth with environmental protection and the global balance of ecosystems. According to this concept, the increase in agricultural production can only occur through the increase of resource productivity and thus the introduction of technologies that simultaneously protect resources and maintain their high quality for future generations. The need to develop environmentally friendly agriculture is increasingly accepted in Poland. An important element of it is the change of some agrotechnical methods due to the need to protect ground and surface waters. Restricting the use of chemicals is justified because they cause a number of negative effects [42]—they contribute to the emergence of pesticide-resistant breeds of pests, diseases and weeds, destroy beneficial organisms by reducing their numbers and can even lead to their complete elimination from ecosystems, reduce the value of the environment by contaminating soil, surface and groundwater, lead to a decrease in the nutritional value of plant and animal products as a result of persistence of their residues or metabolites, negatively affect human health.

Ensuring the quality of spraying is a key process in plant protection, which is to ensure high yields while minimizing the adverse impact on the environment [43]. Proper spraying of field crops is influenced by many factors that can be grouped into three categories—1. Equipment and techniques used; 2. Spray quality. 3. Operator skills [44]. All leading atomizer manufacturers offer "low-drift" nozzles that produce droplets less prone to drift [44]. Currently produced atomizers and in particular Venturi nozzles, are designed to produce larger diameter droplets. These designs were created because droplets with a smaller diameter (<200 μm) are more likely to drift [45]. The quality of the plant protection treatment is influenced by the height of the boom position above the sprayed surface [45].

One of the methods used for improvement of the properties of liquids used for plant protection is usage of adjuvants. They improve spraying efficiency by equalizing droplet size and reducing wind drift. They reduce the level of the surface tension of liquid, so they enable coverage more surface of the leaf blade and at the same time they increase absorbing properties of plants. An additional benefit of adjuvants is availability of combining various types of plant protection agents and their use during one treatment [46].

According to Directive 2009/128/EC of the European Parliament establishing a framework for the sustainable use of pesticides, ways to reduce the impact of pesticides on the environment are determined and thus also maintaining the safety of field crops. Particularly great importance is attached to the precise use of pesticides on protected plants. An important issue is the application of the pesticide in such a way that it is effective in use and at the same time safe for the environment. The safety of pesticide use in agriculture depends on the techniques used. At present, great importance is attached to eliminating drift of liquid used during plant protection treatment. The quality of the plant protection treatment depends, among others on atomizer type, working pressure and working speed. The type of atomizer used and the operating pressure determine the droplet size produced by the atomizer. Sprayers wear out during operation and the parameters of their work change. These changes are expressed by the increase in the flow rate of the liquid as well as the spectrum of droplets produced (VMD) [46].

The research topics discussed in the papers make it important to determine the most precise determination of the distribution of droplets generated by sprayers. Because the statistical probability distribution for droplet trace diameters is not fully specified, as presented in the introduction, computer methods become helpful when analysing these problems. Their example are the bootstrap methods, which allows, for example, to generate confidence intervals for a specific parameter or statistics. As the simulation studies show, they do not always give the right results. Results are influenced by among others size of the random sample or too large variability of the examined feature. However, simulation studies have shown that under certain conditions, these methods give reliable results and allow the study properties of variables with asymmetrical statistical distributions, which is confirmed by Reference [27].

Despite some advantages, bootstrap confidence intervals may be heavily distorted and therefore may be misleading. This problem is especially noticeable for simple bootstrap methods and small samples. This is confirmed by the results of other authors who also received inaccurate estimates of confidence intervals in their studies [47]. In addition, bootstrap methods are time consuming, and not always more complex methods give better results, and significantly increase the cost of calculations [48]. Despite these drawbacks, bootstrap methods are a powerful tool for creating confidence intervals. They do not need sophisticated mathematical models and are often used in the absence of information about statistical distributions of the studied features. In addition, for samples of over 200 elements, all examined bootstrap methods give similar results, so in such cases the percentile method seems sufficient. It is the least time-consuming than other more advanced bootstrap methods. For smaller samples studentized and double bootstrap gives more proper estimates.

The applied method of assessing the size of sprayed droplets obtained from sprayers by means of measuring the size of traces left on the surface is burdened with a certain error. Some drops fall in the same place or close to each other, as a result of which some traces are enlarged compared to real ones. For larger drops, the most commonly left traces do not have a circular shape, which causes additional inaccuracies in the estimation of the size of traces left by the computer image analysis program. However, this method, apart from determining the droplet size distribution, also allows the surface coverage of falling drops to be assessed.

In future research, the authors intend to focus on further analysis of spray droplet distribution using more accurate droplet size measurement methods (using in-flight laser measurement). Additionally, the use of the bootstrap method to estimate other parameters of the droplet size distribution obtained from the sprayer will allow for a more precise assessment of sprayer nozzle wear and unevenness of the generated droplets.

5. Conclusions

The comparison of the empirical probabilities of coverage of the expected value by the examined six bootstrap confidence intervals used for estimation of the mean droplet stain diameter following the spraying process performed with the Lechler 110-03 sprayer nozzle leads to the following conclusions:

- The size distribution of traces of droplets obtained from sprayers is asymmetrical, similar to the log-normal distribution.
- The six bootstrap methods compared give confidence intervals that in general do not hold the assumed confidence level especially for small samples.
- The bootstrap methods are generally useful for constructing confidence intervals for the expected value of droplet diameter but not all methods always give confidence intervals that in general hold the assumed confidence level.
- The simulation studies conducted here can be used in practice with the interval estimation of the expected value of droplet stain diameters. An adequate method should be selected depending on the sample size, in particular in the case of smaller sizes (below 100) and depending on the variability of data.
- The studentized and double bootstrap methods allow obtaining less distinct coverage than the assumed confidence level, compared with the other four methods.
- For small sample sizes, the lengths of the confidence intervals obtained using the studentized and double bootstrap methods are similar, greater than the intervals from the other four methods.
- Although the confidence intervals obtained by the studentized and double bootstrap methods maintain the assumed coverage level, however, for small samples the estimated confidence intervals are too wide, which makes it impossible to use them in practice. It is recommended to use samples with at least 100-200 elements for usefull confidence intervals.
- The coverage of intervals obtained using the studentized and double bootstrap methods is less sensitive to the variability of the feature, compared with the basic and percentile methods. For small sample sizes, a confidence level of 93% or more can be obtained if the coefficient of variation does not exceed certain values. For example, if the sample size is ca. 30 elements, the standard deviation of the sample cannot be higher than 22%–33%, 25%–36%, 25%–34%, 23%–33%, 109%–123% and 110%–123% of the sample mean for the basic, percentile, bias-corrected, bias-corrected and accelerated, studentized and double bootstrap methods, respectively.
- Determining the confidence interval for the expected value of droplet size using bootstrap methods requires knowledge of the exact diameters of individual traces, that is, using image analysis or another measurement method, such as laser droplet size measurement. However, the method presented in this work allows the analysis of surface coverage.

Author Contributions: Conceptualization, A.B. and I.K-.B.; methodology, A.B. and I.K-.B.; software, A.B..; validation, A.B., P.A.K., I.K-.B. and M.K.; formal analysis, P.A.K. and I.K-.B.; investigation, M.K..; resources, M.K..; data curation, P.A.K. and M.K.; writing—original draft preparation, A.B., P.A.K., I.K-.B. and M.K; writing—review and editing, A.B., P.A.K., I.K-.B. and M.K; visualization, A.B. and P.A.K.; supervision, A.B. and I.K-.B.; project administration, A.B. and I.K-.B.

References

1. Cheng, X.; Shuai, C.; Liu, J.; Wang, J.; Liu, Y.; Li, W.; Shuai, J. Modelling environment and poverty factors for sustainable agriculture in the Three Gorges Reservoir of China. *Land Degrad. Dev.* **2018**, *29*, 3940–3953. [CrossRef]
2. Kiełbasa, B.; Pietrzak, S.; Ulén, B.; Drangert, J.O.; Tonderski, K. Sustainable agriculture: The study on farmers' perception and practices regarding nutrient management and limiting losses. *J. Water Land Dev.* **2018**, *36*, 67–75, [CrossRef]
3. Öhlund, E.; Zurek, K.; Hammer, M. Towards sustainable agriculture? The EU framework and local adaptation in Sweden and Poland. *Environ. Policy Gov.* **2015**, *25*, 270–287, [CrossRef]
4. Piesik, D. Biologiczna walka z chwastami na przykładzie Rumex confertus Willd. *Postępy Nauk Rolniczych* **2001**, *3*, 85–98.
5. Biziuk, M.; Hupka, J.; Warendecki, W.; Zygmunt, B.; Siłwoiecki, A.; Zelechowska, A.; Dąbrowski, Ł.; Wiergowski, M.; Zaleska, A.; Tyszkiewicz, H. *Pestycydy: Występowanie, Oznaczanie i Unieszkodliwianie*; Wydawnictwo Naukowo-Techniczne: Warszawa, Poland, 2001.
6. Lipa, J.J. Zwalczanie szkodników, chwastów i patogenów. *Zesz. Probl. Postęp. Nauk Rol., Rol. Ekol.* **1987**, *324*, 131–155.
7. Miłkowski, P.; Woźnica, Z. Zachowanie się kropel opryskowych na powierzchni roślin a skuteczność chwastobójcza Glifosatu. *Pr. Kom. Nauk Roln. Kom. Nauk Lesn., Poznan. Tow. Przyj. Nauk* **2001**, *91*, 77–85.
8. Sawa, J.; Huyghebaert, B.; Koszel, M. Parametry pracy opryskiwaczy a jakość oprysku. In Proceedings of 4th Conference: "Racjonalna Technika Ochrony Roślin", Skierniewice, Poland, 15–16 October 2003; pp. 84–89.
9. Miczulski B. *Podstawy Praktycznej Ochrony Roślin*; Wydawnictwo Akademii Rolniczej w Lublinie: Lublin, Poland, 1991.
10. Rogalski, L.; Konopka, W. Wybrane charakterystyki opryskiwania pszenicy w zależności od wartości dawki cieczy użytkowej. *Zesz. Probl. Postęp. Nauk Rol.* **2002**, *486*, 367–373. Available online: http://agro.icm.edu.pl/agro/element/bwmeta1.element.agro-c9aa0840-51cb-4418-a22e-1e912970f2f1/c/367-373.pdf (accessed on 25 October 2019).
11. Rogalski, L.; Konopka, W. Bilansowanie rozchodu masy oprysku w łanie pszenicy w zależności od rodzaju cieczy użytkowej. *Zesz. Probl. Postęp. Nauk Rol.* **2002**, *486*, 375–380.
12. Nilars, T.; Taylor, B.; Kappel, D. Wpływ rozpylaczy na jakość i bezpieczeństwo opryskiwania. In Proceeding of 3rd Conference: "Racjonalna Technika Ochrony Roślin". Skierniewice, Poland, 16–17 October 2002; pp. 121–134.
13. Marshall, J.S.; Palmer, W.M.K. The distribution of raindrops with size. *J. Meteorol.* **1948**, *5*, 165–166. [CrossRef]
14. Kozu, T.; Nakamura, K. Rainfall parameter estimation from dual-radar measurements combining reflectivity profiles and path-integrated attenuation. *J. Atmos. Ocean. Technol.* **1991**, *8*, 259–270, [CrossRef]
15. Su, C.L.; Chu, Y.H. Analysis of terminal velocity and VHF backscatter of precipitation particles using Chung-Li VHF radar combined with ground-based disdrometer. *Terr. Atmos. Ocean. Sci.* **2007**, *18*, 97–116. [CrossRef]
16. Ulbrich, C.W. Natural variations in the analytical form of the raindrop size distribution. *J. Appl. Meteorol.* **1983**, *22*, 1764–1775. [CrossRef]
17. Maguire, W.B.; Avery, S.K. Retrieval of raindrop size distribution using two Doppler wind profilers: Model sensitivity testing. *J. Appl. Meteorol. Climatol.* **1994**, *33*, 1623–1635. [CrossRef]
18. Feingold, G.; Levin, Z. Application of the lognormal raindrop distribution to differential reflectivity radar measurement (ZDR). *J. Atmos. Ocean. Technol.* **1987**, *4*, 377–382. [CrossRef]
19. Meneghini, R.; Rincon, R.; Liao, L. On the use of the lognormal particle size distribution to characterize global rain. In Proceedings of the International Geoscience and Remote Sensing Symposium, Toulouse, France, 21–25 July 2003; pp. 1707–1709, [CrossRef]
20. Kuna-Broniowski, M.; Kuna-Broniowska, I. The use of comparators for automatic classification of the splashed rain drops. *Electron. J. Pol. Agric. Univ. Agric. Eng.* **2001**, *4*, #08. Available online: http://www.ejpau.media.pl/volume4/issue2/engineering/art-08.html (accessed on 25 October 2019).
21. Wilks, D.S. Rainfall intensity, the Weibull distribution, and estimation of daily surface runoff. *J. Appl. Meteorol.* **1989**, *28*, 52–58. [CrossRef]

22. Joss, J.; Waldvogel, A. Raindrop size distribution and sampling size errors. *J. Atmos. Sci.* **1969**, *26*, 566–569. [CrossRef]
23. Efron, B.; Tibshirani R.J. Statistical data analysis in the computer age. *Science* **1991**, *253*, 390–395. [CrossRef]
24. Efron, B.; Tibshirani R.J. *An introduction to the Bootstrap*; Chapman & Hall: New York, NY, USA, 1993; [CrossRef]
25. Kamińska, J.; Machowczyk, A.; Szewrański, S. The variation of drop size gamma distribution parameters for different natural rainfall intensity, Wydawnictwo ITP. *Woda-Środ.-Obsz. Wiej.* **2010**, *10*, 95–102. Available online: http://yadda.icm.edu.pl/baztech/element/bwmeta1.element.baztech-article-BATC-0005-0103 (accessed on 25 October 2019).
26. Hall, P. *The Bootstrap and Edgeworth Expansion*; Springer: New York, NY, USA, 1992; [CrossRef]
27. Pekasiewicz, D. Bootstrapowa weryfikacja hipotez o wartości oczekiwanej populacji o rozkładzie asymetrycznym. *Acta Univ. Lodz. Folia Oeconomica* **2012**, *271*, 151–159. Available online: http://cejsh.icm.edu.pl/cejsh/element/bwmeta1.element.hdl_11089_1928/c/Pekasiewicz_151-159.pdf (accessed on 25 October 2019).
28. Zhou, X.; Gao, S. Confidence intervals for the log-normal mean. *Stat. Med.* **1997**, *16*, 783–790. [CrossRef]
29. Land, C.E. An evaluation of approximate confidence interval estimation methods for lognormal means. *Technometrics* **1972**, *14*, 145–158. [CrossRef]
30. Angus, J.E. Inferences on the lognormal mean for complete samples. *Commun. Stat. Simul. Comput.* **1988**, *17*, 1307–1331. [CrossRef]
31. Angus, J.E. Bootstrap one-sided confidence intervals for the lognormal mean. *Statistician* **1994**, *43*, 395–401. [CrossRef]
32. Krishnamoorthy, K.; Mathew, T.P. Inferences on the means of lognormal distributions using generalized p-values and generalized confidence intervals. *J. Stat. Plan. Inference* **2003**, *115*, 103–121. [CrossRef]
33. Efron, B. *The Jackknife, the Bootstrap, and Other Resampling Plans*; Society for Industrial and Applied Mathematics: Philadelphia, PA, USA, 1982. [CrossRef]
34. Ozkan, H.E.; Reichard, D.L.; Ackerman, K.D. Effect of orifice wear on spray patterns from fan nozzles. *Trans. ASAE* **1992**, *35*, 1091–1096. [CrossRef]
35. Efron, B. Better Bootstrap Confidence Intervals. *J. Am. Stat. Assoc.* **1987**, *82*, 171–185. [CrossRef]
36. McCullough, B.D.; Vinod, H.D. Implementing the Double Bootstrap. *Comput. Econ.* **1998**, *12*, 79–95. [CrossRef]
37. Maleksaeidi, H.; Karami, E. Social-ecological resilience and sustainable agriculture under water scaricity. *Agroecol. Sustain. Food Syst.* **2013**, *37*, 262–290. [CrossRef]
38. Tikhonovich, I.A.; Provorov, N.A. Microbiology Is a basis of sustainable agriculture: An opinion. *Ann. Appl. Biol.* **2011**, *159*, 155–168. [CrossRef]
39. McNeill, D. The contested discourse of sustainable agriculture. *Glob. Policy* **2019**, *10* (Suppl. 1). 1758–5899. [CrossRef]
40. Domardzki, K.; Rola, H. Efektywność stosowania niższych dawek herbicydów w zbożach. *Pamiętnik Puławski Materiały Konferencji* **2000**, *120*, 53–63.
41. Głazek, M.; Mrówczyński, M. Łączne stosowanie agrochemikaliów w nowoczesnej technologii produkcji zbóż. *Pamiętnik Puławski Materiały Konferencji* **1999**, *114*, 119–126.
42. Szymańska, E. Zużycie chemicznych środków ochrony roślin i możliwości jego ograniczenia w zrównoważonym systemie produkcji zbóż. *Pamiętnik Puławski Materiały Konferencji* **2000**, *120*, 439–444.
43. De Cock, N.; Massinon, M.; Salah, S.O.T.; Lebeau, F. Investigation on optimal spray properties for ground based agricultural applications using deposition and retention models. *Biosyst. Eng.* **2017**, *162*, 99–111. [CrossRef]
44. Arvidsson, T.; Bergström, L.; Kreuger, J. Spray drift as influenced by meteorological and technical factors. *Pest. Manag. Sci* **2011**, *67*, 586–598. [CrossRef]
45. Butts, T.R.; Luck, J.D.; Fritz, B.K.; Hofmann, W.C.; Kruger, G.R.. Evaluation of spray pattern uniformity using three unique analyses as impacted by nozzle, pressure, and pulse-width modulation duty cycle. *Pest. Manag. Sci.* **2019**, *75*, 1875–1886. [CrossRef]
46. Parafiniuk, S.; Milanowski, M.; Subr A.K. The influence of the water quality of the droplet spectrum produced by agricultural nozzles. *Agric. Agric. Sci. Procedia* **2015**, *7*, 203–208. [CrossRef]

47. Benkwitz, A.; Lütkepohl, H; Wolters, J. Comparison of bootstrap confidence intervals for impulse responses of German monetary systems. *Macroecon. Dyn.* **2001**, *5*, 81–100. Available online: https://www.researchgate.net/publication/23775280_Comparison_of_Bootstrap_Confidence_Intervals_for_Impulse_Responses_of_German_Monetary_Systems (accessed on 4 December 2019). [CrossRef]
48. Trichakis, I.; Nikolos, I; Karatzas, G.P.; Comparison of bootstrap confidence intervals for an ANN model of a karstic aquifer response. *Hydrol. Process.* **2011**, *25*, 2827–2836. [CrossRef]

17

Ultraviolet Fluorescence in the Assessment of Quality in the Mixing of Granular Material

Dominika Barbara Matuszek

Faculty of Production Engineering, Opole University of Technology, 45-758 Opole, Poland; d.matuszek@po.edu.pl

Abstract: The aim of the study was to determine the possibility of using ultraviolet fluorescence to evaluate the quality of the mixing process of industrial feed. A laboratory funnel-flow mixer was used for the mixing process. The studies were carried out using three different feeds for pigs. A key component in the form of ground grains of yellow maize covered with the fluorescent substance Rhodamine B was introduced into the mixer before the mixing process began. After the illumination of the sample by UV lamps, the images were taken with a digital camera. The images were analyzed in Patan® software. The information obtained on the percentage content of the key component was used for further calculations. At the same time, the tracer content was determined using the control method (weight method). The comparison of the results obtained by the two methods (statistical comparative analysis) did not indicate significant differences. Therefore, the usefulness of the proposed method to track the share of the key component by inducing it to glow in the ultraviolet light has been proven. The introduced tracer is also one of the components of the feed, which translates into the possibility of observing the material having the characteristics of a mixed material.

Keywords: mixing of granular materials; fluorescence; tracer; industrial feed; image analysis

---

1. Introduction

Mixing granular materials is a difficult and multifaceted process which, although it was already described in the 1940s, is still a current research problem. This is due, among other things, to the development and availability of new tools, thanks to which it is possible to precisely observe the process itself, as well as to describe the behavior of selected components of the mixed bed. The degree of mixing of granular components is an important quality parameter of the final product in the context of e.g., nutrition [1–3]. It seems to be particularly important to conduct research on the process of mixing components with considerable fineness, such as powders. This is due, among other things, to the scale of production of products in the form of granules or powders, which reaches up to one billion pounds a year [4–7] and the aspects of safety [8].

The correct level of mixing ingredients in the mixture is important to obtain satisfactory results (mixture of good quality) in the production of farm animals; the preparation of feed in accordance with the given recipe. Incomplete mixing may result in negative nutritional effects. The appropriate uniformity of a mixture is important also for the granulation process [1,8,9]. In industrial conditions, ensuring safe feed production requires the implementation of Quality Management Systems such as Good Hygienic Practice—GHP, Good Manufacturing Practice—GMP, and Hazard Analysis and Critical Control Points—HACCP [10]. Quality control in feed production consists of verifying the correctness of the key stages and is carried out by eliminating possible biological, chemical and physical hazards. In the HACCP system, the mixing stage is often indicated as a critical control point. The assessment of homogeneity is an important Critical-To-Quality (CTQ) factor and it must be contolled periodically. This allows us, inter alia, to determine the required mixing time to obtain a high-quality product and involves the homogeneity analysis of samples taken from the mixer or from bags containing the finished mix. Determining the exact mixing time is important due to the fact that both too long and too short mixing can cause incorrect homogeneity. Moreover, the mixing time recommended by the mixers' manufacturers may not give the declared results in industrial conditions. This is due to the multi-threading of the process of mixing granular materials, which underlies research problems. Of course, determining the effective mixing time translates into production costs. Optimizing the work of mixing devices requires quantitative means to evaluate mixing [8,11]. Moreover, even a slight change in the parameters of the mixing process, such as humidity of the mixed material, degree of filling of the mixer and others, affects the degree of mixing [5,12,13]. The quality of the feed's raw materials is another important issue. Due to its properties (biologically active material), the feed may undergo contamination with pathogenic microflora [8,14]. However, as shown by some research carried out in Poland, this type of contaminant does not exceed accepted standards, and the use of thermal treatment such as drying allows for the additional reduction of this risk [14–16].

The degree of mixing of granular materials is described by various mathematical relationships in which the main element is the content of a given ingredient in the mixed bed. The determination of the state of the mixture is usually based on its deviation from the state of segregation or the state of perfect mixture. When mixing a two-component material, the determination of the state of the mixture is quite simple and is based on the analysis of sample composition [2]. When mixing heterogeneous multicomponent systems, is often impossible to analyze the composition of the samples accurately. There are only a few items in the literature where the authors have evaluated the mixing process based on the analysis of the content of each component of the mixture. However, methods of this type are very labor- and time-consuming [17,18]. A new method for determining the mixing index based on determining the distance from the adjacent particle and coordination number concept in multiple-spouted bed was proposed by Chen et al. [19].

However, most often the condition of a multicomponent granular mixture is determined on the basis of the content of the selected key component in the taken samples. The key ingredient can be a deliberately introduced element like microtracer—colored uniformly sized iron particles or microgrits [20–24]. This solution allows the quick assessment of the homogeneity of grain mixtures in industrial conditions. However, iron filings have different characteristics from mixed components, e.g., feed components or biologically active materials. This does not allow us to use them to track the behavior of actually mixed materials.

A number of methods for assessing the homogeneity of granular mixtures (indicator methods) are based on the determination of the content of specially dyed particles or on the use of the fluorescence emission phenomenon. Many researchers have used laser-induced fluorescence (LIF method) for this purpose. Lai et al. [25,26] used it to describe the kinetics of the pharmaceutical powder mixing process. A similar methodology was also used by Karumanchi et al. [27] for determining the dead zones in the mixer during powder mixing. Meanwhile, Durao et al. [28] determined in this way the concentration of five different substances in a multicomponent vitamin preparation. The phenomenon of X-ray fluorescence (XRF) was used, among others, in the assessment of sulphur and chlorine content in feeds for ruminants. The research was conducted for the following feeds: grass, grass silage and maize silage [29]. The use of a laser does not require physical sampling, which is the main advantage of this solution. However, it is necessary to prepare the mixer properly (window to take the sample) and the possible falsification of the results with the contaminants present. In addition, the window must be kept sufficiently clean, which may not be possible if the mixed components are very fine [30].

The analysis of the distribution of the stained components is often based on computer image analysis of the mixed material. Realpe et al. [31] used image analysis to evaluate the homogeneity of colored powders mixed in a cylindrical device. Berthiaux et al. [32] used this tool to assess the homogeneity of the mixture flowing out of the mixer. The homogeneity of the mixture at the mixer outlet was tested by Muerza et al. [33] based on on-line image analysis, and Dal Grande et al. [34] proved the usefulness of the HSV color model description for the analysis of binary grain mixture images. Computer image analysis to assess the impact of vibration intensity on the segregation of a bed consisting of steel and glass balls was used by Yang [35] and then by Tai et al. [36]. Hu et al. [37] analyzed images of samples taken during mixing in a rotating conical mixer with and without a mixing blade. Liu at al. [38] used an interesting optical technique consisting of the simultaneous acquisition of images with an infrared thermal camera and an ordinary RGB camera. Red (heated to 40 and 80 °C) and white balls (at room temperature) were used in this test. Techniques of this type are a relatively cheap and good tool, especially for analyzing the surface of a mixed material or its flow, and are most often used in laboratory conditions [30,39]. The limitation of these methods results from certain difficulties in obtaining information, e.g., a lack of access to the free space of the mixed bed, covering of the tracer by loose material [30].

In the research carried out by the authors of this work, an optical method was used that combines the fluorescence of a tracer in ultraviolet light with the acquisition of images by means of a digital camera with a standard lens. The usefulness of the proposed method in the mixing of whole grains in two- and multicomponent systems was confirmed, and the results obtained were described in the works of Matuszek et al. [40–42]. Referring to the commonness of powders in industrial practice and the limitations of computer image analysis in the evaluation of their mixing [4–6,30], the developed method was tested in the mixing of multicomponent feed consisting of ground grains and additives in the form of powders—so-called micronutrients. The aim of the study is to determine the possibility of using the method based on UV-induced fluorescence to assess the mixing of multicomponent industrial ground feed. The tests were carried out under laboratory conditions; however, the mixtures were obtained from feed factories. This makes the work more functional.

2. Materials and Methods

The mixtures used for the research consisted of components used for feed production, i.e., cereal grains and feed additives aimed at improving the nutritional value or sensory attractiveness (flavors). To assess the homogeneity of this type of mixture, indicator methods based on the assessment of the content of the key component (tracer) in feed samples were used. Therefore, there was the question of choosing the component that would properly represent the state of the mixed bed. Based on the previous experience of the author of this study and the literature information, it was decided that this ingredient may be maize. The choice was dictated, among others, by the widespread use of this ingredient in feeding farm animals and the ease of grinding or coating by coloring matter. Determining the size and number of samples is another research problem. On the one hand, the sample may consist of a single grain; on the other hand, the whole mixed system. The most important thing for the sample is to be representative. In the tests, 10 samples were taken due to the construction of the mixer. The mass of a single sample was determined experimentally. Finally, one of the most frequently appearing problems, i.e., how to assess the content of a particular ingredient against the background of many components that differ in color or fineness. The authors, using a known and proven tool, namely image analysis, attempted to eliminate this disturbance–for this, the fluorescent phenomenon was used.

Feed mixes obtained from a local feed factory were used in the study. These were one of the most commonly used mixtures intended for pigs. The mixtures were multicomponent systems (10-, 13- and 14-component systems) consisting of ground products of plant origin, such as cereal grains and minerals, and chemical additives. The types and proportions of the individual components and the degree of fineness (assessment by sieve analysis) of the tested feed are shown in Table 1.

Table 1. Characteristics of the tested feed mixture.

Type of Component	Percentage of Component [%]		
	Mixture 1	Mixture 2	Mixture 3
Fodder chalk	9	1.5	7.1
Barley	-	30	-
Maize	8	9	7
Triticale	-	20	-
Wheat	-	20	-
Soya meal	65.55	12	72
Rape meal	-	5	-
Dry maize decoction	5.45	-	4.3
Sodium chloride	2.5	0.5	2
Phosphate	3.5	1	2.8
Premix	2.5	1	2
Lysine	2.5	-	1.8
Methionine	0.5	-	0.4
Threonine	0.35	-	0.3
Phytase	0.05	-	0.05
Grindazyn	0.05	-	0.05
Luctarom (aroma)	0.05	-	0.05
Neubaciol	-	-	0.15
Number of components	13	10	14
Fineness degree M [mm]	0.64	0.61	0.63

A tracer was introduced to the mixing process (before it started). The seeds of yellow maize were used as the tracer—they were subjected to grinding and then wet treated with 0.01% Rhodamine B solution ($C_{28}H_{31}ClN_2O_3$, red-violet powder, wavelength 627 nm, excitation area 553 nm). The dyeing was carried out in laboratory conditions and the obtained materials (after drying) were stored in identical conditions in tightly closed packages, protecting against light. Two types of tracer were obtained in this way—maize of average particle size d_1 = 2.00 mm and d_2 = 1.25 mm. Before starting the mixing process, the tracer was placed in the upper part of the mixer (ring 10) in the amount of 100 g (10% of the mixture). The remaining part (rings 1–9) was an industrial feed mix of 900 g (90% of the mixture). After filling the mixer, the mixing process started. Mixing was carried out in a laboratory funnel-flow mixer by means of subsequent flows from 1 to 10. The characteristics of the mixer were presented in detail in another work by the authors [43]. The mixing was completed after the 10th flow and sampling were carried out. Single samples of 10 g were taken from ten mixer locations (levels). This was possible by the special design of the mixing tanks—10 removable rings.

The samples on the Petri dishes (120 mm × 20 mm) were placed in a chamber equipped with ultraviolet fluorescent lamps (2 lamps with 15 W each, mounted inside in the upper part of chamber). The chamber was made of black material, which limited the influence of light from outside. The sample was placed on a pull-out drawer. The chamber was then closed and the lighting controlled from outside. In the upper part of the chamber was a hole for a digital camera lens. A digital camera with a standard lens, 20.1 Mpix resolution and 35 mm focal length was used. When taking pictures, the exposure correction was set to −0.5 EV. After the illumination of the sample, images with a resolution of 1600 × 1200 pixels were obtained. The images obtained were analyzed in Patan®. This program analyzes the image based on the RGB 256 scale. After the loading of the image, three areas (designation of surface fragments, assignment of pixels) responsible for individual classes were marked in the examined area (circular area): 1—tracer (fluorescent maize) and 2 and 3—background (industrial mix). Thanks to this, it was possible to capture the desired information (the tracer's share) against a multicolored sample. The results obtained referr to the percentage content of particular classes. For further research, data on the percentage share of the first class, i.e., tracer, were used. In the next stage, the manual separation of the taken samples was performed and the separated tracer was weighed on an analytical balance with accuracy of ±0.01 g. In this way, the share of the tracer was estimated using two methods: (1) computer image analysis, (2) weighing method. The second method was used as a tool to verify the results obtained by method one. The selected stages of the methodology are schematically presented in Figure 1. On the basis of the results, the arithmetic mean, the standard deviation, the difference between the results obtained by the two methods and the coefficient of variation were calculated. The coefficient of variation (CV) was treated as a parameter indicating the degree of homogeneity of the obtained mixture. The calculations were made according to the instructions of the National Veterinary Institute [44]. The coefficient of variation was in the range from 0% to 100%. According to these instructions, the mixture is considered homogeneous when the CV ≤ 15%. In this range, a good mix of the tracer in the mixture is obtained. CV > 15% means bad homogeneity of the mixture, meaning there is poor mixing of the tracer in the mixture.

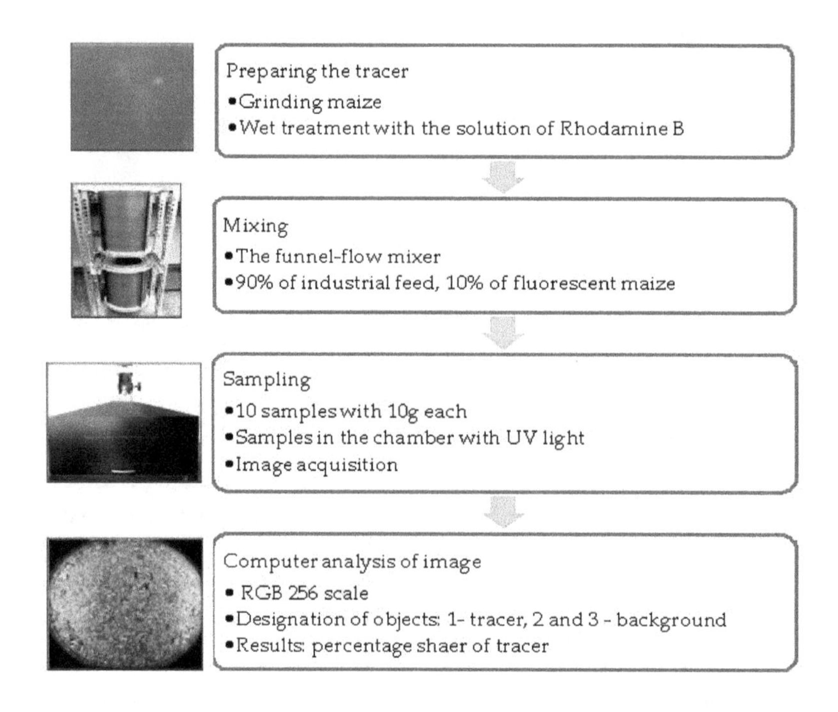

Figure 1. Schematic diagram of the methodology for the determination of fluorescent maize in the feed.

In order to verify the results obtained by the two methods, statistical comparative analysis was carried out. After checking the normality of distribution (Shapiro–Wilk test) and homogeneity of variance (Levene test), appropriate parametric (Student *t*-test) and non-parametric (U Mann-Whitney test) tests were used. The significance level $\alpha = 0.05$ was assumed for calculations. The null hypothesis regarding no differences in results obtained by the two methods was tested compared to the alternative hypothesis about the occurrence of such differences.

3. Results

3.1. Results Obtained by Means of Computer Image Analysis and the Weighing Method

The share of the tracer and the difference in results obtained by the two methods and selected statistical parameters (arithmetic mean, standard deviation, coefficient of variation) are presented in Tables 2 and 3. Graphical interpretations of the results are presented in diagrams (Figures 2–4).

Table 2. Results of the tracer's share (maize d = 2.00 mm).

Series of Tests	Method 1 [1]	Method 2 [2]	Difference [3]
	Mixture 1		
1	8.00 ± 0.88	8.12 ± 0.94	0.30 ± 0.20
2	9.04 ± 0.79	9.05 ± 0.73	0.33 ± 0.18
3	8.98 ± 1.14	9.25 ± 1.06	0.41 ± 0.17
mean, %	8.67 ± 1.08	8.81 ± 1.06	0.34 ± 0.19
CV, %	10.83 ± 1.62	10.37 ± 1.66	0.55 ± 0.09
	Mixture 2		
1	10.04 ± 1.20	10.03 ± 1.21	0.27 ± 0.17
2	10.38 ± 1.59	10.24 ± 1.58	0.26 ± 0.07
3	9.90 ± 1.27	9.85 ± 1.18	0.26 ± 0.09
mean, %	10.10 ± 1.41	10.04 ± 1.37	0.26 ± 0.12
CV, %	13.46 ± 1.32	13.17 ± 1.60	0.41 ± 0.16
	Mixture 3		
1	9.34 ± 0.93	9.51 ± 1.04	0.41 ± 0.23
2	8.66 ± 0.92	8.84 ± 1.05	0.43 ± 0.23
3	9.45 ± 0.76	9.35 ± 0.91	0.29 ± 0.16
mean, %	9.15 ± 0.96	9.23 ± 1.06	0.38 ± 0.22
CV, %	9.55 ± 1.10	10.87 ± 0.87	0.56 ± 0.01

[1] arithmetic mean of the tracer's share obtained by computer image analysis ± standard deviation; [2] arithmetic mean of the tracer's share obtained by weighing ± standard deviation; [3] difference of results obtained by two methods ± standard deviation.

Table 3. Results of the tracer's share (maize d = 1.25 mm).

Series of Tests	Method 1 [1]	Method 2 [2]	Difference [3]
	Mixture 1		
1	9.13 ± 0.87	9.37 ± 0.78	0.53 ± 0.20
2	9.49 ± 0.93	9.42 ± 0.96	052 ± 0.30
3	8.94 ± 1.09	8.89 ± 0.97	0.44 ± 0.17
mean, %	9.18 ± 1.01	9.23 ± 0.96	0.50 ± 0.23
CV, %	10.50 ± 1.19	9.83 ± 1.07	0.44 ± 0.09
	Mixture 2		
1	9.39 ± 0.76	9.13 ± 0.79	0.38 ± 0.26
2	9.22 ± 0.75	9.21 ± 0.84	0.42 ± 0.26
3	9.56 ± 0.81	9.53 ± 0.89	0.33 ± 0.14
mean, %	9.39 ± 0.80	9.29 ± 0.87	0.38 ± 0.23
CV, %	8.25 ± 0.16	9.04 ± 0.28	0.58 ± 0.11
	Mixture 3		
1	9.35 ± 0.75	9.61 ± 0.85	0.62 ± 0.26
2	9.09 ± 1.02	9.21 ± 0.94	0.64 ± 0.26
3	9.13 ± 0.92	9.19 ± 0.77	0.50 ± 0.25
mean, %	9.19 ± 0.92	9.33 ± 0.89	0.59 ± 0.30
CV, %	9.75 ± 1.33	9.15 ± 0.76	0.49 ± 0.06

[1] arithmetic mean of the tracer's share obtained by computer image analysis ± standard deviation; [2] arithmetic mean of the tracer's share obtained by weighing ± standard deviation; [3] difference of results obtained by two methods ± standard deviation.

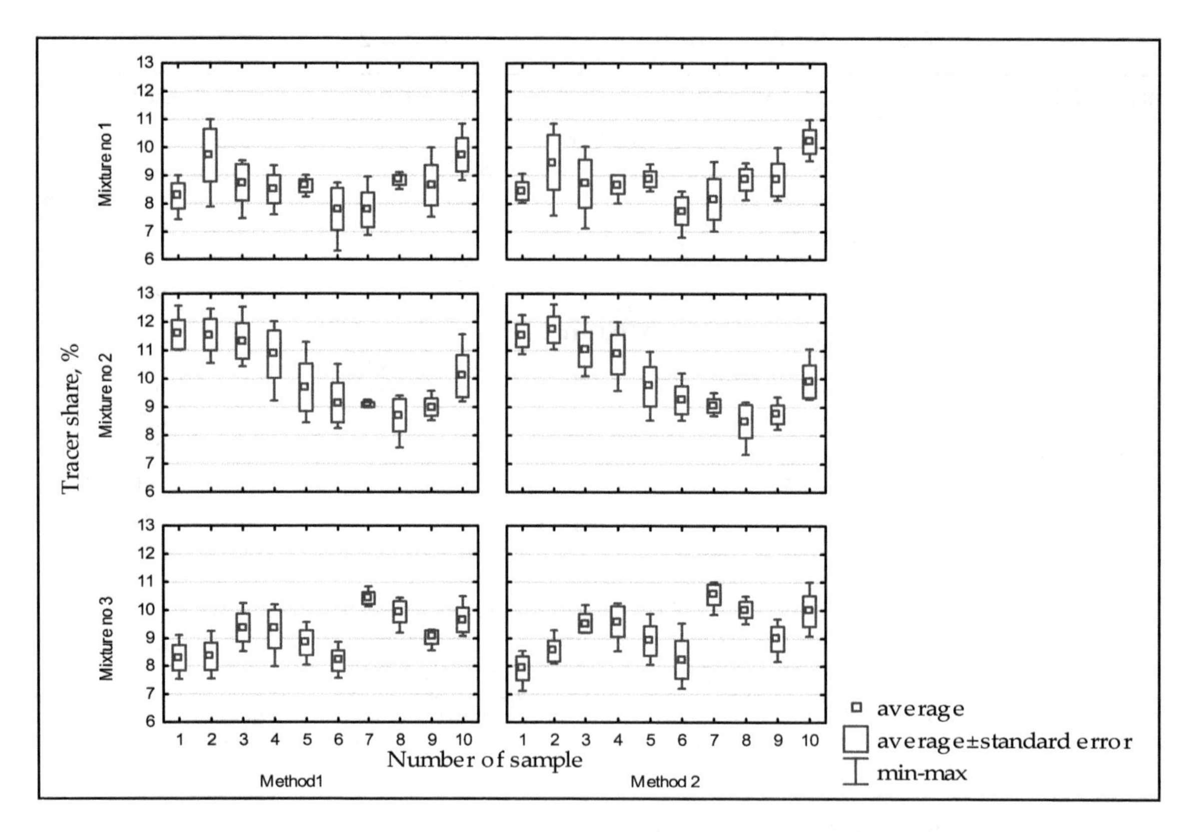

Figure 2. Results of the tracer's share (2.0 mm maize) in individual samples obtained by two methods for three compound feed mixes.

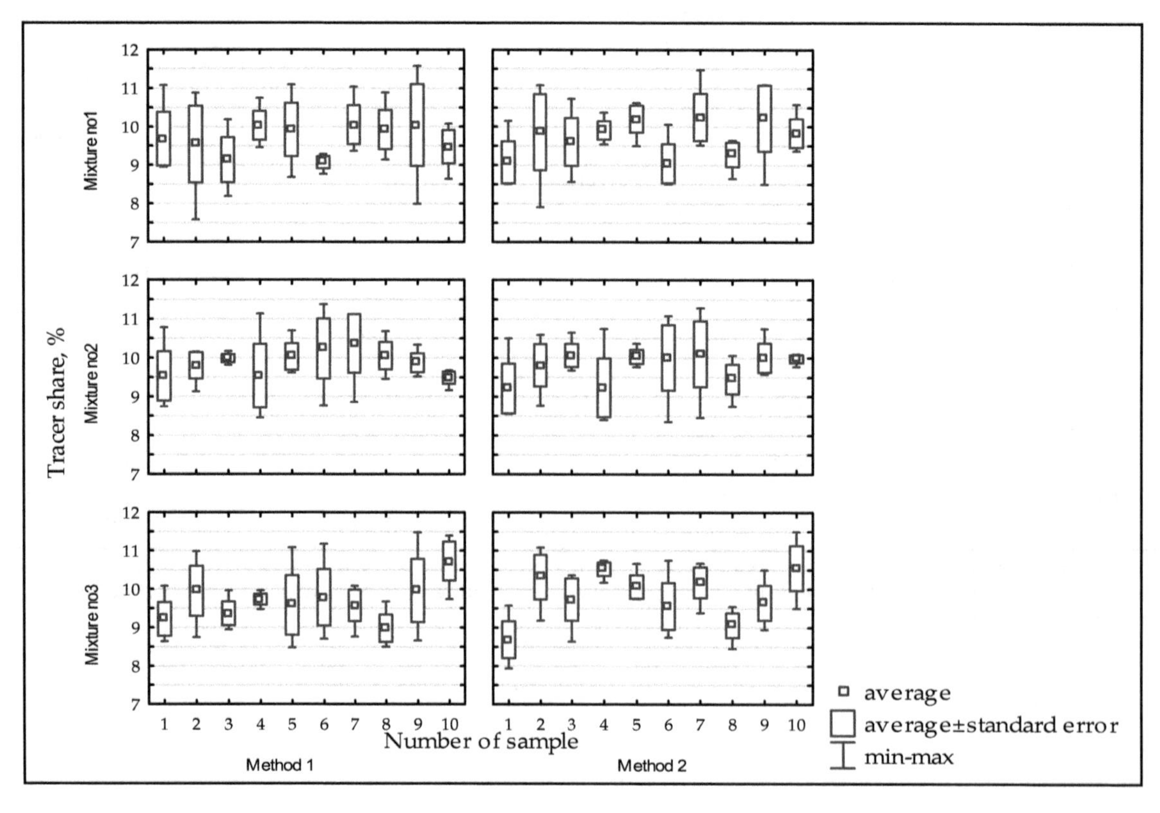

Figure 3. Results of the tracer's share (1.25 mm maize) in individual samples obtained by two methods for three compound feed mixes.

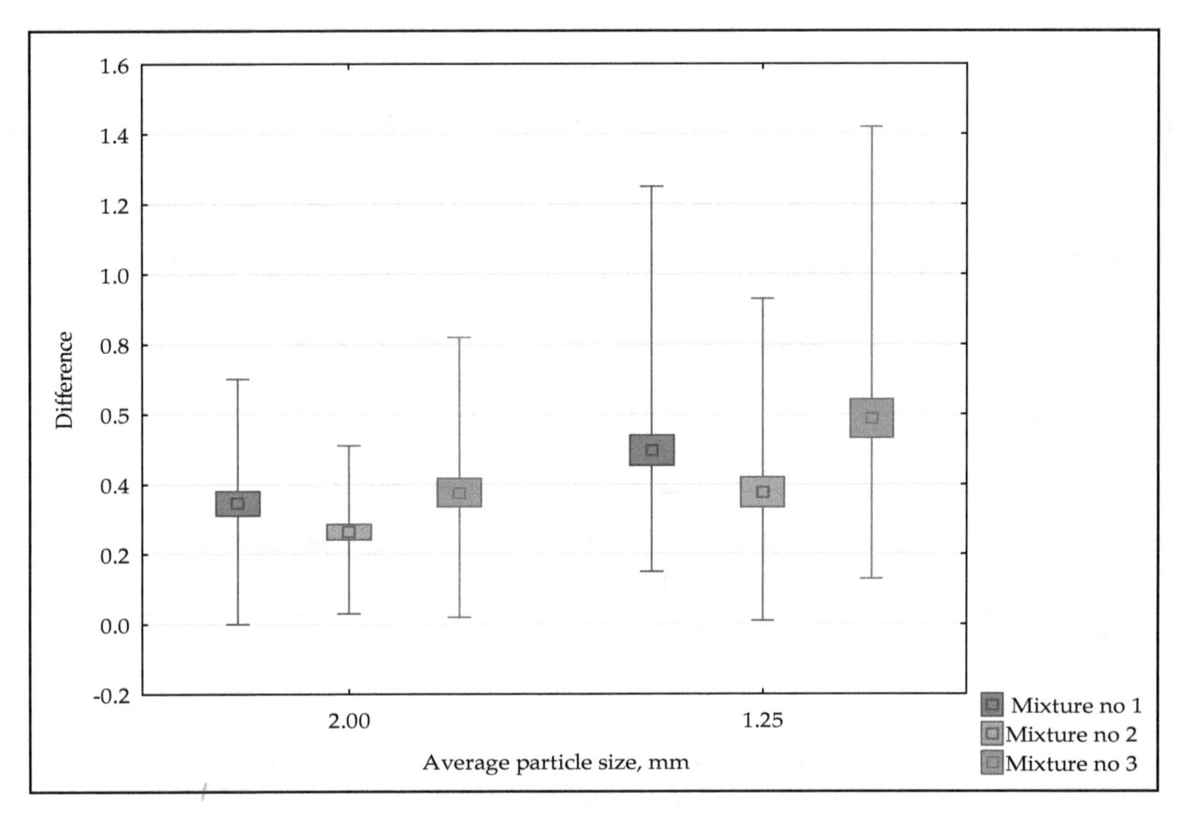

Figure 4. Box plot graph (mean, standard error, min-max) of the difference between the results obtained by the two methods for the mean tracer dimension $d_1 = 2.00$ mm and $d_2 = 1.25$ mm.

The results presented in Tables 2 and 3 show the great similarity of the results obtained by the two methods. The mean difference of the results obtained by the two methods for the tracer with mean particle size d = 2.00 ranged from 0.3 to 0.41 for mixture 1, 0.26–0.27 for mixture 2 and 0.29–0.43 for mixture 3 (Table 2). In the case of mixing with a smaller average particle size (d = 1.25 mm), slightly larger differences were obtained: 0.44–0.53 for mix 1, 0.33–0.42 for mix 2 and 0.50–0.64 for mix 3 (Table 3). In one case, a zero difference was obtained and this concerned the results of mixing feed 1 with the maize of larger average particle size. However, the highest level of the difference of 1.42 was recorded for mixing feed 3 with the d = 1.25 mm tracer. Similar values of the tracer's share obtained by method 1 and 2 in particular segments (samples) of the mixer after the mixing process was completed are also visible on the diagrams (Figures 2 and 3). By analyzing these charts, you can also track the tracer's distribution in the individual mixing zones (rings). For example, when mixing feed 2 with 2.00 mm maize, the tracer gathered in the lower part of the mixer (rings 1 to 3). Then, there is a decrease in its share (rings 4–8) and again a slight increase (rings 9 and 10). The graphical interpretation also shows that when using a tracer with an average particle size (d = 1.25 mm), larger deviations from the average value (average ± standard error) were obtained than in the case of a tracer with a smaller degree of fineness. Similarly, by analysing Figure 4, lower values of the differences obtained in mixing with the d = 2.00 mm tracer can be observed. In this case, lower mean values were obtained (0.34%, 0.26%, 0.43% respectively for mixtures 1, 2 and 3) than in tests with finer d = 1.25 mm (mean values 0.50%, 0.38%, 0.59% respectively for mixtures 1, 2 and 3). Additionally, in the same case, the values are more favorably (lower values) distributed in the min-max range (0–0.7, 0.03–0.51, 0.02–0.82 respectively for mixtures 1, 2 and 3). The values within the same range, for a series of tests with a d = 1.25 mm tracer, are 0.15–1.25, 0.01–0.93 and 0.13–1.42, respectively.

Referring to the results of the coefficient of variation, each of the mixtures achieved good homogeneity, and thus good mixing of the tracer. In each case, a coefficient of variation at $CV \leq 15\%$ was obtained. This observation concerns the results obtained with two methods and with the use of two different tracers of different degrees of fineness.

3.2. Results of Statistical Comparative Analysis

The results of the statistical analysis are presented in Tables 4 and 5 and their graphical interpretation in Figures 5 and 6.

Table 4. Results of statistical comparative analysis (Student's *t*-test) of the results obtained by the two methods using the tracer d_1 = 2.00 mm.

Mixture No.	*t*	*p*
1	−0.49398	0.62318
2	0.18586	0.85320
3	−0.33460	0.73913

Table 5. Results of statistical comparative analysis (Student's *t*-test and U Mann-Whitney's test) of the results obtained by the two methods using the tracer d_2 = 1.25 mm.

Mixture No.	Statistical Test Result	*p*
1 [1]	−0.16586	0.86884
2 [2]	0.34004	0.73383
3 [2]	−0.73183	0.46427

[1] parametric Student's *t*-test. Normal distribution of variables, homogeneity of variance; [2] non-parametric U Mann-Whitney test. No normal distribution of variables, homogeneity of variance.

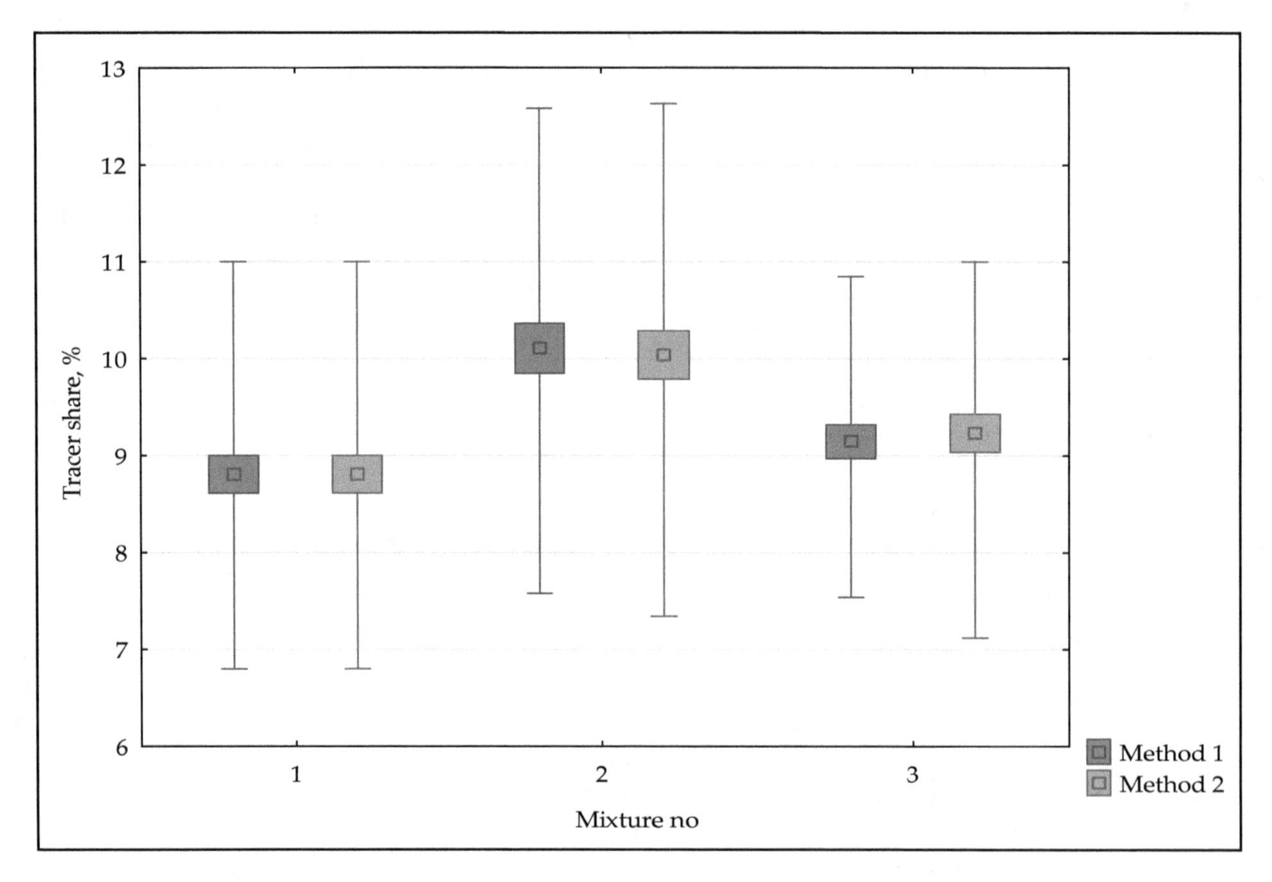

Figure 5. Box plot graph of the share of the tracer (maize d = 2.0 mm) obtained by two methods for the three analysed mixtures. Point: average, box: average ± standard error, whiskers: min-max.

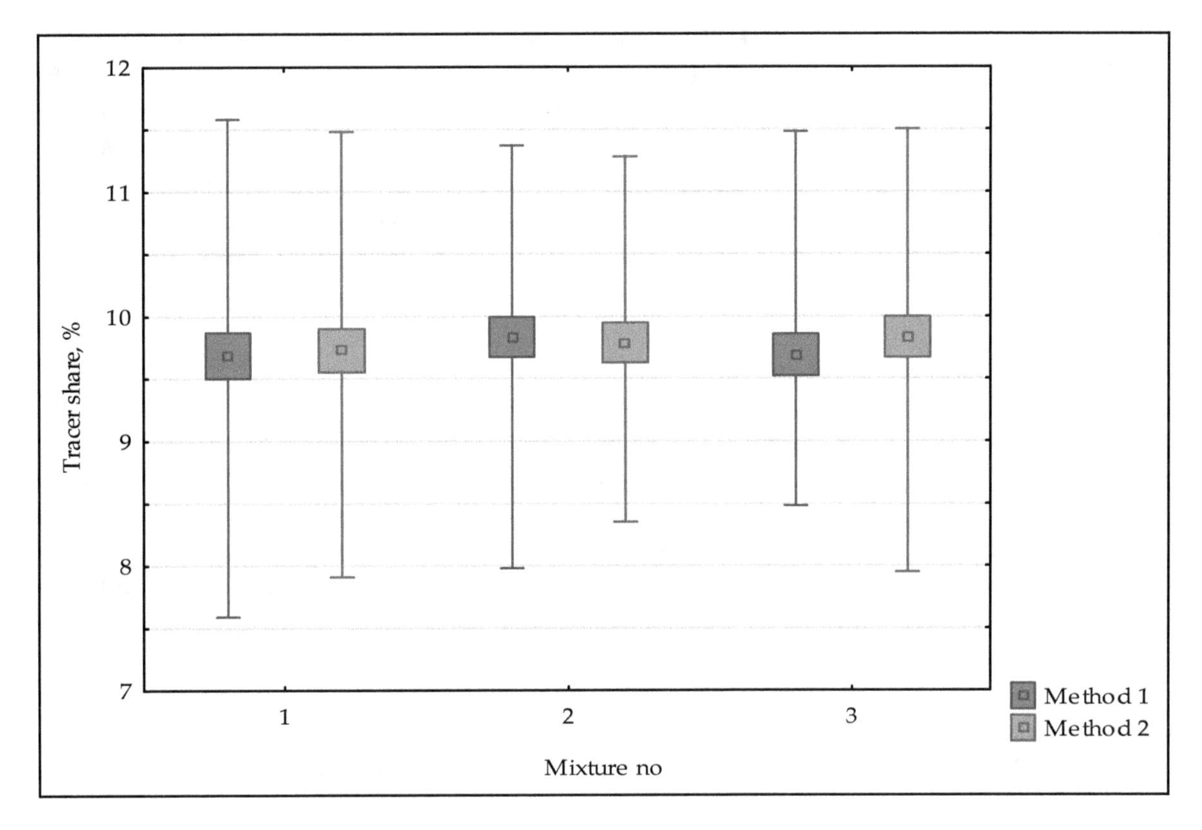

Figure 6. Box plot graph of the share of the tracer (maize d = 1.25 mm) obtained by two methods for the three analysed mixtures. Point: average, box: average ± standard error, whiskers: min-max.

The results obtained (Tables 4 and 5) do not allow us to reject the zero hypothesis about the lack of statistically significant differences in the results of the tracer's share obtained with both methods (no significant statistical differences between two methods: image analysis and weighing method). This statement applies to both tests with fluorescent maize with an average particle size of $d_1 = 2.00$ mm and $d_2 = 1.25$ mm. This confirms the preliminary observations presented in Section 3.1. The high similarity of the results obtained by the two methods is also confirmed by the graphical interpretation of statistical analysis (Figures 5 and 6).

4. Discussion

The obtained results confirm the possibility of using fluorescence induced by ultraviolet radiation to analyze the mixing process of multi-component heterogeneous feed mixtures. This method gives highly reliable results compared to the control method (weighing method), which was confirmed by appropriate statistical tests (Student's *t*-test and U Mann–Whitney test). It was observed that the differences in the results obtained by the two methods using a tracer with an average particle size of $d_1 = 2.00$ mm are smaller than in the case of a tracer with an average particle size of $d_2 = 1.25$ mm. However, in both cases, statistical comparative analysis did not indicate the significance of these differences. In addition, a mixture with an appropriate level of homogeneity was obtained in each test series.

In the proposed method, the key component (dyed maize with 0.01% Rhodamine B solution) was observed on the basis of computer image analysis (software working on the basis of RGB 256 scale) of particular mixer segments. The special design of the tank made it possible to take significant number (N = 10) of samples and thus to track the tracer distribution in the entire volume of the mixed bed. In addition, the phenomenon of fluorescence induced by ultraviolet radiation was used, which allowed us to eliminate interference during image acquisition. The main disadvantage of computer image analysis in the evaluation of mixing of granular components subjected to the milling process is the covering of the tracer by a loose material [30]. In such a situation, image analysis may give incomplete and unreliable information. The light of the tracer allowed for its "singling out" against a multicolored sample. The disadvantage presented has therefore been eliminated. Another limitation of optical methods in the mixing of granular systems is the necessity to take samples, i.e., to interfere with the mixed bed. The proposed methodology is also based on sampling. This does not explicitly exclude its use to track the tracer's behavior on line. Of course, this requires careful analysis and testing. It is worth noting, however, that non-invasive methods to characterize grain mixing require more advanced technology and can therefore be more expensive.

The authors have not yet come across a study in which a similar solution has been applied in the evaluation of the mixing of multi-component ground grain systems (degree of fineness M = 0.61–0.64 mm). Most of the studies using image analysis concerned the mixing of two-component systems differing significantly in colour, e.g., white and red balls [4,34–36,45–48]. What is more, the research carried out by the authors made use of feed mixes that are produced in a feed factory and the tracer introduced was one of the components of these mixes. Therefore, the proposed method is more likely to be used under industrial conditions. In the next stage of the research, the authors plan to verify the usefulness of the developed method by conducting tests with additional feed mixes differing in the degree of fineness and composition.

The proposed method uses a well-known computer image analysis tool. However, the use of an additional stage (illuminating the sample with ultraviolet light) allows the elimination of disturbances in the case of multi-component systems subjected to a milling process. What is more, the method allows us to assess the homogeneity of the mixture based on the ingredient used to produce feed, i.e., maize. This is important because of the properties of biologically active materials. Their parameters, such as humidity, brittleness, the ability to agglomerate and many others, can change significantly in response to the conditions of the mixing process. Materials often used by other authors (presented in introduction and discussion chapters) like steel, glass or plastic balls have definitely different properties. As presented in the introduction, slight changes in the parameters of a mixed system or mixer affect the degree of mixing or segregation. The described method is not a tool without limits. This is due to the Rhodamine B coloring agent used. This substance is a chemical reagent that has irritating properties. In laboratory conditions, when applying general precautions, it does not matter much. However, in industrial conditions, the use of this method would require appropriate handling of mixed feed (disposal). Therefore, this method is applicable only in laboratory conditions, for now.

References

1. Neumann, K.D. Work of the Mixer is crucial for additives. *Feed Mag.* **2000**, *10*, 371–383.
2. Tasirin, S.M.; Kamarudin, S.K.; Hweage, A.M.A. Mixing behavior of binary polymer particles in bubbling fluidized bed. *J. Phys. Sci.* **2008**, *19*, 13–29.
3. Jarray, A.; Shi, H.; Scheper, B.J.; Habibi, M.; Luding, S. Cohesion-driven mixing and segregation of dry granular media. *Sci. Rep.* **2019**, *9*, 13480. [CrossRef] [PubMed]
4. Weineköttter, R.; Gericke, H. *Mixing of Solids: Particle Technology*, 1st ed.; Springer: Berlin, Germany, 2000; ISBN 978-94-015-9580-3.
5. Cleary, P.W.; Sinnott, M.D. Assessing mixing characteristics of particle-mixing and granulation devices. *Particuol. Simul. Modeling Part. Syst.* **2008**, *6*, 419–444. [CrossRef]
6. Sarkar, A.; Wassgren, C.R. Simulation of a continuous granular mixer: Effect of operating conditions on flow and mixing. *Chem. Eng. Sci.* **2009**, *64*, 2672–2682. [CrossRef]
7. Królczyk, J.B. Metrological changes in the surface morphology of cereal grains in the mixing process. *Int. Agrophys.* **2016**, *30*, 193–202. [CrossRef]
8. Matuszek, D.; Królczyk, J.B. Aspects of safety in production of feeds—A review. *Anim. Nutr. Feed Technol.* **2017**, *17*, 367–385. [CrossRef]

9. Walczyński, S.; Korol, W. Long-term monitoring of homogeneity of compound feed in the government supervision. *Krmiva Zagreb* **2008**, *50*, 311–317.
10. Zawiślak, K.; Sobczak, P.; Wełdycz, A. Mixing as CCP in the production of industrial feed. *J. Cent. Eur. Agric.* **2012**, *13*, 554–562. [CrossRef]
11. Bothe, D. Evaluating the quality of a mixture: Degree of homogeneity and scale of segregation. In *Micro and Macro Mixing. Heat and Mass Transfer*; Bockhorn, H., Mewes, D., Peukert, W., Warnecke, H.J., Eds.; Springer: Berlin, Germany, 2010; pp. 17–35. [CrossRef]
12. Liu, P.Y.; Yang, R.Y.; Yu, A.B. Self-diffusion of wet particles in rotating drums. *Phys. Fluids* **2013**, *25*, 063301. [CrossRef]
13. Li, H.; McCarthy, J.J. Controlling Cohesive Particle Mixing and Segregation. *Phys. Rev. Lett.* **2003**, *90*, 184301. [CrossRef] [PubMed]
14. Cegielska-Radziejewska, R.; Stupe, K.; Szablewsk, T. Microflora and mycotoxin contamination in poultry feed mixtures from western Poland. *Ann. Agric. Environ. Med.* **2013**, *20*, 30–35. [PubMed]
15. Sobczak, P.; Zawiślak, Z.; Żukiewicz-Sobczak, W.; Mazur, J.; Nadulski, R.; Kozak, M. The assessment of microbiological purity of selected components of animal feeds and mixtures which underwent thermal processing. *J. Cent. Eur. Agric.* **2016**, *17*, 303–314. [CrossRef]
16. Kwiatek, K.; Kukier, E. Microbiological contamination of animal feed. *Vet. Med.* **2008**, *64*, 24–26.
17. Królczyk, J.; Matuszek, D.; Tukiendorf, M. Modelling of quality changes in a multicomponent granular mixture during mixing. *Electon. J. Pol. Agric. Univ.* **2010**, *13*.
18. Królczyk, J.B. An attempt to predict quality changes in a ten-component granular system. *Teh. Vjesn.* **2014**, *21*, 255–261.
19. Chen, M.; Liu, M.; Li, T.; Tang, Y.; Liu, R.; Wen, Y.; Liu, B.; Shao, Y. A novel mixing index and its application in particle mixing behavior study in multiple-spouted bed. *Powder Technol.* **2018**, *339*, 167–181. [CrossRef]
20. Eisenberg, S.; Eisenberg, D. Closer to perfection. *Feed Manag.* **1992**, *11*, 8–20.
21. Zawiślak, K.; Grochowicz, J.; Sobczak, P. The Analysis of mixing degree of granular products with the use of Microtracers. *TEKA Kom. Mot. Energ. Roln. OL Pan* **2011**, *11*, 335–342.
22. Matuszek, D. The analysis of homogeneity of industrial fodder for cattle. *J. Res. Appl. Agric. Eng.* **2013**, *58*, 118–121.
23. Królczyk, J.B. Industrial conditions of the granular material manufacturing process. *App Mech. Mater.* **2014**, *693*, 267–272. [CrossRef]
24. Królczyk, J.B. Analysis of kinetics of multicomponent, heterogeneous granular mixtures—laminar and turbulent flow approach. *Chem. Process. Eng.* **2016**, *37*, 161–173. [CrossRef]
25. Lai, C.K.; Holt, D.; Leung, J.C.; Cooney, C.L.; Raju, G.K.; Hansen, P. Real time and noninvasive monitoring of dry powder blend homogeneity. *AIChE J.* **2001**, *47*, 2618–2622. [CrossRef]
26. Lai, C.K.; Cooney, C.L. Application of a fluorescence sensor for miniscale on-line monitoring of powder mixing kinetics. *J. Pharm. Sci.* **2004**, *93*, 60–70. [CrossRef] [PubMed]
27. Karumanchi, V.; Taylor, M.K.; Ely, K.J.; Stagner, W.C. Monitoring powder blend homogeneity using light-induced fluorescence. *AAPS Pharmscitech.* **2011**, *12*, 1031–1037. [CrossRef]
28. Durão, P.; Fauteux-Lefebvre, C.; Guay, J.M.; Batzoglou, N.; Gosselin, R. Using multiple process analytical technology probes to monitor multivitamin blends in a tableting feed frame. *Talanta* **2017**, *164*, 7–15. [CrossRef]
29. Necemer, M.; Kump, P.; Rajcevic, M.; Jacimovic, R.; Budic, B.; Ponikvar, M. Determination of sulfur and chlorine in fodder by X-ray fluorescence spectral analysis and comparison with other analytical methods. *Spectrochim. Acta* **2003**, *58*, 1367–1373. [CrossRef]
30. Nadeem, H.; Heindel, T.J. Review of noninvasive methods to characterize granular mixing. *Powder Technol.* **2018**, *332*, 331–350. [CrossRef]
31. Realpe, A.; Barrios, K.; Rozo, M. Assessment of homogenization degree of powder mixing in a cylinder rotating under cascading regime. *Int. J. Eng. Technol.* **2015**, *7*, 394–404.
32. Berthiaux, H.; Mosorov, V.; Tomczak, L.; Gatumel, C.; Demeyre, J.F. Principal component analysis for characterising homogeneity in powder mixing using image processing techniques. *Chem. Eng. Process. Process. Intensif.* **2006**, *45*, 397–403. [CrossRef]
33. Muerza, S.; Berthiaux, H.; Massol-Chaudeur, S.; Thomas, G. A dynamic study of static mixing using on-line image analysis. *Powder Technol.* **2002**, *128*, 195–204. [CrossRef]

34. Dal Grande, F.; Santomaso, A.; Canu, P. Improving local composition measurements of binary mixtures by image analysis. *Powder Technol.* **2008**, *187*, 205–213. [CrossRef]
35. Yang, S.C. Density effect on mixing and segregation processes in a vibrated binary granular mixture. *Powder Technol.* **2006**, *164*, 65–74. [CrossRef]
36. Tai, C.H.; Hsiau, S.S.; Kruelle, C.A. Density segregation in a vertically vibrated granular bed. *Powder Technol.* **2010**, *204*, 255–262. [CrossRef]
37. Hu, G.; Gong, X.; Huang, H.; Li, Y. Effects of geometric parameters and operating conditions on granular flow in a modified rotating cone. *Ind. Eng. Chem. Res.* **2007**, *46*, 9263–9268. [CrossRef]
38. Liu, X.; Gong, J.; Zhang, Z.; Wu, W. An image analysis technique for the particle mixing and heat transfer process in a pan coater. *Powder Technol.* **2016**, *295*, 161–166. [CrossRef]
39. Kingston, T.A.; Heindel, T.J. Granular mixing optimization and the influence of operating conditions in a double screw mixer. *Powder Technol.* **2014**, *266*, 144–155. [CrossRef]
40. Matuszek, D.; Wojtkiewicz, K. Application of fluorescent markers for homogeneity assessment of grain mixtures based on maize content. *Chem. Process. Eng.-Inz.* **2017**, *38*, 505–512. [CrossRef]
41. Matuszek, D.; Biłos, Ł. Use of fluorescent tracers for the assessment of the homogeneity of multicomponent granular feed mixtures. *Przem. Chem.* **2017**, *96*, 2356–2359. [CrossRef]
42. Matuszek, B.D. The use of UV-induced fluorescence for the assessment of homogeneity of granular mixtures. *Open Chem.* **2019**, *17*, 485–491. [CrossRef]
43. Matuszek, D. Modelling selected parameters of granular elements in the mixing process. *Int. Agrophys.* **2015**, *29*, 75–81. [CrossRef]
44. Kwiatek, K.; Przeniosło-Siwczyńska, M. *Instructions for Testing the Homogeneity of Medicated Feeds and Intermediate Products on the Basis of Testing the Degree of Mixing of the Active Substance*; Państwowy Instytut Weterynaryjny: Puławy, Poland, 2007.
45. Weinekötter, R.; Reh, L. Characterization of particulate mixtures by in-line measurments. *Part. Part. Syst. Char.* **1994**, *11*, 284–290. [CrossRef]
46. Boss, J.; Krótkiewicz, M.; Tukiendorf, M. The application of picture analysis as a method to evaluate the quality of granular mixture during funnel-flow mixing. *Agric. Eng.* **2002**, *4*, 27–32.
47. Arratia, P.E.; Duong, N.H.; Muzzio, F.J.; Godpole, P.; Lange, A.; Reynolds, S. Characterizing mixing and lubrication in the Bohle bin blender. *Powder Technol.* **2006**, *161*, 202–208. [CrossRef]
48. Brone, D.; Muzzio, F.J. Enhanced mixing in double-cone blenders. *Powder Technol.* **2010**, *110*, 179–189. [CrossRef]

Permissions

All chapters in this book were first published in MDPI; hereby published with permission under the Creative Commons Attribution License or equivalent. Every chapter published in this book has been scrutinized by our experts. Their significance has been extensively debated. The topics covered herein carry significant findings which will fuel the growth of the discipline. They may even be implemented as practical applications or may be referred to as a beginning point for another development.

The contributors of this book come from diverse backgrounds, making this book a truly international effort. This book will bring forth new frontiers with its revolutionizing research information and detailed analysis of the nascent developments around the world.

We would like to thank all the contributing authors for lending their expertise to make the book truly unique. They have played a crucial role in the development of this book. Without their invaluable contributions this book wouldn't have been possible. They have made vital efforts to compile up to date information on the varied aspects of this subject to make this book a valuable addition to the collection of many professionals and students.

This book was conceptualized with the vision of imparting up-to-date information and advanced data in this field. To ensure the same, a matchless editorial board was set up. Every individual on the board went through rigorous rounds of assessment to prove their worth. After which they invested a large part of their time researching and compiling the most relevant data for our readers.

The editorial board has been involved in producing this book since its inception. They have spent rigorous hours researching and exploring the diverse topics which have resulted in the successful publishing of this book. They have passed on their knowledge of decades through this book. To expedite this challenging task, the publisher supported the team at every step. A small team of assistant editors was also appointed to further simplify the editing procedure and attain best results for the readers.

Apart from the editorial board, the designing team has also invested a significant amount of their time in understanding the subject and creating the most relevant covers. They scrutinized every image to scout for the most suitable representation of the subject and create an appropriate cover for the book.

The publishing team has been an ardent support to the editorial, designing and production team. Their endless efforts to recruit the best for this project, has resulted in the accomplishment of this book. They are a veteran in the field of academics and their pool of knowledge is as vast as their experience in printing. Their expertise and guidance has proved useful at every step. Their uncompromising quality standards have made this book an exceptional effort. Their encouragement from time to time has been an inspiration for everyone.

The publisher and the editorial board hope that this book will prove to be a valuable piece of knowledge for researchers, students, practitioners and scholars across the globe.

List of Contributors

Ariovaldo Luchiari Júnior, Daniel de Castro Victoria and Célia Regina Grego
Embrapa Informática Agropecuária, Brazilian Agricultural Research Corporation, Campinas 13083-886, Brazil

Édson Luis Bolfe
Embrapa Informática Agropecuária, Brazilian Agricultural Research Corporation, Campinas 13083-886, Brazil
Department of Geography, Graduate Program in Geography, University of Campinas (Unicamp), Campinas 13083-885, Brazil

Lúcio André de Castro Jorge, Cinthia Cabral da Costa and Ricardo Yassushi Inamasu
Embrapa Instrumentação, Brazilian Agricultural Research Corporation, São Carlos 13560-970, Brazil

Ieda Del'Arco Sanches
Divisão de Sensoriamento Remoto, National Institute for Space Research (INPE), São José dos Campos 12227-010, Brazil

Victor Rodrigues Ferreira and Andrea Restrepo Ramirez
Unidade de Competitividade do Sebrae Nacional, Brazilian Micro and Small Business Support Service (Sebrae), Brasília 70770-900, Brazil

Juan Fernández-Peláez and Manuel Gómez
Food Technology Area, College of Agricultural Engineering, University of Valladolid, 34004 Palencia, Spain

Candela Paesani
Food Technology Area, College of Agricultural Engineering, University of Valladolid, 34004 Palencia, Spain
Área de Mejora Nutricional y Alimentos Nutricionales, Instituto de Ciencia y Tecnología de Alimentos Córdoba (ICYTAC-CONICET-UNC), X5016BMB Córdoba, Argentina

Svetlana R. Derkach, Nikolay G. Voron'ko, Yuliya A. Kuchina and Daria S. Kolotova
Department of Chemistry, Murmansk State Technical University, 183010 Murmansk, Russia

Iwona Gruss and Jacek P. Twardowski
Department of Plant Protection, Wroclaw University of Environmental and Life Sciences, 50-363 Wrocław, Poland

Agnieszka Latawiec
Institute of Agricultural Engineering and Informatics, University of Agriculture in Kraków, 30-149 Kraków, Poland
Department of Geography and the Environment, Rio Conservation and Sustainability Science Centre, Pontifícia Universidade Católica, Rio de Janeiro 22453-900, Brazil

Jolanta Królczyk
Department of Manufacturing Engineering and Production Automation, Faculty of Mechanical Engineering, Opole University of Technology, 45-271 Opole, Poland

Agnieszka Medyńska-Juraszek
Institute of Soil Sciences and Environmental Protection, Wroclaw University of Environmental and Life Sciences, 53, 50-357 Wrocław, Poland

Joanna Karcz
Faculty of Production and Power Engineering, University of Agriculture in Krakow, ul. Balicka 116B, 30-149 Kraków, Poland

Marcin Niemiec
Faculty of Faculty of Agriculture and Economics, University of Agriculture in Krakow, al. Mickiewicza 21, 31-121 Kraków, Poland
Department of Agricultural and Environmental Chemistry, University of Agriculture in Krakow, 31-120 Krakow, Poland

Zofia Gródek-Szostak
Department of Economics and Enterprise Organization, Cracow University of Economics,ul. Rakowicka 27, 31-510 Krakow, Poland

Monika Komorowska
Faculty of Biotechnology and Horticulture, University of Agriculture in Krakow, al. Mickiewicza 21, 31-121 Kraków, Poland

Katarzyna Anna Koryś
International Institute for Sustainability, Estrada Dona Castorina 124, Rio de Janeiro 22460-320, Brazil
Rio Conservation and Sustainability Science Centre, Department of Geography and the Environment, Pontifícia Universidade Católica, Rio de Janeiro 22453900, Brazil

Katarzyna Grotkiewicz
Department of Production Engineering, Logistic and Applied Computer Sciences, University of Agriculture in Kraków, Balicka 116B, 30-149 Kraków, Poland

Agnieszka Ewa Latawiec
International Institute for Sustainability, Estrada Dona Castorina 124, Rio de Janeiro 22460-320, Brazil
Rio Conservation and Sustainability Science Centre, Department of Geography and the Environment, Pontifícia Universidade Católica, Rio de Janeiro 22453900, Brazil
Department of Production Engineering, Logistic and Applied Computer Sciences, University of Agriculture in Kraków, Balicka 116B, 30-149 Kraków, Poland
School of Environmental Science, University of East Anglia, Norwich NR4 7TJ, UK

Maciej Kuboń
Department of Production Engineering, Logistic and Applied Computer Sciences, University of Agriculture in Kraków, Balicka 116B, 30-149 Kraków, Poland
Institute of Technical Sciences, State Vocational East European Higher School in Przemyśl, Książąt Lubomirskich 6, 37-700 Przemyśl, Poland
Institute of Agricultural Engineering and Informatics, University of Agriculture in Krakow, 30-149 Krakow, Poland

Andrzej Marczuk and Jacek Caban
Department of Agricultural, Forestry and Transport Machines, University of Life Sciences in Lublin, 20-612 Lublin, Poland

Alexey V. Aleshkin, Petr A. Savinykh and Alexey Y. Isupov
N.V. Rudnitsky North-East Agricultural Research Institute, Kirov 610007, Russia

Ilya I. Ivanov
FSBEI HE Vologda State Dairy Farming Academy (DSFA) named after N.V. Vereshchagin, Vologda—Molochnoye 160555, Russia

Jawad Kadhim Al Aridhee
College of Agriculture, Al Muthanna University, Al Muthanna 66001, Iraq

Grzegorz Łysiak, Ryszard Kulig, Monika Wójcik and Marian Panasiewicz
Department of Food Engineering and Machines, University of Life Sciences in Lublin, 20-033 Lublin, Poland

Agnieszka Kubik-Komar
Department of Applied Mathematics and Computer Science, University of Life Sciences in Lublin, 20-612 Lublin, Poland

Izabela Kuna-Broniowska
Department of Applied Mathematics and Computer Science, University of Life Sciences in Lublin, 20-612 Lublin, Poland
Department of Applied Mathematics and Computer Science, University of Life Sciences in Lublin, 20-950 Lublin, Poland

Beata Ślaska-Grzywna and Leszek Rydzak
Department of Biological Bases of Food and Feed Technologies, University of Life Sciences in Lublin, 20-612 Lublin, Poland

Dariusz Andrejko
Department of Biological Bases of Food and Feed Technologies, University of Life Sciences in Lublin, 20-612 Lublin, Poland
Department of Biological Bases of Food and Feed Technologies, University of Life Sciences in Lublin, Akademicka 13, 20-612 Lublin, Poland

Zbigniew Kobus and Anna Pecyna
Department of Technology Fundamentals, University of Life Sciences in Lublin, 20-612 Lublin, Poland

Monika Stoma
Department of Power Engineering and Transportation, University of Life Sciences in Lublin, 20-612 Lublin, Poland

Agata Blicharz-Kania and Marek Szmigielski
Department of Biological Bases of Food and Feed Technologies, University of Life Sciences in Lublin, 20-612 Lublin, Poland

Paweł Sobczak and Kazimierz Zawiślak
Department of Food Engineering and Machines, University of Life Sciences in Lublin, Akademicka 13, 20-612 Lublin, Poland

Kamil Wilczyński
Department of Food Engineering and Machines, University of Life Sciences postcode, 20-612 Lublin, Poland

Dariusz Dziki
Department of Thermal Technology and Food Process Engineering, University of Life Sciences, 20-612 Lublin, Poland

Andrzej Anders, Ewelina Kolankowska, Dariusz Jan Choszcz, Stanisław Konopka and Zdzisław Kaliniewicz
Department of Heavy Duty Machines and Research Methodology, University of Warmia and Mazury in Olsztyn, 10-719 Olsztyn, Poland

Bartosz Kulczyński and Anna Gramza-Michałowska
Department of Gastronomy Sciences and Functional Foods, Faculty of Food Science and Nutrition, Poznań University of Life Sciences, 31Wojska Polskiego Str., 60–624 Poznań, Poland

Jolanta B. Królczyk
Department of Manufacturing Engineering and Production Automation, Faculty of Mechanical Engineering, Opole University of Technology, ul. Prószkowska 76, 45–758 Opole, Poland

Joanna Kapusta-Duch, Teresa Leszczyńska and Barbara Borczak
Department of Human Nutrition, Faculty of Food Technology, University of Agriculture in Krakow, 30-149 Krakow, Poland

Anna Szeląg-Sikora and Jakub Sikora
Institute of Agricultural Engineering and Informatics, University of Agriculture in Krakow, 30-149 Krakow, Poland
Faculty of Production and Power Engineering, University of Agriculture in Krakow, ul. Balicka 116B, 30-149 Kraków, Poland

Agnieszka Starek, Agnieszka Sagan and Beata Zdybel
Department of Biological Bases of Food and Feed Technologies, University of Life Sciences in Lublin, Akademicka 13, 20-612 Lublin, Poland

Wioletta Żukiewicz-Sobczak
Pope John Paul II State School of Higher Education in Biala Podlaska, Sidorska 95/97, 21-500 Biala Podlaska, Poland

Andrzej Bochniak and Paweł Artur Kluza
Department of Applied Mathematics and Computer Science, University of Life Sciences in Lublin, 20-950 Lublin, Poland

Milan Koszel
Department of Machinery Exploitation and Management of Production Processes, University of Life Sciences in Lublin, 20-950 Lublin, Poland

Dominika Barbara Matuszek
Faculty of Production Engineering, Opole University of Technology, 45-758 Opole, Poland

Index

Printed in the USA
CPSIA information can be obtained
at www.ICGtesting.com
JSHW051410091023
49903JS00006B/370